Handbook of Value Addition Processes for Fabrics

Handbook of Value Addition Processes for Fabrics

B. Purushothama

WOODHEAD PUBLISHING INDIA PVT LTD

New Delhi, India

Published by Woodhead Publishing India Pvt. Ltd.
Woodhead Publishing India Pvt. Ltd., 303, Vardaan House, 7/28, Ansari Road,
Daryaganj, New Delhi - 110002, India
www.woodheadpublishingindia.com

First published 2018, Woodhead Publishing India Pvt. Ltd.
© Woodhead Publishing India Pvt. Ltd., 2018
Reprint 2020

Woodhead Publishing India Pvt. Ltd. ISBN: 978-93-85059-44-5
Woodhead Publishing India Pvt. Ltd. Master E-ISBN: 978-93-85059-92-6

Typeset by Versatile PreMedia Services
Digitally Printed and bound by Replika Press Pvt. Ltd.

Contents

Fabrics taken out at loom state cannot fetch good price as they are not attractive and do not have a good hand. Also, they are not suitable for wearing or for any kind of use at all.

The size present in the fabrics need to be removed (desized), if sized, and the impurities in the materials are to be removed by treatments such as scouring and various treatments such as bleaching, dyeing, printing and different finishes are to be given. In a number of processes, chemicals are involved and; hence, people call these processes as chemical processing of textiles. As water is invariably used in number of processes, these processes are also termed as wet processes. There are a number of processes which neither use chemicals nor water and; hence, the terms chemical processing or wet processing might not suit all the time. All these processes add value to the fabric and; hence, term 'value addition processes' is more appropriate.

It is not possible to list all the value addition processes practiced in the world in one book. In this book, an attempt is made to collect details of some of the commonly practiced value addition processes, especially for apparel purposes. The functional treatments given for various technical textiles like medical textiles, protective textiles, industrial textiles, aggrotech materials, geotextiles and sport tech, etc. are not covered in this book.

This book has aimed mainly to guide the new entrants in the textile field, who would like to supervise the processes and manage them. The process parameters and the chemicals used vary depending on the processes that need to be decided by the senior technical person in the section considering the fabric in use, the effect required, the machinery and chemicals available. Therefore, the chemicals and chemical reactions are not discussed in this book. This book gives general guidelines that are applicable for all. This book can be used as a guide for training technical staff.

I hope the readers shall get a fair idea by which they can start their career.

B. Purushothama

Value addition processes and pretreatment

1.1 Various value additions done on fabrics

The fabric coming from the loom does not have properties such as absorbency, softness and so forth, and the appearance of the fabric is dirty or pale yellow. They are not undesirable and cannot be used directly for making apparels or clothing. Therefore, it is necessary to follow some processing of the material to make it wearable. These processes add value to the fabrics and; hence, are referred as *value addition processes*.

There are innumerable value addition processes in textile industry. The purposes are different for each value addition process. They are all done as per the requirement of the customers. Although the value addition processes done on fabrics have different purposes and the processes, they can be broadly grouped as follows:

Pretreatments	• Singeing • Desizing
Scouring/ bleaching/ mercerizing	• Kier boiling • Bleaching • Mercerizing • Scouring and milling of wool fabrics • Rotary drum washing
Dyeing	• Padding process • Jigger dyeing • Jet dyeing
Drying	• Drying ranges—cylinder drying • Hot-air drying. • Air-flow processing of fabrics • Hot-air stenters • Radio frequency dryers • Relax dryers

Shrink-proof finishing	• Suprema KD Biella • Formula-1, KD Biella • Sanforizing • Zero–zero finish • Comfit • Steam relaxing • Jet air • Relax shrinking
Special effects	• Weight reduction processes for polyester • Peaching or sueding of fabrics • Raising operation on fabric • Fabric shearing • Decatizing of wool • Calendaring
Printing styles	• Direct printing • High-density printing of polyester • Overprinting • Mordant printing • Resist dyeing • Discharge printing • Resist printing • Pigment printing • Emboss printing—pub printing • Glitter printing • Inkjet printing
Printing methods	• Block printing • Burnout printing • Blotch printing • Digital printing • Duplex printing • Engraved roller printing • Electrostatic printing • Flock printing • Inkjet printing • Jet spray printing • Photo printing • Photographic printing

	• Screen printing—flat screen • Rotary screen printing • Stencil printing • Spray printing • Transfer printing • Warp printing • Special methods—tie and dye, Batik printing, space dyeing
Garment washing	• Wet process/chemical process • Normal wash/garment wash/rinse wash • Pigment wash • Caustic wash • Enzyme wash • Stonewash with or without bleach • Stone enzyme wash • Tinting (tie) and overdyeing (dip dyeing) • Super-white wash • Bleach wash • Acid wash • Silicon wash • Soft wash • Whitening • Metal wash Dry process/mechanical process • Sandblasting • Hand scraping • Overall wrinkles • Permanent wrinkle • Broken and tagging • Grinding and destroy • PP spray, PP sponging and so forth.

It is practically impossible to list all the value addition processes practiced worldwide. We shall be discussing some of the processes, especially those used for apparel purposes. There are a number of value addition processes for garments normally referred as garment washing, which are discussed in brief.

The value addition processes for fabrics to be used for apparels are arranged as per the requirement of the customers. A general flow chart shall be as follows.

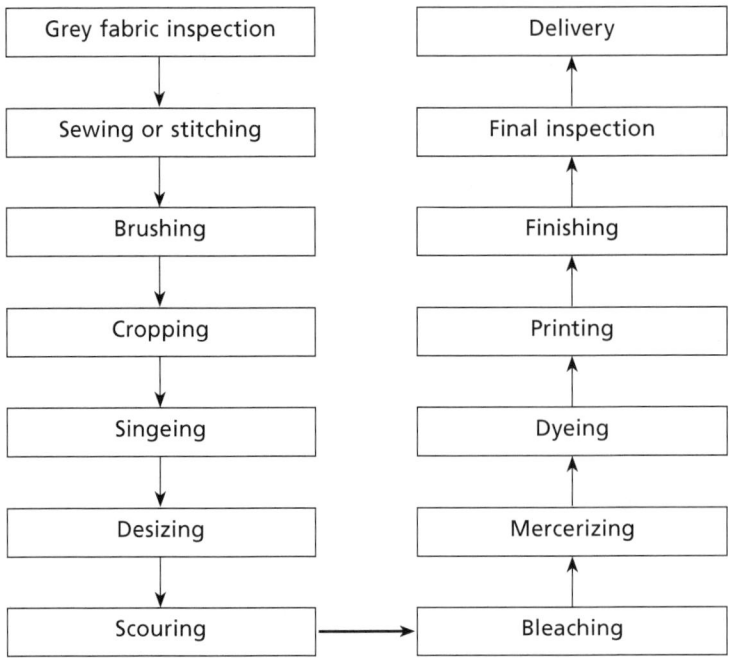

1.1.1　Grey fabric inspection

After manufacturing of grey fabric on loom, it is inspected using an inspection table. During this inspection, defects such as knots, broken and loose warp ends, broken weft ends are removed. In case of holes, either it is mended or the fabric is cut off depending on the position of the hole and its severity.

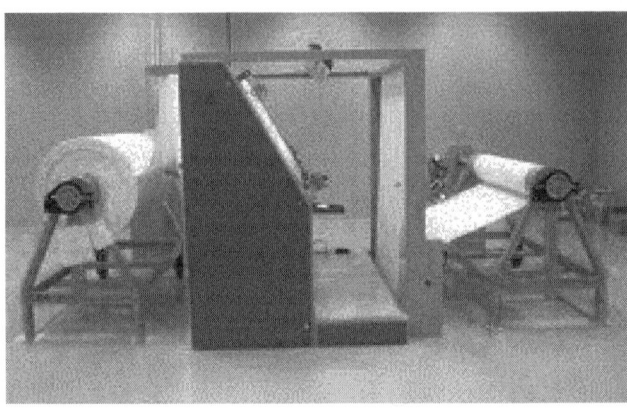

Figure 1.1　Batch inspection machine.

1.1.2 Batching

Batching is the receiving section of grey fabric, where the fabrics which should be dyed and processed for a particular lot of a particular order are grouped, assembled together and the length is increased (Fig. 1.1). From batching section, the grey fabric is sent to the dyeing section for dyeing.

Flow chart of batching section is given as follows:

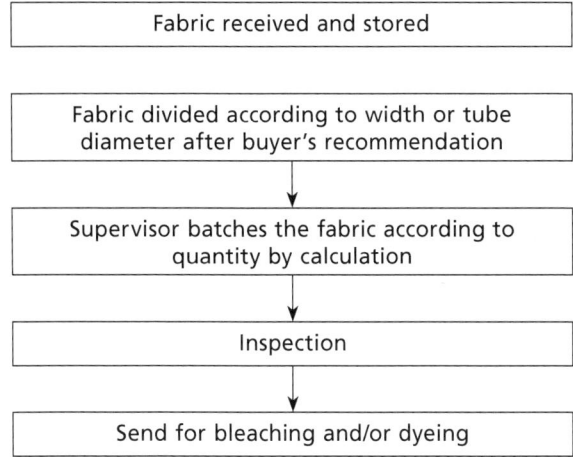

Batching of fabric for dyeing is done according to the following criteria:

- Order sheet (received from the buyer)
- Dyeing shade
- Machine available
- Type of fabrics (100% cotton, polyester cotton, chief value cotton, etc.)
- Emergency

The grey fabric is sent to the dyeing floor with a batch card providing details of the fabric rolls, the processes to be carried out and the sales contract against which the process is taking place. Records are maintained for the same. The criteria for proper batching are:

1. To use maximum capacity of existing dyeing machines.

2. To minimize the washing time or preparation time and machine stoppage time.

3. To keep the number of batches as less as possible for same shade.

4. To use a particular machine for dyeing same shade.

1.1.3 Stitching

Stitching is done of different fabrics of the same variety undergoing same process to increase the length of the fabric, making it suitable for processing. It is done by plain sewing machine.

1.1.4 Brushing

Brushing is done to remove the dirt, dust, loose fibre and loose ends of the warp and weft threads. If brushing is not done, the impurities obstruct the absorption of chemicals in next processes.

1.1.5 Shearing/cropping

The shearing or cropping is the process by which the attached loose ends of the warp and weft thread is removed by cutting by the knives or blades. After this, the fabrics undergo singeing process.

1.1.6 Singeing

In singeing process, the protruding and/or projecting fibres are removed from the fabrics by burning and/or heating to increase the smoothness of the fabric. If required, both sides of the fabric are singed. This process is optional.

1.1.7 Desizing

Desizing is the process in which the sizing materials are removed from the fabric. This must be done before other wet processes of bleaching, mercerizing, dyeing, printing or finishing are carried out.

1.1.8 Scouring

The process by which the natural impurities (oil, wax, fat, etc.) and added/external/adventitious impurities (dirt, dust, etc.) are removed from the fabric is called scouring. It is done by strong alkali like caustic soda (sodium hydroxide).

1.1.9 Souring

Souring is the process in which the alkali is removed from the scoured fabric by using dilute acid solution.

1.1.10 Bleaching

Cotton has some natural colouring matter, which confers a yellowish-brown colour to the fibre. Bleaching is the process by which the natural colours (nitrogenous substance) are removed from the fabric to make the fabric pure and white. It is done by bleaching agents such as hydrogen peroxide or sodium hypochlorite. Although there are different bleaching agents that can be used for bleaching cotton, using hydrogen peroxide (H_2O_2) is most common. In addition to an increase in whiteness, bleaching results in an increase in absorbency, levelness of pretreatment and complete removal of seed husks and trash.

1.1.11 Mercerizing

Mercerizing is the process by which the cellulosic materials are treated with highly concentrated sodium hydroxide (NaOH) to impart properties such as strength, absorbency capacity and lustre. This process is optional.

1.1.12 Dyeing

Dyeing is the process of colouring fibres, yarns or fabrics with either natural or synthetic dyes. Dyeing imparts beauty to the textile by applying various colours and their shades onto a fabric. Dyeing can be done at any stage of the manufacturing of textile such as fibre, yarn, fabric or a finished textile product including garments and apparels. The property of colour fastness depends on two factors, selection of proper dye according to the textile material to be dyed and selection of the method for dyeing.

Colour is applied to the fabric by different methods and at different stages of the textile manufacturing process.

- In stock dyeing, the fire is dyed even before it is spun.

- Top dyeing is the process of dyeing the slivers, that is, the fibre is dyed at the stage just before the appearance of finished yarn. This process is more popular in combed wool and in the production of melange yarns.

- Yarn dyeing may be done in hank form or in package form. The package dyeing may be done for yarn cheeses or yarn beams.

- Space dyeing consists of dyeing the yarn at intervals along its length.

- Piece dyeing consists of dyeing fabrics in small batches according to the demands for a given colour.

- In solution pigmenting or dope dyeing, the dye is added to the solution before it is extruded through the spinnerets for making synthetic filaments.

- In garment dyeing, the dye is applied to finished products such as apparels and garments.

1.1.13 Printing

Printing is the process for producing a pattern on yarns, warp, fabric or carpet by any of a large number of printing methods. The colour or other treating materials, usually in the form of a paste, are deposited onto the fabric which is then usually treated with steam, heat or chemicals for fixation.

1.1.14 Finishing

There are numbers of finishing treatments to give special effects. The finishing treatments are carried out according to buyer requirements, which are followed by folding, packaging and delivery.

1.2 Pretreatment

Satisfactory preparation of the substrate before any value addition process such as bleaching, dyeing, printing and finishing makes major contributions to consistent attainment of the desired end-product quality. In order to get the best results, we should conduct right pretreatment right at first instance.

Successful preparation depends on four factors:

- The amounts of the various impurities present
- The purity of the water supply
- The chemicals used in the various preparation processes
- The machinery available for processing of the goods

It is necessary to remove the unwanted materials from the fabric which hinder the value addition processes. They may be referred to as impurities from the point of view of value addition processes. Let us see the normal unwanted materials in a fabric that hinders value addition processes.

- Cotton impurities and preparation chemicals
- Pectins are polygalacturonic acids and their calcium, magnesium and iron salts
- The inorganic ash containing calcium, magnesium and potassium phosphates and carbonates
- The spin finish and knitting oil containing mineral oil and surfactants applied to decrease friction on machinery parts
- Sizing agents, which are film-forming polymers applied to warp yarns before fabric weaving in order to minimize yarn breakage
- Metallic ion contamination, particularly iron and copper, is of serious concern during oxidative bleaching processes

Important factors to select and use of surfactants in preparation are as follows:

- Wetting agents used in the desizing stage must be compatible with the enzyme preparation.

- Detergents selected for scouring must be stable at the temperature and concentrations of alkali and electrolyte required.

- Surfactants added to bleach liquors must be stable at strong oxidizing conditions.

- Residual surfactants which are retaining must not cause problems in subsequent printing or water-repellent finishing.

- The cloud point of any non-ionic surfactants used must be high enough to avoid impairing the wetting or detergency performance.

- Surfactants must be low-foaming to avoid risks of pump cavitation in circulating liquor systems and loss of traction in conveyor or roller-bed steamers.

- The viscosity of the surfactant solution should allow satisfactory performance in automatic dosing systems.

The various pretreatments are singeing, desizing, scouring, bleaching, mercerizing for cotton fabrics. The degumming is a special pretreatment for natural silk fabrics.

1.3 Machinery for textile value addition processes

Textile machinery used for value addition processes can be classified as batch, semi-continuous and continuous one. While semi-continuous and continuous machineries are generally used for textile fabrics, batch machines are available for fibre, yarn as well as fabrics and garments.

The examples of batch machines used in value addition processes are kiers, jiggers, winch, jet dyeing machine, drum washers, HDHP dyeing machines for yarns and so forth. Semi-continuous and continuous machines include chain mercerizing machine, stenters, J box, sanforizing machines and so forth.

Singeing and desizing of fabrics

2.1 Introduction

The warp yarns, especially when used as single yarns, are normally sized to give sufficient strength and elasticity while weaving. This size hinders the wet processes by forming a film around the warp yarns, which prevent absorption of processing chemicals and dyes. Desizing is the process of removing the sizing materials present in the fabric in order to make fabric suitable for further operations in wet processing.

The hairs on the grey fabrics produce fuzzy effect when dyed by absorbing more dyes at hairy portions. Hence for high-value products, it is suggested to remove the hairs. Singeing is the process of removing protruding fibres from the fabric by burning them out (Fig. 2.1).

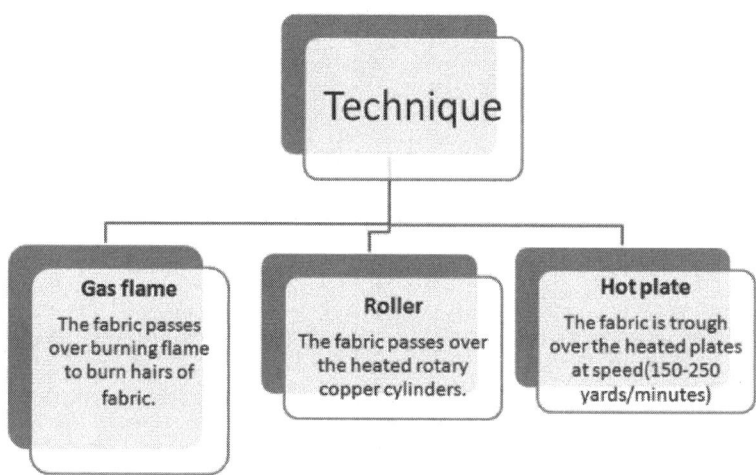

Figure 2.1 Techniques of singeing.

Singeing machine without   Singeing cum desizing machine
desizing (Osthoff)   (Osthoff)

Figure 2.2   Singeing machines.

Singeing and desizing can be done on a single machine or on two different machines depending on the requirement (Fig. 2.2).

There are three types of singeing machines available for fabric singeing. They are:

a)   Plate singeing machine

b)   Roller singeing machine

c)   Gas singeing machine

Gas singeing is normally practiced in majority of the mills as flames can penetrate the gaps in the fabric and singeing can be more effective.

There are three types of techniques in desizing and four methods of desizing that has been used in wet processing. Desizing techniques are different depending on the kind of sizing agent to be removed.

Natural starch, starch ethers, cellulose ethers and polyacrylates are usually used as sizes. Economical starch-based formulations are effective for cotton yarns. Polyvinyl alcohol is often preferred for sizing polyester/cotton blends. Waxy plasticizers may be added to the size mix. Considering the sizing chemicals used, the desizing chemicals are to be decided.

Factors to be considered for removing the size are:

• Concentration and viscosity of the size formulation

• Nature and amount of plasticizer present

• Fabric construction

• Ease of dissolution of the size

• Washing-off procedure and temperature

Desizing involves impregnation of the fabric with the desizing agent, allowing the desizing agent to degrade or solubilize the size material and finally to wash out the degradation products. The major desizing processes are:

- Enzymatic desizing of starches on cotton fabrics
- Oxidative desizing
- Acid desizing
- Removal of water-soluble sizes
- Bioscouring

Enzymatic desizing: Enzymatic desizing is the process of degrading starch size on cotton fabrics using enzymes. Enzymes are complex organic, soluble biocatalysts formed by living organisms that catalyze chemical reaction in biological processes. Enzymes are quite specific in their action on a particular substance. A small quantity of enzyme is able to decompose a large quantity of the substance it acts upon. Enzymes are usually named by the kind of substance degraded in the reaction it catalyzes.

Amylases are the enzymes that hydrolyze and reduce the molecular weight of amylose and amylopectin molecules in starch, rendering it water soluble enough to be washed off the fabric. Amylase enzymes are highly effective catalysts for the hydrolysis of the amylose and amylopectin components of the starch.

Effective enzymatic desizing requires strict control of pH, temperature, water hardness, electrolyte addition and choice of surfactant. Normal types are applied for several hours at 65–70°C.

Thermostable hydrolytic enzymes have been introduced, allowing brief dwell times at temperatures up to 120°C. Common salt and calcium ions increase the rate of hydrolysis but amylase is deactivated by copper or zinc ions, as well as most anionic surfactants.

There is some interest in the use of pectinases as scouring agents and lignases to degrade the lignin in bast fibres, but as yet no commercial processes have been developed.

Oxidative desizing: In oxidative desizing, the risk of damage to the cellulose fibre is very high, and, hence, its use for desizing is rare. Oxidative desizing uses potassium or sodium persulphate or sodium bromite as an oxidizing agent.

Oxidative desizing reduces the number of fabric preparation stages minimizing the overall energy consumption. The oxidant can be added to the hot caustic scour liquor and little or no magnesium silicate or organic stabilizer is needed. Rapid desizing treatments require more critical control of alkali and oxidant concentrations. Increased oxidant above the quantity necessary for effective desizing and increasing the alkalinity for a given oxidant concentration both tend to increase the degree of chemical damage. Persulphate promotes desizing rather than bleaching and requires more critical control of concentration than does hydrogen peroxide.

Acid desizing: Cold solutions of dilute sulphuric or hydrochloric acids are used to hydrolyze the starch; however, this has the disadvantage as it also affects the cellulose fibre in cotton fabrics.

Removal of water-soluble sizes: Fabrics containing water-soluble sizes can be desized by washing using hot water containing wetting agents (surfactants) and a mild alkali. The water replaces the size on the outer surface of the fibre, and get absorbs within the fibre to remove any fabric residue.

Bioscouring: Bioscouring is a process in which alkali-stable pectinase as the enzyme is applied to selectively remove pectin and waxes from cotton fibres. By hydrolyzing the pectin material between the waxes and the fibre surface, the enzyme exposes the waxes to emulsification when the scouring bath temperature exceeds their melting range. Bioscouring does not eliminate motes (cottonseed fragments) or the natural colour of the cotton, which can be beneficial when scouring for a natural look.

2.2 Purpose of singeing the fabrics

The basic purpose of singeing the fabrics is to remove the hairs projecting out of the fabric by burning them out and making the fabric surface smooth. This helps in improving lustre of the fabric and also in getting good printing effect. The fabrics may be singed either on one side or on both the sides.

Singeing is an optional process. It is performed only on the fabrics where the presence of protruding hairs is a hindrance for processing and is not attractive to the customer.

2.3 Purpose of desizing

The size applied during sizing is required only while weaving, and has no function once the fabric is woven. The size, if remains in the fabric, does not allow the dyes and chemicals to penetrate inside the fibres. Hence the purpose of desizing is to remove the size applied to the warp yarn and make the fabric suitable for further wet processing such as scouring, bleaching, dyeing, printing and/or finishing.

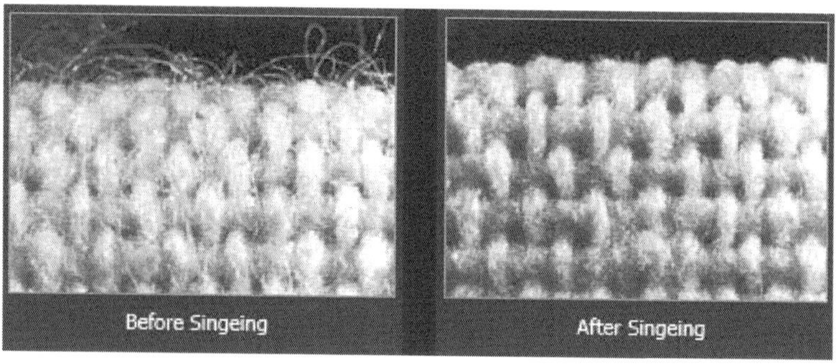

Figure 2.3 Before and after singeing.

2.4 What singeing should and should not do?

2.4.1 Should do

a) Singeing should remove the protruding fibres from the surface of fabric (Fig. 2.3).

b) Singeing should make the fabric surface smooth.

2.4.2 Should not do

a) Singeing should not burn the fabric.

b) Singeing should not make the fabric weak.

c) Singeing should not blacken the fabrics.

d) Singeing should not lead to fire accidents in the factory.

2.5 What desizing should and should not do?

2.5.1 Should do

a) Desizing should break and remove the size particles from the warp yarns.

b) Desizing should make the fabric suitable for undergoing wet processes by making fabric capable of absorbing water and other chemicals used in wet processing.

2.5.2 Should not do

a) Desizing should not degrade the yarns or fabrics by damaging the fibres.

b) Desizing should not make the fabric weak.

c) Desizing should not result in water pollution.

d) Desizing should not result in uneven patches on fabrics.

2.6 General activities in singeing and desizing of fabrics

The general activities in singeing and desizing can be listed as follows. Gas singeing operation (Figs. 2.4 and 2.5) is considered in the illustration.

a) The fabrics to be singed are opened and prepared so that they can be fed to singeing machine without any hindrances. Normally the fabrics are batched and rolled.

b) The fabrics are fed in open width without any slackness or wrinkles. Singeing cannot be done in rope form.

c) The cloth is exposed to flame from a twin gas burner where the gas comes through the filter at a pressure of 100–140 mbar.

d) The ratio of gas to air mixture is maintained at 1:4. The fabric is brushed using special brush roller. Different types of brush rollers are developed for different type of fabrics.

e) As the cloth runs at a high speed, only protruding fibres are burnt out. The cloth passes in water after burning to avoid chances of fire.

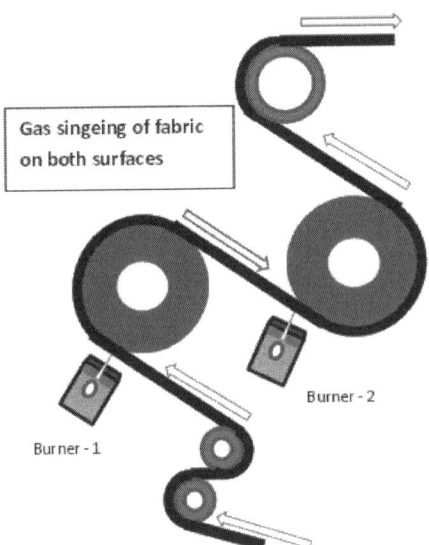

Gas singeing of fabric on both surfaces

Burner - 1

Burner - 2

Figure 2.4 Gas singeing of fabric.

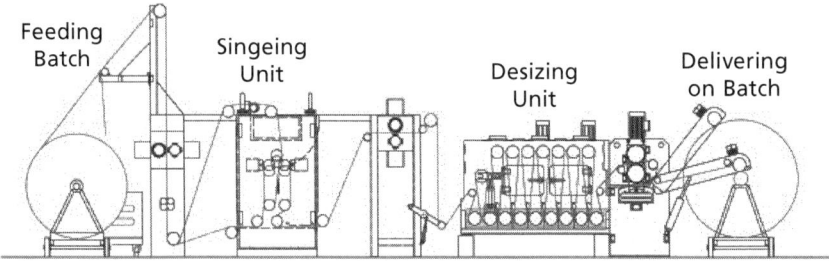

Figure 2.5   Schematic diagram of singeing machine with desizing attachment.

f)   In singeing cum desizing machine, the singeing is done first followed by desizing. The desized materials are wound on the batch, and kept for rotating slowly. Minimum 8 h rotation is needed to complete desizing activity (breaking of sizing materials to microfine level so that they can be washed off in the next process) and to get the uniform effect so that the further processes give uniform results.

g)   There are three different positions to get varied effect of singeing (Fig. 2.6).

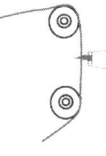 i.   The singeing flame meets the free-guided fabric at right angles as the fabric passes between two guide rollers. This position is recommended for materials made of 100% natural fibres and for blended fabrics which have been thoroughly beaten, with weights over 125 g/m^2.

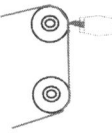 ii.   The singeing flame meets the fabric at right angles as it is bent over a water-cooled roller. The choice of this position is recommended for qualities of fabric composed of temperature sensitive fibres, open-weave blended fabrics and those with weights of less than 125 g/m^2.

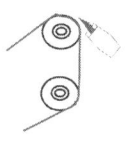 iii.   The singeing flame passes close to the fabric with the jet direction being at a tangent to the fabric surface. This singeing process is recommended for all materials which cannot tolerate direct exposure to flame. Tangential singeing can also equalize protruding fibres and repair filament breaks. However, at present, this facility is not available on our machines.

Figure 2.6   Different positions of flame to get different singeing effect.

2.7 Operating instructions for operating a combined gas singeing and desizing machine. (example—Osthoff singeing cum desizing machine)

Following is a typical operating instruction for working on Osthoff singeing cum desizing machine.

1. Clean the machine with compressed air and wipe with clean cloth before starting

2. Clean the desizing chamber with pressure water

3. Clean the brushing units and remove the threads

4. Clean the double jet burners by using the gauge provided with the machine

5. Clean the ducting unit and dust collecting airbags once in a shift or whenever the bag becomes half full

6. Clean all the chemical preparation tanks

7. Bring the material to be fed to the machine and keep at the feeding place

8. Prepare the desizing recipe as needed. Take the required chemicals by weighing, mix them appropriately and fill them in the overhead tank.

9. Open the air valve, water supply valve and gas supply valve

10. Following is an example of recipe for desizing

    • Water—1000 L

    • Bactasol PHC 3 g/L

    • Acetic Acid 1 g/L

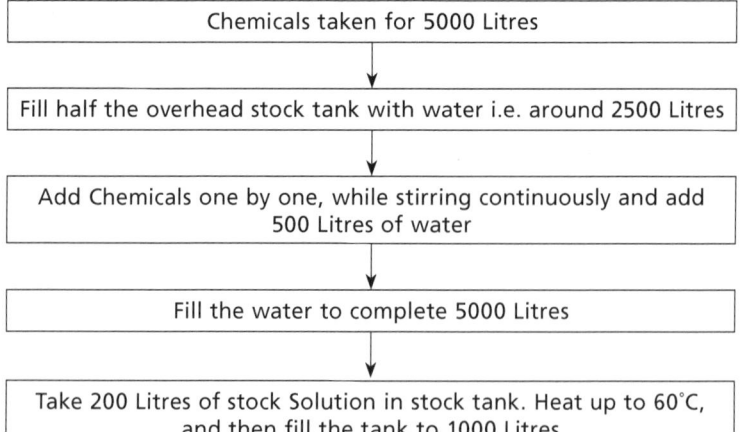

11. Maintain pH at 5.5–6.0 and saturator temperature at 60°C

12. Check the pickup per cent of mangles

13. Discuss with process house head and set the parameters such as speed, flame intensity, as per the fabric requirement and the intensity of singeing required

14. When singeing is done onto a cooled roller, verify and ensure that inlet and outlet of water is working and water is flowing smoothly

15. Open the manual gas stop valve and allow the gas

16. Ensure that the brush rollers are set properly

17. Start the Mangle motor, brush motor and batcher motor

18. Run the machine at the desired speed

19. Take the burner at its positions

20. Check the working of discharge fan and ensure that all smoke and residual fuel are discharged out safely in time

21. While desizing check the pH from lot-to-lot

22. Separate the designs by putting a polythene sheet at the end of each design to prevent colour migration from one design to another

23. While taking the processed material out of the machine to the batcher, put polythene sheet on the fabric

24. Keep the singed and desized material at rotating station (Fig. 2.7) as instructed by the supervisor.

25. While working if you find any quality problem in the cloth, inform the superiors immediately

Figure 2.7   Rotating station.

26. If the quality problem is coming from the singeing machine, remove the materials immediately, and restart only after rectification

27. Enter the details of the materials worked, the lot number, the quantity produced in each lot, batch number in the production book

28. Enter the completion time of desizing on the display board hanged on the batcher kept for rotating

2.8 Precautions to be taken for singeing and desizing

Following precautions are to be taken while working on a singeing and desizing machine.

a) If the gas line is filled with air, the ignition takes time. Do not allow the air to enter the gas line

b) Fabric should be dry when fed for singeing

c) Fabric component should be clean when fed to singeing

d) Speed should not alter when the singeing operation is on

e) There should not be any power failure while the singeing operation is on. Hence, install uninterrupted power supply system

f) It is needed to install the brush that is suitable for the type of material being singed

2.9 Control points and checkpoints

It is essential to have clarity on the points to be controlled to achieve the targets agreed for quality and production and those to be checked to ensure the process in control. These points need to be reviewed from time to time and modified to suit the requirements of individual companies and their targets. Each mill should prepare its own "control points and check points" and display them in the work area so that the people can refer and follow. Following is an illustration.

2.9.1 Control points

a) Deciding on the fabrics needing singeing; all fabrics cannot be singed

b) Deciding the flame height and speed of singeing depending on the fabric structure, that is, the count of warp and weft, and the weave

c) Deciding on the flame angle and the position of the burner

d) Deciding on the enzymes to be used depending on the sizing chemicals applied

e) Deciding on the process parameters for desizing

2.9.2 Checkpoints

Material related

a) Verify whether the fabrics received are matching with the plan, that is, the design number, the metres batched for singeing

b) Check for the proper batching with selvedge exactly in one plain

Machine related

a) The condition of the gas pipes, joints and the burners

b) The functioning of emergency stop motion to prevent fabric from burning off

c) The condition of water pipes and joints

d) The functioning of water drain

e) The availability of gas in the gas cylinders

f) The condition of brush rollers

g) The condition of fire bricks

Setting related

a) The alignment of feeding batch to the burners

b) The flame height and the angle

c) Positioning of the flame in relation to the fabric density

d) The speed of the machine

e) The desizing recipe, that is, the chemicals planned and prepared

f) The batching parameters set at delivery

Performance related

a) The quality of singeing, that is, the hairs burnt and uniformity of singeing

b) The quality of desizing

Work practice related

a) Cleaning of the machine before starting the work

b) Housekeeping around the machine

c) Putting polythene sheet after every design and fabric roll

Documentation and records

a) The design numbers and the number of meters singed in each design

b) Number of gas cylinders replaced in case of using gas cylinder.

c) The meter reading of the gas meter in case of direct pipeline connection to singeing machine

Logbook related

a) The design number wise singeing and desizing done

b) Consumption of gas

c) Consumption of chemicals for desizing

d) Instructions for next shift

Management information system related

a) Design number

b) Lot number

c) Number of metres

d) Quantity of gas consumed

e) Desizing chemicals used and their quantity

General

a) The condition of firefighting system near to the singeing system

b) The number of batches in rotating station

c) Timings written on each batch

2.10 Normal problems in singeing and desizing of fabrics

a) Uneven flame heights due to burnt fibres chocking the burners. Thorough and periodic cleaning of burners can solve the problem.

b) Fluffs those have fallen on the fabric before singeing leading to excess burning. As the fluff fallen on the fabric are loose and have no control, catches fire and starts burning. This can damage the fabric and also result in the uneven shade.

c) Improper brushing leads to improper removing of burnt particles and fluff.

d) Improper desizing leads to uneven bleaching and dyeing.

2.11 Dos and don'ts for singeing and desizing of fabrics

It is necessary to clearly understand what is supposed to be done without fail and what should not be done at any cost. Each market complaint and observations by quality control can add points to the list of 'Dos and Don'ts'. Some examples are given below.

2.11.1 Dos

a) Get the fabric cleaned by blowing compressed air before batching for feeding into singeing machine and remove superfluous fluff to avoid problems in singeing.

b) Have periodic preventive maintenance of burning elements.

c) Ensure that the hot air is exhausted out along with burnt fibres.

d) Check and ensure that the fabrics are not wet before feeding to singeing.

e) Check the flame height and the colour of the flame before starting the singeing operation. Constantly watch the flame height and its uniformity.

f) Prepare recipe as per the instructions without making any change.

g) Check the pilling of the fabric after singeing.

h) Clean the flame nozzles thoroughly with 0.15-mm gauge before starting the machine (Fig. 2.8).

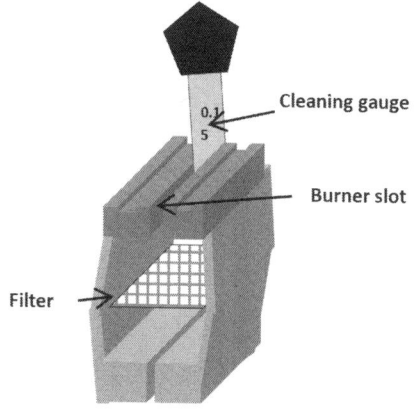

Figure 2.8   Cleaning the burner slot.

i) Keep the batch for rotating after desizing. Display the time of desizing on the rotating batch.

j) Ensure that fire extinguishers are kept near the machine and are maintained up to date.

k) Check and ensure that burner auto rotate is working properly.

2.11.2 Don'ts

a) Do not close the water inlet or outlet while singeing onto a water cooled roller.

b) Do not put all chemicals at once while preparing the recipe.

c) Do not allow dust to accumulate especially near the fire nozzles.

d) Do not clean the ramp of the burner with water (Fig. 2.9). It may burst.

e) Do not take the fabric for singeing if the fabric width is more than the burner width.

f) Do not run the machine with higher speed than the specified speed. The maximum limit is 150 m/min for majority of machines.

g) Do not replace single worn ramp bodies. Always replace the complete set.

h) Do not keep any material by the side of singeing machine.

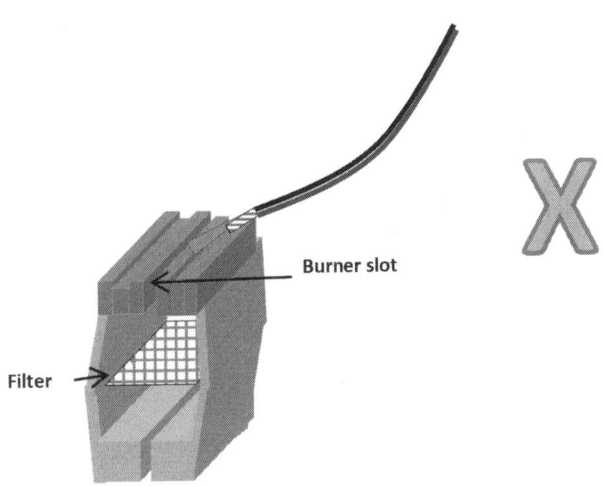

Figure 2.9 Do not clean ramp of burner with water.

2.12   Responsibilities and authorities of supervisor in singeing and desizing

2.12.1   Responsibilities

Following are the normal responsibilities of a supervisor in singeing and desizing area apart from regular supervisory responsibilities.

a) Ensuring that the required quantity of material is singed, desized and supplied in time

b) Planning the operations

c) Planning the parameters such as flame height and angle, speed and recipe depending on the materials being scoured

d) Ensuring that the desized material is batched and kept in rotating station for the required time (minimum 8 h) before sending out for next process

e) Monitoring the flame height

f) Monitoring the consumption of chemicals

g) Ensuring all safety precautions while handling chemicals and running the machines

h) Checking the quality of effluents being discharged and controlling them

2.12.2   Authorities

Following authorities are needed to be given to a supervisor to perform effectively in singeing and desizing area. The management should ensure that adequate authorities are given.

a) Authorized to stop the process and remove the materials out in case of any problem in the running machine

b) Authorized to stop feeding the materials if batching is not done well

c) Authorized to reject the chemicals if they are found contaminated or weak

d) Authorized to question the operators when the quality or productivity is not as expected

e) Authorized to stop the process if gas is found leaking

Scouring and washing

3.1 Purpose of scouring

Textile materials contain various impurities that may be natural impurities or impurities that resulted from processing. These impurities are to be removed from the fabric before taking them for further processes such as bleaching, dyeing, printing or finishing. Boiling off or kier boiling removes the natural and other impurities resulting from processes such as soil, grease, stains, wax and so forth.

If the process is performed at boiling temperature, it is known as boiling off; if it is performed under pressure, it is called kier boiling. Moreover, the scouring process can be performed as a continuous operation by incorporating a padding process followed by J-box and washing.

3.2 Kier boiling

Kier boiling is often referred as scouring which is the hot alkaline process necessary to remove the non-cellulosic impurities. The main effects of this treatment are a 5–10% loss in mass and an improvement in wettability and absorbency. The change in mass results from degradation of protein to amino acids, conversion of pectin to soluble sodium salts, hydrolytic dissolution of hemicellulose and a limited amount of oxidative degradation of cellulose. Saponification of the cotton wax is incomplete but it becomes no longer capable of forming a continuous film over the fibre surface.

A kier or keeve (or similar spellings) is a large circular boiler or vat used in bleaching or scouring cotton fabric (Fig. 3.1). They are also used for processing paper pulp. When in use, they are continuously rotated by an engine, steam being supplied through a rotating joint in the axle. They are usually spherical, sometimes cylindrical, and some were recycled from old boiler shells.

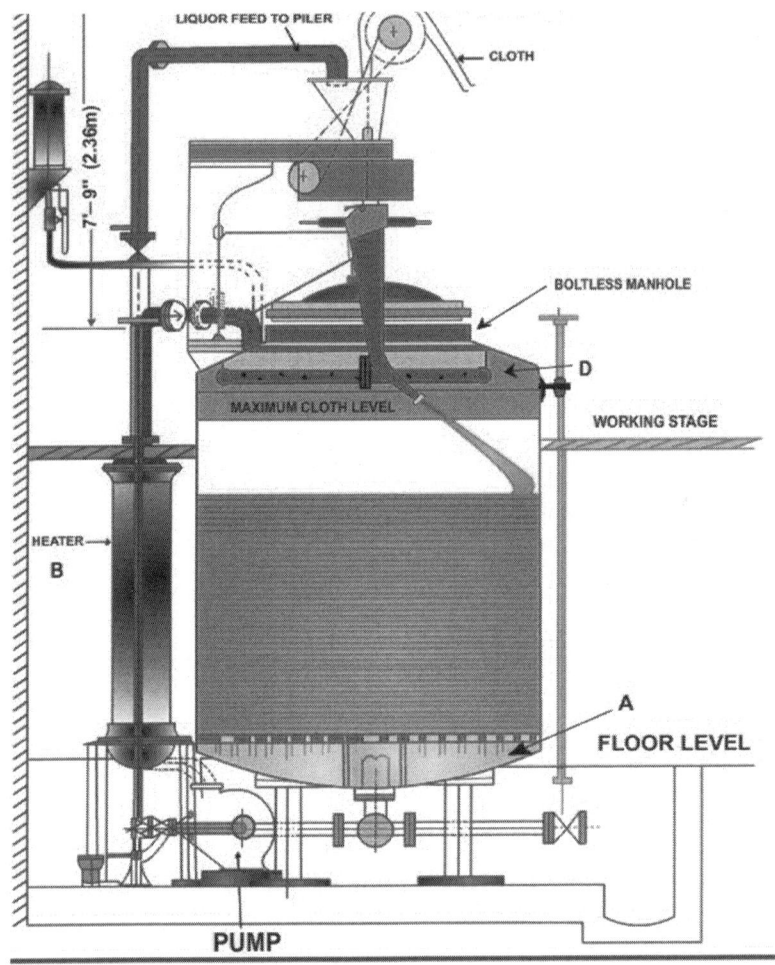

Figure 3.1   Schematic diagram of a kier.

Kier is a cylindrical vessel designed to carry out scouring of cotton fabrics in rope form. The kiers may either be open where scouring is carried out at boil at atmosphere pressure or be closed to carry out scouring at high temperature and pressure. The material used for the construction is generally steel or cast iron in which the inner side is rendered passive by a coat of sodium silicate, cement and magnesium sulphate. The kier has a perforated false bottom, located about 18 in above the base. The fabric is piled on it either manually or by a mechanical piler. It is important to lay the fabric in an even manner or else the channels of least resistance may lead to form an uneven scouring. The movement of fabric during boiling can be avoided by placing heavy weights over the fabric pile.

The capacity of a kier normally varies from 0.5 to 5 t, whereas 2-t capacity kiers are more common.

The circulation of scouring liquor is maintained by passing a stream in the kier by a central piper (puffer) through a hole in the perforated bottom. The steam draws liquor with it which hits a curved baffle plate above the goods and is distributed over the surface of the fabric. The liquor gradually percolates down the pile below the perforated bottom from where it is again forced through the buffer pipe by the steam pressure. The use of steam to circulate liquor as well as heat it is not very efficient and economical. Moreover, steam condensation may result in an increase in material to liquor ratio (M:L ratio) which is generally maintained at around 1:3. Many modern kiers use external heat exchangers and the circulation is maintained by a powerful centrifugal pump.

For scouring of cotton in kiers, it is important that all the air is forced out from the system as cotton can be oxidized into oxycellulose in the presence of air in hot alkaline liquors. This is ensured by heating the liquor to 90°C and then heating and circulation is stopped for 10–15 min. The steam and air is allowed to escape from a valve. The liquor is heated again and air is allowed to escape periodically till only steam emerges out from the valve. After this, the valve can be closed and scouring is continued for desired time.

3.2.1   What kier boiling should and should not do?

Kier boiling should help in scouring of the textile material by the following.

a)   Saponification of fats into water-soluble soap and water miscible glycerine under alkaline conditions

b)   Hydrolysis of proteins into water-soluble degradation products

c)   Dissolution of amino compounds

d)   Solubilization of pectose and pectins by converting into soluble salts

e)   Dissolution and extraction of mineral matter

f)   Emulsification and solubilization of natural oils and waxes

g)   Removal and dispersion of dirt particles and kitty by the action of alkali and detergent

Kier boiling should not result in loss of strength of the fabrics by overreaction and should not end up with accidents causing loss to the humans as well as to the company.

3.2.2 General activities in kier boiling

The activities of kier boiling includes obtaining the materials such as cotton, yarn or fabric for kier boiling, loading in the kiers, wetting and then boiling with mild alkali solution. After kier boiling, the materials are washed and then taken out for drying.

Examples of recipe for kier boiling are given below.

	Recipe 1	Recipe 2
Caustic soda	0.5–3 g/L	2–5 g/L
Soda ash	0.5–1 g/L	To adjust pH at 10.5
Wetting agent	0.5–1 g/L	1 g/L
Sequestering agent	0.5–1 g/L	1 g/L
Detergent	Nil	1–2 g/L
M:L ratio	1:10	1:10
Temperature	100–120°C	100–120°C
Material form	Open form	
Time	6 h (close vessel) and 8 h (open vessel)	
Note: Water level should be kept 6–8 in above the fabric.		

a) Kier boiler is provided with one manhole for loading and unloading the material.

b) The liquor is prepared into the mixing tank by chemicals and then brought into the preheater and heated by the steam.

c) In case of kier boiling of cottons, the cotton is filled in stainless steel cages and the cages are inserted in the kier using a chain block. The yarns are normally handled in hank form and the fabrics in the rope form.

d) The fabric is loaded in the machine by manhole and kept in the rope form.

e) The hot liquor is pumped and sprayed by spider plate onto the fabric which is packed into the kier.

f) The temperature of the kier is maintained at about 100°C and boiled for 8 h.

g) After scouring, the material is washed at 80°C to remove the impurities. Then the material is neutralized with 0.1% acetic acid and then cold wash is carried out.

3.2.3 Knowledge required for kier boiling

The person looking after kier operations should have knowledge of following:

a) Basic knowledge of chemistry, especially the reaction of caustic soda with cottons, the reaction of acids while neutralizing, the role of pressure and temperature while kiering

b) Safety precautions to be taken while handling chemicals

c) Safety systems in kier; controlling and monitoring pressure, the safety valves, controlling of the temperature, safe draining of used chemicals and water

3.2.4 Significance of different steps in kier boiling cycle

• Cleaning the kiers before starting the work to ensure that no residual chemicals of previous lot are remaining is very important to get the consistency in quality.

• Loading the materials of predetermined quantity is very important as the water and chemicals are worked out on that basis and fed to the machine.

• Loading the materials uniformly in the kier so that the chemicals can move freely and uniformly at all places is very important to get uniform kiering effect.

• Wetting the materials before kier boiling ensures uniform reaction.

• The temperature and pressure enhances the kiering operation.

• Neutralizing with acetic acid prevents cellulose from getting damaged due to overreaction of alkalis.

• Washing after kiering removes the impurities and makes the materials clean.

3.2.5 Precautions to be taken for kier boiling

a) Kier should be cleaned by washing with water before loading new lot.

b) Material should be packed evenly so that chemicals can move freely and uniformly and the materials can have uniform treatment at all places.

c) The material such as cotton, yarn or fabric should be immersed in liquor completely.

d)   Kier works with high pressure and high temperature. It is very essential to ensure that the joints are in good condition and there are no leaks or else there are chances of explosion and accidents.

e)   Before scouring, the fabric should be starch free.

f)   The kiers need to be maintained well and periodically get inspected by a competent authority for its safety systems. It is mandatory to display the inspection certificate issued by the competent authority.

3.2.6   Control points and checkpoints for kier boiling

It is essential to have clarity on the points to be controlled to achieve the targets and those to be checked to ensure the process in control. These points need to be reviewed from time to time and modified to suit the requirements of individual companies and their targets. Each mill should prepare its own 'control points and checkpoints' and display them in the work area so that the people refer and follow. Following are some of the common points.

Control points

1. Lot size: the maximum quantity that depends on the capacity of the kier

2. Kier boiling parameters such as

   - Temperature
   - Pressure
   - Time
   - Liquor ratio
   - Wetting agents to be used
   - Concentration of caustic solution
   - Neutralizer
   - Number of washes after kiering
   - Draining out the water

3. Persons to be employed

Checkpoints

1. **Material related**

   a)   Whether the quantity of cotton filled in cages is as per the predetermined quantity in case of boiling cotton

b) Whether the predetermined quantity of yarn hanks is laid uniformly in the kier in case of yarn boiling

c) Whether the fabric loaded is weighed and ensured that the quantity loaded is as per the requirement

d) Whether the fabrics are laid uniformly

e) Whether the lots are properly identified

2. **Machine related**

a) The condition of lid and gaskets

b) Locking system of kier door

c) Operations of valves

d) Working and calibration of the time, temperature and pressure monitoring systems

e) Check for leakages of steam and water

f) Condition of the hoist

3. **Setting related**

a) Timings set for different operations

b) Temperatures set for various operations such as boiling, neutralizing, washing and so forth

c) The incoming steam pressure

d) The steam pressure in the kier

e) The temperature displayed

f) The quantity of water filled and water drained

4. **Performance related**

a) The pH of caustic solution

b) The pH after neutralizing

c) The pH of water after washing

d) The pH of water drained out

5. **Documentation related**

a) The lot number and the quantity of material kiered

b) The quantity of chemicals used

c) Quantity of water consumed

6. **Work practice related**

   a) Whether the vessel was cleaned thoroughly before starting?

   b) Whether the temperature, pressure and time are monitored by the workmen as per requirement?

   c) Whether the operators have taken all safety precautions?

   d) Whether the unloading was done in clean containers?

7. **Logbook related**

   a) Lots kiered and waiting for kiering

   b) Lots running in the machines and the probable time of completion of lots

   c) Chemicals consumed and chemicals in stock

   d) Special instructions if any

8. **Management information system related**

   a) The kier number

   b) The lot number of the material kiered

   c) The quantity kiered

   d) Starting time

   e) Ending time

   f) Quantity of caustic soda used

   g) Quantity of acetic acid used

   h) Men employed

9. **General**

   a) Whether the kier was inspected by a competent authority and certified as fit for use

   b) Whether the certificate issued is displayed prominently in the work area

   c) Check the validity of the certificate displayed

   d) Whether the drains are maintained clean all the time

3.2.7 Normal problems in kier boiling

Following are the normal problems experienced in kier boiling.

a) Improper loading of materials with more height at some place, and hence chemicals not reaching to that level

b) Inadequate wetting leading to improper scouring

c) Low temperature and pressure due to shortage or leakage of steam

d) Improper neutralization leading to tendering of cellulose fibres

e) Insufficient washing leaving traces of acids on material gradually damaging them

f) Improper closing of lid leading to blast and injury to persons operating

g) Worn-out gasket leading to steam leakage

3.2.8 Dos and don'ts for kier boiling

It is necessary to understand clearly what are supposed to be done without fail and what should not be done at any cost. Some examples are given below.

Dos

a) Thoroughly clean the kier before loading new material.

b) Weigh the materials and load exactly the predetermined quantity in the kier.

c) Ensure complete wetting of the material before starting the operation of kier boiling.

d) Maintain the required liquor ratio.

e) Check all the valves and gaskets periodically and ensure that they are in good condition all the time.

f) Check the purity of caustic lye before preparing the solution.

g) Wear gloves, goggles and gum shoes while handling chemicals.

h) Always check the quality of water for its softness and contents of iron oxides.

i) Wet the materials thoroughly before starting kier boiling operations.

j) Be precise in following the process as per predetermined process sequence.

k) Wear helmet if overhead cranes are in operation in the section.

l) Use only soft water for kier boiling.

Don'ts

a) Do not allow a worn out gasket to work.

b) Do not allow leaking valves to work.

c) Do not increase the steam pressure to enhance kier boiling.

d) Do not load the materials more than the specified limit of the kier in use.

e) Do not fill the kiers partially as it increases cost of operation.

f) Do not use water if it is having traces of iron oxide.

3.2.9 Responsibilities and authorities of supervisor in kier boiling

Responsibilities

The supervisor handling kier operations shall be responsible for the following.

a) Ensuring that the required quantity of material is kier boiled and supplied in time

b) Planning the operations, timings and recipe depending on the materials being scoured

c) Ensuring that the kier boiled material is hydro-extracted before sending out for next process

d) Monitoring the consumption of chemicals

e) Ensuring all safety precautions while handling chemicals and running the machines

f) Checking the quality of effluents being discharged and controlling them

Authorities

The supervisor handling kier boiling operations should have the following authorities.

a) To stop loading the materials if kiers are not cleaned well

b) To reject the chemicals if they are found contaminated or weak

c) To question the operators when the quality or productivity is not as expected

d) To stop a kier if gasket is found hard or steam is leaking from the lid

e) To stop a kier if the door locks are not secured well

3.3   Continuous scouring using J-box

Continuous scouring operation consists of padding, J-box treatment followed by washing. Padding is used for application of chemicals to the fabric in a uniform manner in open-width form. The completion of the reaction is carried out by batching, steaming, curing and so forth. After application of liquor by padding, the required time for chemicals to act on impurities is provided in a J-box or batching tray or steamer. We shall discuss padding operation in detail in a separate chapter.

3.3.1   J-box

J-box is one of the popular machines generally used in scouring of cotton fabrics (Fig. 3.2). It is essentially a J letter-shaped stainless steel chute with large fabrics holding capacity. The fabric is fed from one end and taken out of the other (first in first out) in contrast to a kier where it is first in last out. It has a polished inner side and it is insulated to minimize heat losses.

The fabric after saturation with liquor (from a saturator located just in front of J-box) is fed from the top and taken out from the lower end. Depending on fabric speed (150–300 m/min) and the capacity (10 000 m), a residence time of 60–90 min can be provided. The fabric may be heated at the entry to the

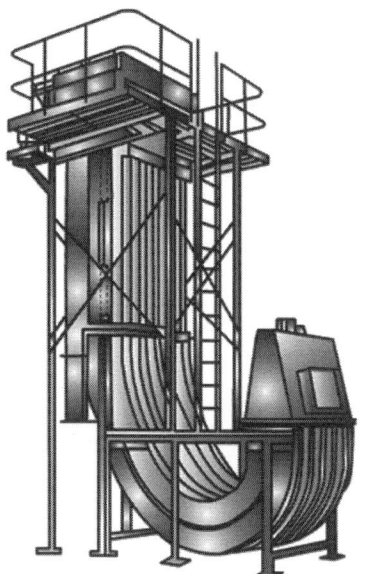

Figure 3.2   J-box.

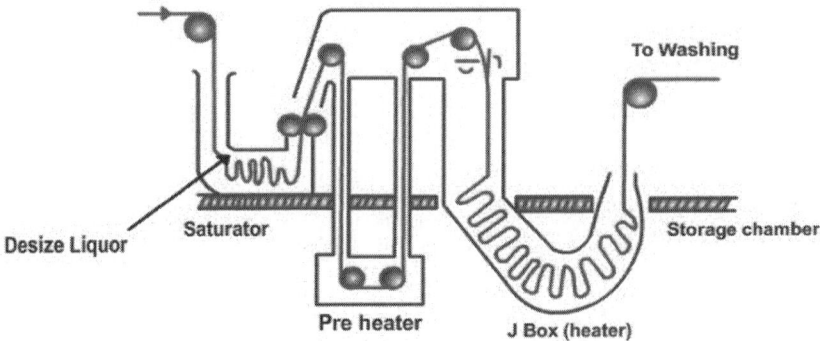

Figure 3.3   Desizing in a J-box.

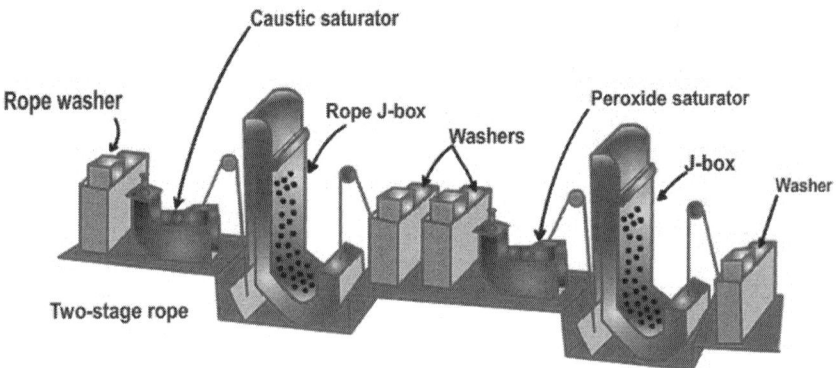

Figure 3.4   Dupont two-stage continuous scouring and bleaching range.

J-box and may retain the temperature due to insulation inside the chamber. M:L ratio is 1:1. After recovery from the J-box, the fabric is sent to a washer.

Alternately, after saturating the fabric with the liquor, it can be sent to a steamer. The speed is about 60 m/min, temperature 100–110°C and residence time 30 sec to 5 min. The J-boxes have large capacity as the fabric is stored in the rope form and moves down due to gravity. In steamers, the capacity is much lower as fabric is processed in open-width form (Fig. 3.3).

In continuous processing, various processes are combined to save time, water, energy, manpower and so forth. In case of cotton, combining desizing and scouring is possible by using oxidative desizing agents which can work in alkaline medium. Similarly, scouring and bleaching can be carried out in two continuous stages in one operation (Fig. 3.4).

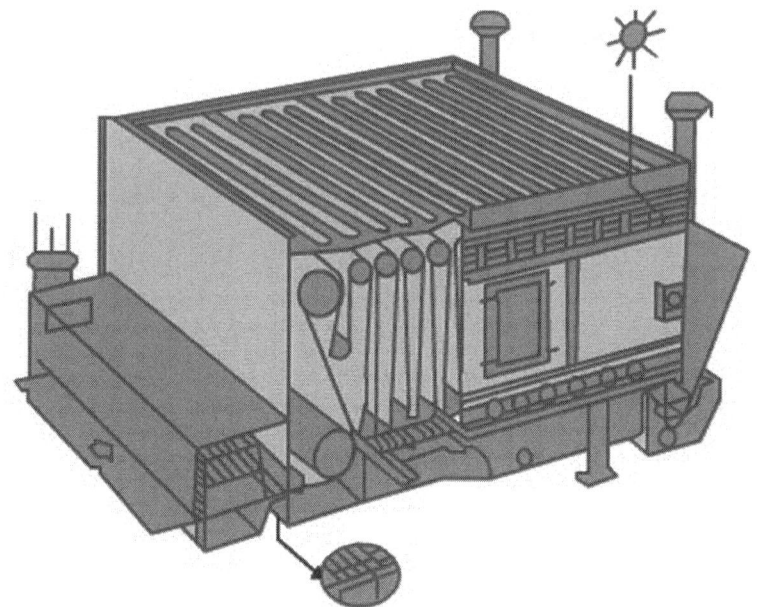

Figure 3.5   Steamer.

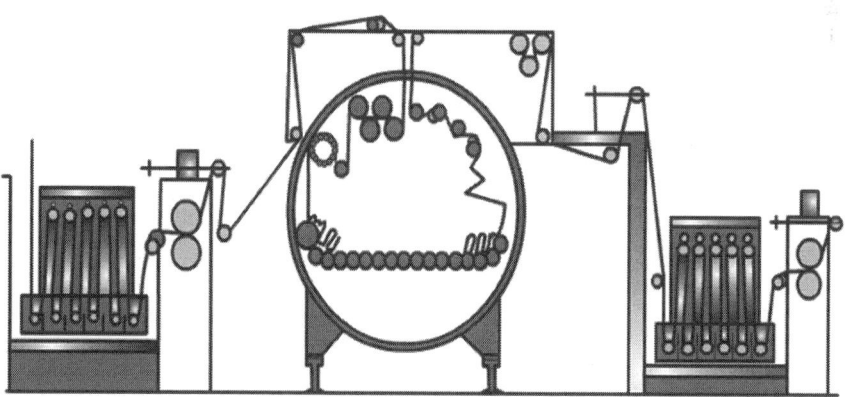

Figure 3.6   Mather and Platt continuous open-width pressure reaction chamber
            (vapour lock).

Continuous machines generally do not operate at high pressure. However, there is one high pressure reaction chamber that can operate at high pressures and hence can reduce reaction time significantly. This is known as vapour lock unit.

In vapour lock unit, which is essentially a high pressure steamer (Figs. 3.5 and 3.6), fabric entry and exit is through hydraulic sealing heads. These heads consists of polytetrafluoroethylene diaphragms where pressure is higher than the chamber pressure. The pressure in the unit can be 2 atm and temperature 134°C. The reaction time is only 90–120 s. The capacity of the machine is around 200 m which allows fabric speed of about 120 m/min.

3.4 Bleaching

Cotton is not pure white in colour; it has some natural colouring matter having slightly yellowish brown colour. The purpose of bleaching is to remove this colouring material and to confer a white appearance to the fibre. In addition to an increase in whiteness, bleaching results in an increase in absorbency, levelness of pretreatment, and complete removal of seed husks and trash.

In the case of the production of full white finished materials, the degree of whiteness is the main requirement. The amount of residual soil is also taken into consideration because of the possibility of later yellowing of the material.

In the case of pretreatment for dyeing, the degree of whiteness is not as important as the cleanliness of the material, especially the metal content. Similar demands refer to the production of medical articles. In this case, too, the metal content as well as the ash content are important factors.

If whiteness is of primary importance, it requires a relatively large amount of bleaching agent as well as a high operating temperature and a long dwell time. Accurate regulation of the bleaching bath is a further obligatory requirement. Where the destruction of trash, removal of seed husks and an increase in absorbency are prime necessities, a high degree of alkalinity is important. It is, however, not the alkali alone that is responsible for these effects. The levelness of pretreatment can only be guaranteed if cotton of the same or equal origin is processed in each bath.

Suitable pretreatment has to be undertaken to obtain the required uniformity. A pretreatment with acid and/or a chelating agent will even out varying quantities of catalytic metallic compounds.

Although there are different bleaching agents that can be used for bleaching cotton, hydrogen peroxide is, by far, the most commonly used bleaching agent today.

The mere formulation of the correct initial bath concentration is not sufficient to ensure a controlled bleaching process. There is a need to have regular checks of the bath composition during the operation. Such checks contribute to an economic bleaching operation, and also allow an early tracing of the defects and failures of the system. The important parameters for bleaching with hydrogen peroxide are as follows:

- Concentration of hydrogen peroxide
- Concentration of alkali
- pH
- Temperature
- Time
- Nature and quality of the goods
- Water hardness and other impurities
- Types and concentration of auxiliaries
- Desired bleaching effect
- Available equipment and stabilizer system employed

Most of these factors are interrelated, and all have a direct bearing on the production rate, the cost and the bleaching quality.

The concentration of bleaching liquor is considered in two ways: (a) that based on the weight of the goods and (b) that based on the weight of the solution. All other factors being equal, the concentration on the weight of the goods determines the final degree of whiteness.

In order to get adequate bleach there must be enough peroxide present from the start. The peroxide concentration based on the weight of the solution will determine the bleaching rate—the greater the solution concentration, the faster the bleaching. The alkalinity in the system is primarily responsible for producing the desired scour properties and maintaining a reasonably constant pH at the desired level throughout the bleaching cycle.

3.5 Washing

Washing is an important stage in all stages of value addition, and is more important in pretreatment. The aim of washing is to remove the added or natural impurities, residual auxiliaries and so forth. Since washing is done with water, the impurities must be either water-soluble or emulsifiable.

Modern washing practice aims to improve washing efficiency by utilizing minimum water to remove impurities; in conventional systems, it is about 4–6 L water per kilogram of goods.

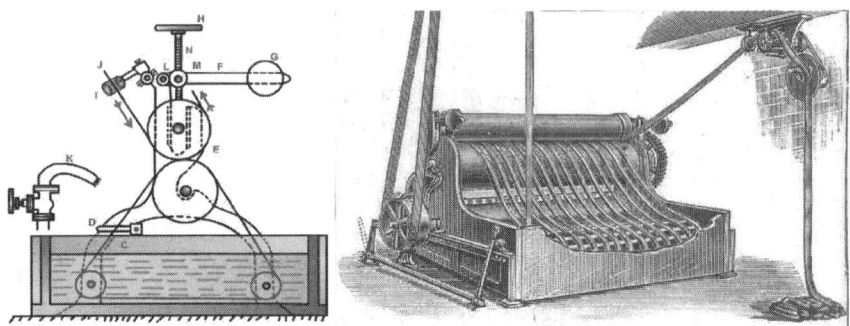

Figure 3.7   Tight rope washing machine.

Washing can be done either in rope form or in open width. There are three types of washing machines in rope form, namely (a) tight rope washing machine, (b) slack rope washing machine and (c) rotary drum washing machine. In open-width washing, the machines can be of (a) horizontal type or (b) vertical type.

3.5.1 Rope washing machine

3.5.1.1 Tight rope washing machine

This machine maintains high tension on the fabric during washing and hence is used for medium to heavy fabrics (Fig. 3.7).

The fabric in rope form passes through the nip of squeezing rolls in to the water trough and again into the nip. Thus, many loops of the fabrics are made which are separated by the pegs below the bowls that allow the rope to move in a spiral fashion. Fresh water is continuously fed to the tank. The rollers are heavy, with a wooden lower roller and the upper ones are made from highly compressed cotton. Squeezing is provided by levers and weights.

The main disadvantage of tight rope washing machine is fast contamination of the water in the box. Feeding fresh water and removing soiled water helps to some extent. However, the problem has been tackled to a large extent by placing a small trough below the squeeze rollers. Fresh water is fed at the exit end and the soiled squeezed water is collected in the smaller trough which is then directed to a drain. This reduces contamination of the water in the bigger trough considerably.

3.5.1.2 Slack rope washing machine

Slack rope washing machine is similar to the tight rope machine with slight difference (Fig. 3.8). The tight rope washing machine relies on squeezing

Figure 3.8 Slack rope washing machine.

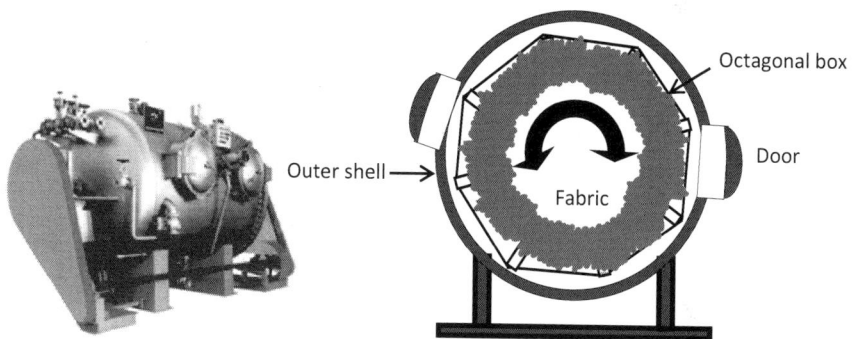

Figure 3.9 Drum washer.

action for liquor interchange. In slack rope washing machine low tension is exercise. It also consists of two hard wooden rollers with lower roller driven and the upper one friction driven. Beneath these rollers is situated a large wooden tank with slanting bottom.

The fabric enters the nip of the rollers, into the tank where it slides down the bottom of the tank. From the deeper section of the tank, the rope is led over a freely rotating roller, into the nip and finally to the exit. A small wooden box is situated just below the bowls to collect soiled water squeezed out from the bowls to minimize the contamination of the tank.

3.5.1.3 *Rotary drum washing machine*

In rotary drum washing machine (Fig. 3.9), the fabric is rotated inside a perforated drum along with washing liquors. The washing ensures bulkiness of the fabrics being washed. It is used for washing 100% synthetic fabrics and also for garments. We shall discuss this in detail in a separate chapter.

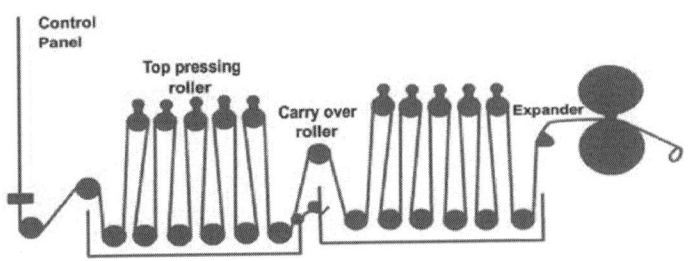

Figure 3.10 Continuous washing unit Aquatex.

3.5.2 Open-width continuous washing machine

Open-width continuous washing machines are used whenever the fabric is being processed on a continuous machine in open-width crease-free form. Generally, the liquor is stationary and the fabric movement through the liquor provides some mechanical agitation and liquor interchange.

Mostly, the washing machines are a series of washing boxes/chambers working on the principle of countercurrent flow of fresh washing liquor. The fabric and the liquor movement from one box to the other is designed to maximize interchange of liquor and minimize the contamination of subsequent chamber. The process is aided by frequent squeezing of fabric. The basic processes involved are squeezing, diluting and diffusing.

One of the machines working on this concept is Mather and Platt Aquatex unit (Fig. 3.10). The fabric passes between a set of top and bottom parallel rollers such that the bottom rollers are immersed in liquor. Each compartment or chamber may consist of five or six such top/bottom rollers. As the fabric moves from the bottom rollers upwards to the upper roller, some carry over liquor flows down back. Additionally, pressure rollers above the top rollers help to squeeze more contaminated liquor back. The bottom rollers are separated by metal plates to avoid mixing of liquor as fabric moves forward. Countercurrent flow of wash liquor is used which ensures that as the fabric moves forward it comes in contact with fresh liquor. As the liquor moves against the fabric it gets dirtier and finally exits from the fabric entry point.

The need to save space has led to double threading with additional rollers in vertical direction (refer Benninger Becoflex; Fig. 3.11). This increases fabric capacity and number of immersions at a given speed. However, threading becomes complicated and costs also go up.

The temperature of washing liquor affects washing efficiency at higher temperature, the viscosity of liquor goes down, the solubility of impurities

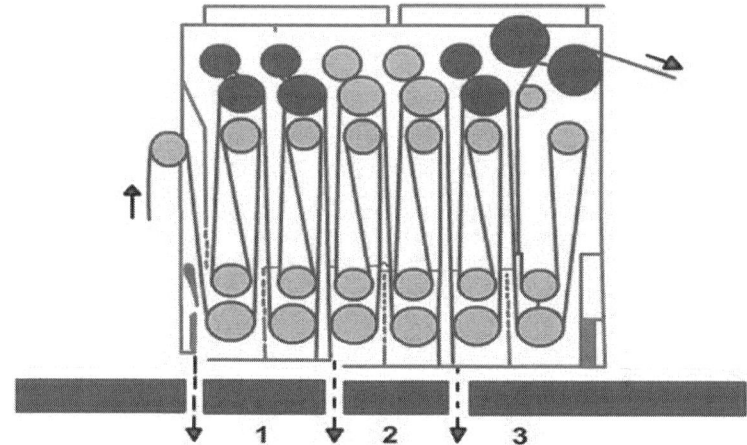

Figure 3.11   Benninger Becoflex washing compartment.

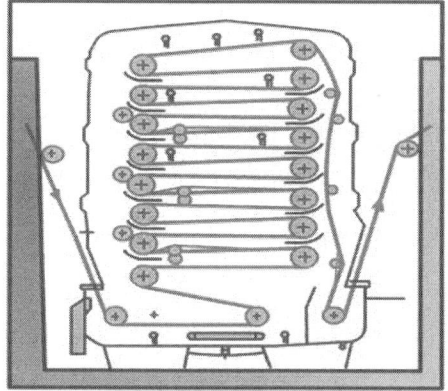

Figure 3.12   Horizontal washer.

goes up and the rate of diffusion also increases. In general, a temperature of 95°C seems to be optimum.

A continuous washing range often consists of a number of compartments or chambers. The fabric movement from one to the next is facilitated by a carryover roller or nip mangle to reduce liquor carryover.

3.5.3   Horizontal washing machines

In a horizontal washing machine (Figs. 3.12), the fabric moves horizontally between two sets of vertical rollers and gradually moves up. The fresh liquor is fed from the top and moves down. The water coming in contact of the fabric may either pass through it or spill over the selvedge.

The horizontal washing machine is more used to save the space, where the space is very costly. Otherwise, horizontal washing units are more popular as it is easy to handle.

The selection of washing machine should consider the availability of space, washing requirements, availability of water and the water quality.

Mercerizing of fabrics

4.1 Mercerization purpose and systems

Mercerization, discovered and developed by John Mercer, is a treatment of cotton with a strong alkaline solution which improves the lustre, strength, hand and other properties. It leads to a number of changes in fibre and fabric properties as follows:

a) A more circular fibre cross section

b) Increased lustre

c) Increased tensile strength

d) Increased apparent colour depth after dyeing

e) Improved dyeability of immature cotton and making dyeing more even

f) Increase in moisture regain of the fibres

g) Increase in water sorption

h) Improved dimensional stability

Mercerization can be done either in yarn stage or in fabric stage, either knitted or woven. We are discussing mercerization of woven fabrics in this chapter.

There are three systems of mercerization for cotton fabrics:

a) Chainless mercerization (roller mercerization)

b) Chain mercerization (stenter mercerization)

c) Batch-up mercerization

Mercerization can be classified depending on the conditions adapted as follows:

a) Water content: dry mercerization and wet mercerization

b) Tension: fixed length mercerization, tension mercerization and tensionless mercerization

c) Alkaline concentration: low-concentration alkaline mercerization, high-concentration alkaline mercerization and two-step mercerization

d) Temperature: ambient-temperature mercerization, high-temperature mercerization and low-temperature mercerization

e) Timing: grey mercerization, pre-dyeing mercerization and post-dyeing mercerization

f) Type of alkali used: caustic soda mercerization and liquid ammonia (LA) mercerization

g) Alkali pad: dry method and alkali pad-steam method

4.2 Chain mercerizing using caustic soda

In this method, a stenter is used to regulate the tension in the fabric while it is treated with strong alkali (Fig. 4.1). For the sake of understanding, Kyoto clip-type mercerizing range is taken as an example.

The Kyoto clip-type mercerizing range has an entrance section with a fabric reserve 'J', impregnation section with provision for 3 impregnations, stretching and stabilization with chain and clips, steam recuperation zone with 5 spraying and recuperating points, washing section with 9 wash boxes, each with 5 pairs of guide rollers, and drying section with 42 drying cylinders (Fig. 4.2). In the first 3 washing units, there are 15 motors to control the speed and stretch.

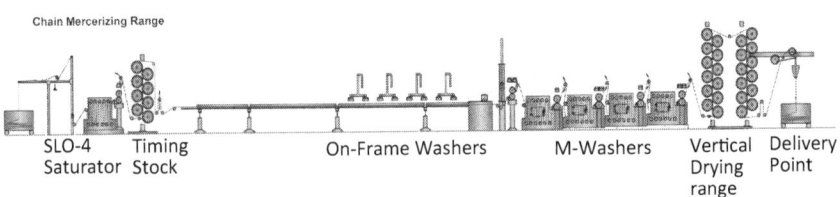

Figure 4.1   Chain mercerizing range.

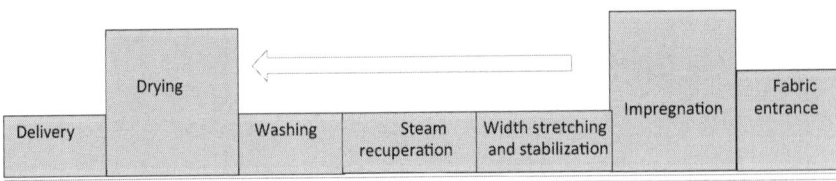

Figure 4.2   Basic parts of a chain mercerizing machine.

4.2.1   What chain mercerization should and should not do?

- **Should do**

  a)   Treating cotton fabrics in open width with caustic soda solution under predetermined tension and width control

  b)   Recovering excess caustic soda by steam recuperation

  c)   Washing and neutralizing the fabric after the treatment is over

  d)   Drying the fabric using vertical drying range (VDR)

  e)   Improving strength, lustre, dyeability and dimensional stability of fabric

- **Should not do**

  a)   Should not damage cotton fabric by long exposure to alkali.

  b)   Fabric should not get tendered by the action of acid while neutralizing after caustic treatment.

  c)   Treated water with chemicals should not be discharged without treating and neutralizing.

  d)   Fabric should not shrink unevenly.

  e)   Mercerization should not cause variations in fabric width.

4.2.2   Quality of mercerizing

There are four different methods of measuring the quality of mercerizing. They are explained in brief as follows:

a)   Quality of mercerizing is generally measured by barium activity number. Mercerized sample absorbs barium hydroxide (alkali) to a greater degree than sodium hydroxide. The absorption rate of barium hydroxide is easier to estimate and is normally practiced in mills. The ratio of uptake of this reagent has been referred as barium activity number.

Two grams of mercerized and unmercerized samples are placed separately in two conical flasks containing 30 mL of N/4 barium hydroxide and kept for 2 h or preferably overnight. A total of 10 mL of clear solution is withdrawn and titrated against N/10 HCl using phenolphthalein as indicator. A blank titration is also carried out on the measured barium hydroxide solution using methyl red as an indicator.

$$\text{Barium activity number} = (b - s) \times 100/(b - u)$$

where

b=mL required for blank test

s=mL required for mercerized cotton

u=mL required for unmercerized cotton

For exact estimation, correction should be made for moisture regain of the sample.

Barium activity number of unmercerized cotton is considered as 100 and semi-mercerized cotton ranges from 115 to 130 and for completely mercerized cotton, it is about 155.

b) Pulfrich photometer, the Gorez Glarimeter comparative glass method and microscopic examination of cross section of fibre are qualitative methods for degree of mercerization. This is not practiced in the textile mills.

c) X-ray photograph of native cellulose (unmercerized) reveals the presence of two arcs close together and inside the prominent 002 arc. Further analysis can be made to measure the level of mercerization but normally not practiced in textile mills.

d) Determination of infrared crystallinity at different wavelengths is another method of measuring mercerization. This method is also not practiced in the mills.

4.2.3 General activities in fabric mercerizing using caustic soda

Following are the general activities in a fabric mercerizing section:

a) Getting requirement of fabrics to be mercerized against various orders and their priority

b) Taking fabric off from its plaited condition or from a batch through a set of automatic cloth guiders, guide rollers and tension bars

c) Impregnation of fabric with the mercerizing lye in suitably dimensioned padding mangles

d) Width stretching and stabilizing using a set of rollers and stenter clips

e) Weak lye of caustic soda is sprayed on the fabric while it is being stretched in the stenter frame

f) After squeezing the fabric at the end of the stenter through the squeezing mangle, the fabric is introduced into the lye recuperation section

g) In the recuperation washing compartment with top and bottom rollers, the steam-heated water recovers the major quantity of lye

h) The remaining portion of the lye is washed out of the cloth in the washing section having the requisite number of washing compartments

i) Fabric neutralizing using acetic acid

j) Drying using VDR

k) Delivering the mercerized fabrics in batches or in containers as practiced

4.2.4 Typical operating procedure of a chain mercerizing machine using caustic soda

Following are the steps in a typical operating procedure; however, each mill has to write their own procedure depending on the type of machine they have, the concentration of chemicals they are using, the type of fabric they are mercerizing, the normal grams per square metre (GSM) of the fabrics they mercerize, the degree of mercerization required and so on.

i. Take the programme from the shift officer and prepare the batch in the sequence of the designs aligning the production requirements.

ii. Drain out all the previous chemicals and wash with water jet using a hosepipe.

iii. Clean the machine with water and then wipe with a clean cloth before starting the work.

iv. Drain the water from compressor pipes.

v. Drain the steam condensates from steam pipes.

vi. Open the compressed air valve.

vii. Check the condition of the washer liquor; if it is coloured, replace the water.

viii. Check the guide rollers for chemical depositions and wipe with a wet cloth with clean water and then with a dry clean cloth.

ix. Fill the wash water tank manually.

x. Open the steam valve manually of VDR and washer.

xi. Check the stock level of caustic by means of a dipstick at the beginning of the shift, while starting a new programme and at the end of each programme and inform the supervisor.

xii. Prepare caustic solution for mercerizing by taking guidance from supervisor depending on the type of material being mercerized. If used caustic soda solution is added, check its concentration and develop the solution to get the required twaddle. Generally, the chemical tank is provided underground and its capacity is 2000 L. This may vary from mill to mill.

xiii. Start the circulation of caustic from pit tank to impregnation tank and back to pit tank in a cycle.

xiv. Start the rotary filter pump for caustic after cleaning it.

xv. The chain and clips are to be aligned properly ensured proper gripping of fabric.

xvi. Adjust the width of the clip unit in line with the fabric width throughout the clip unit (Fig. 4.3).

xvii. The width of the cloth should be continuously monitored while the machine is running.

xviii. In case any damaged material is observed, the speed of the machine is to be reduced and the defect to be corrected while the machine is working. The machine should not be stopped fully as it results in an uneven reaction.

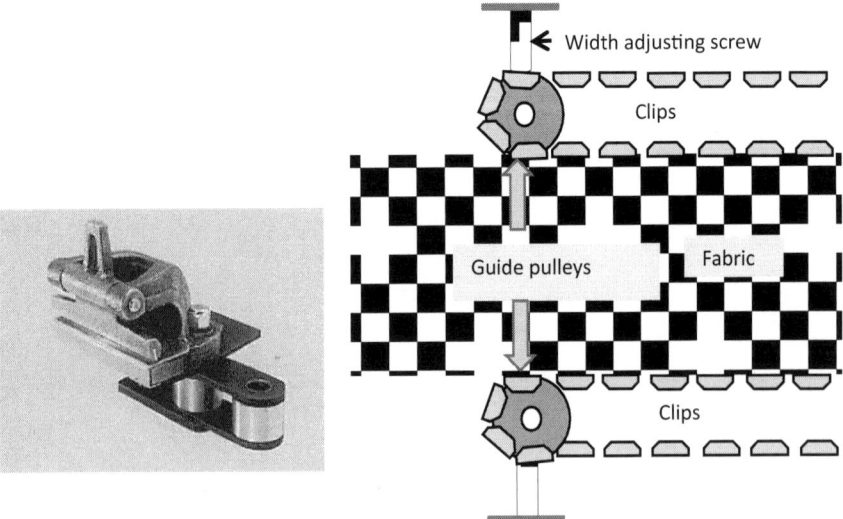

Figure 4.3 Adjusting width of fabric using clips.

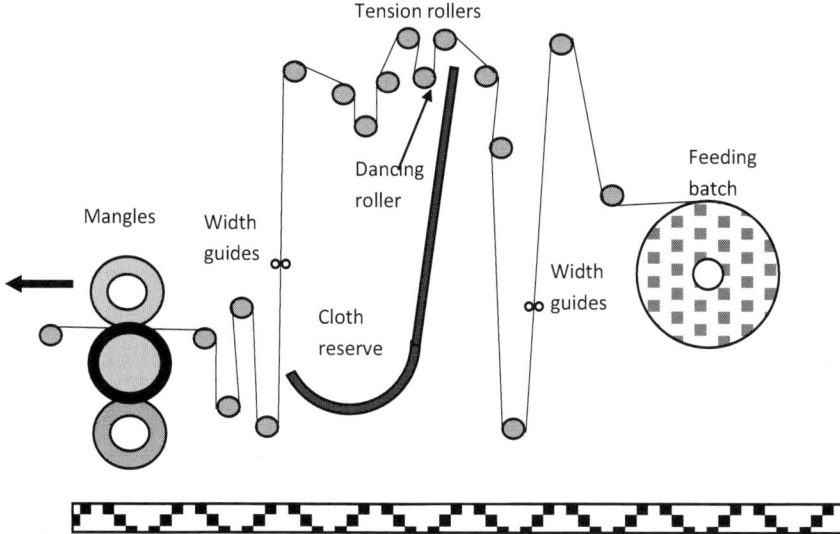

Figure 4.4   The 'J' scray.

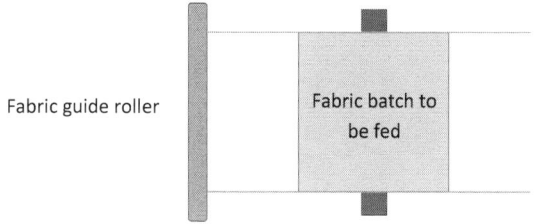

Figure 4.5   Alignment of the batch exactly at the centre.

xix. The 'J' scray should be set to keep the material in reserve for stitching of the batch (Fig. 4.4).

xx. The feeding batch should be exactly in the centre (Fig. 4.5).

xxi. By using the pair of width guides, the fabric is kept exactly at the centre.

xxii. Check and ensure that the width guide pairs (Fig. 4.6) are gripping the fabric properly and are not creating any impression because of damaged or rough surfaces.

xxiii. The speed is set at 65 ± 5 m/min for fabrics going for piece dyeing and 75 ± 5 m/min for yarn-dyed fabrics. The maximum speed can be taken at 80 m/min for large batches with dyed yarns.

xxiv. Before starting the programme, the machine shall run at slow speed till the leader cloth is completely out of the machine while keeping

Figure 4.6   Width guides.

Figure 4.7   Impregnation section.

only one shower working for lubricating the clip chain. Then, the trough is taken up.

xxv. Generally, three pairs of mangles in the impregnation zone (Fig. 4.7) are provided so that the machine can handle different types of fabrics with varying densities (Fig. 4.8).

xxvi. Set the air pressure at mangles at around at 0.3 MPa.

xxvii. Check the tensions in different drives of the machine and ensure that all dancing rollers are free while working.

xxviii. In case of mercerizing yarn-dyed fabrics, the caustic solution is to be chilled to 18°C temperature, whereas for fabric going for piece dyeing, room temperature is maintained (Fig. 4.9).

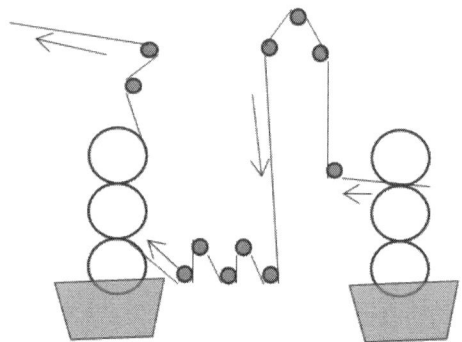

Figure 4.8 Passage of fabrics to the first impregnation mangle.

Figure 4.9 View of chain and clip controlling the width.

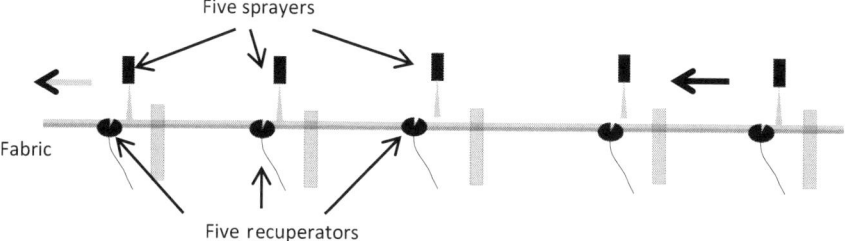

Figure 4.10 Sprayers and recuperators.

xxix. There are three pumps in each recuperating unit, one for hot-water spray, one for recuperation and third for pumping the collected water (minimum 15°Tw) to storage tank for recirculation, which has 30-t capacity (Fig. 4.10).

xxx. The speed and tension are controlled by monitoring the individual motors at metters in the first three washing units (Fig. 4.11).

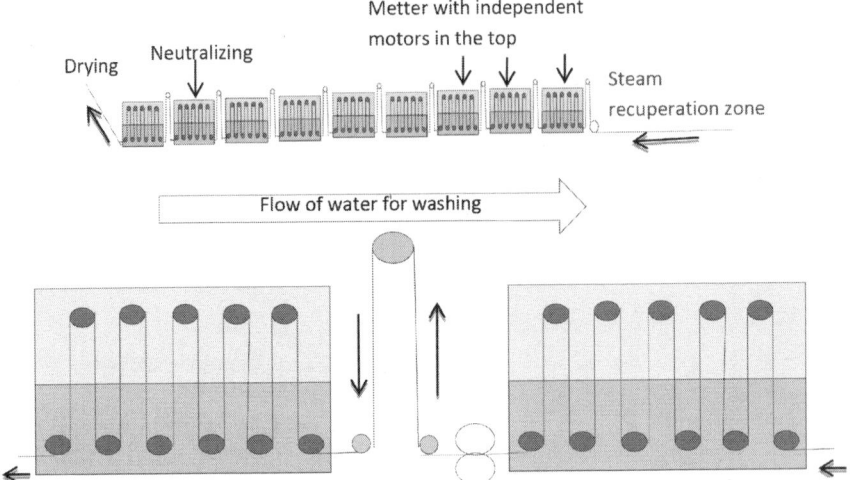

Figure 4.11   Washing units.

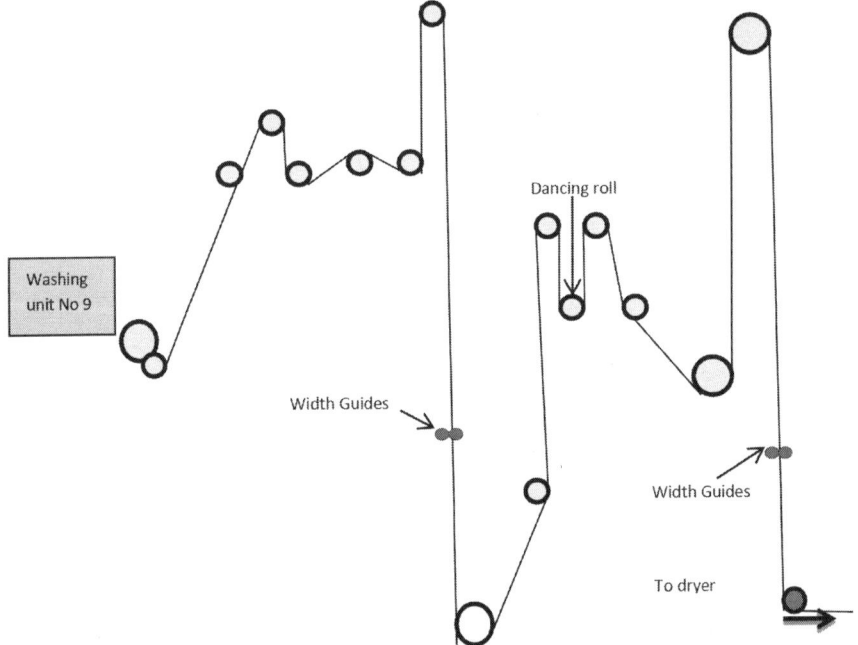

Figure 4.12   Passage between last washing units and dryer.

xxxi. The eighth washing unit (the fifth after metters) is meant for neutralizing. Acetic acid is used for neutralizing. The pH is maintained slightly acidic by setting in the control panel.

xxxii. The temperature in the panel is set to control the temperature of water at the washers.

xxxiii. Adjust the width and tension after washing before drying. There are two pairs of width guides provided between washing and drying sections (Fig. 4.12).

xxxiv. The cloth is then taken over the drying cylinders without any fold or crease as shown in the diagram. There are 42 drying cylinders arranged in 3 groups each of 14 (Fig. 4.13). Steam pressure of 0.16 MPa is kept for the first two groups and 0.14 MPa for the third group. The exhaust fan is provided over the VDR drying range to remove the hot moist air continuously. The moisture content in the fabric is monitored after drying and not allowed to become over dry.

xxxv. The details of the batch mercerized such as the batch number, the design numbers, the quality numbers and total weight of chemicals

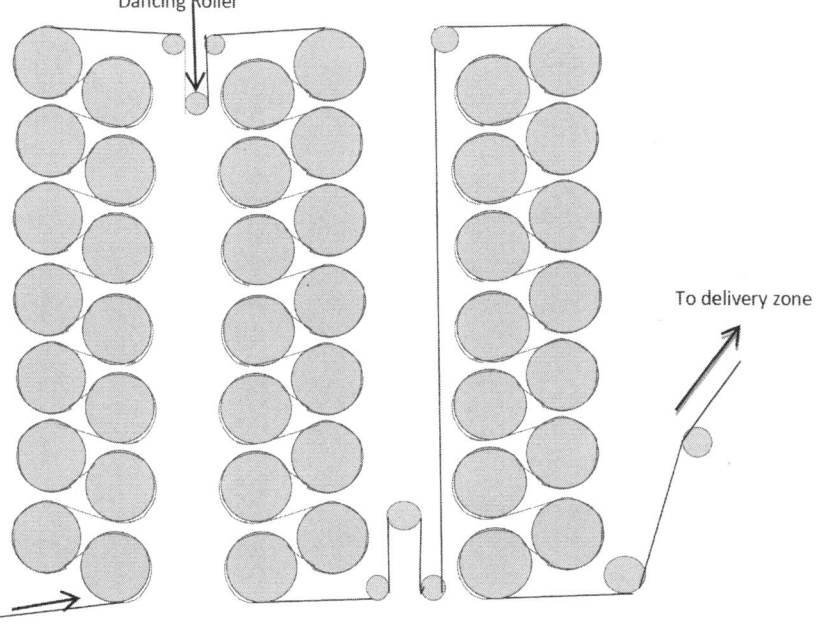

Figure 4.13   Forty-two drying cylinders.

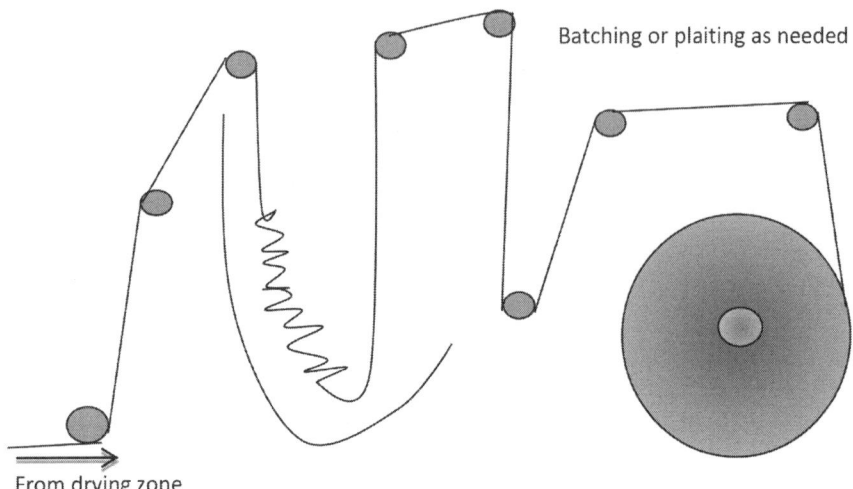

Batching or plaiting as needed

From drying zone

Figure 4.14   Delivery zone.

consumed, total metres of fabric mercerized, the stoppages and quality problems faced, if any, are entered in the register (Fig. 4.14).

xxxvi. At the end of the programme, leader fabric is stitched after checking that the fabric is not worn out in between. It is necessary to ensure that clips, chain and width sensors are cleaned before a leader cloth is entered.

xxxvii. The impregnation bath is taken down to pit tank after completing the programme.

4.2.5   Precautions to be taken for mercerizing using caustic soda

a)   Wearing safety gadgets such as gloves, aprons and gum shoes while handling chemicals

b)   Wearing a mask when the chemicals are active reacting with cellulose

c)   Not storing of alkali and acid side by side

d)   Not to keep the chemical containers open to air

4.2.6   Control points and checkpoints in chain mercerizing using caustic soda

It is essential to have clarity on the points to be controlled to achieve the targets and those to be checked to ensure the mercerizing process in control. These points need to be reviewed from time to time and modified to suit the requirements of individual companies and their targets. Each mill should prepare its own 'control points and checkpoints' and display them in the work area so that the people refer and follow. Following is just an example.

- **Control points**
  a) Deciding and selection of process parameters, namely machine setting, speed, mercerizing time and temperature, stretch, caustic lye concentration, washing and neutralizing sequences
  b) Deciding the acceptance criteria for degree of mercerization
  c) Deciding on chemicals and their quality to be procured
  d) Employing trained and qualified employees
  e) Evolving production norms
  f) Evolving norms for consumption of chemicals, water and steam
  g) Checkpoints

- **Material related**
  a) The fabrics received against the design number and lot numbers specified in the programme
  b) The total quantity (weight) of fabric received for the processing
  c) Checking the chemicals and auxiliaries and ensuring for meeting the acceptance criteria.
  d) Quantity of chemicals received and to be used in the run.

- **Mercerizing machine related**
  a) Condition of the mangles and trough
  b) Alignment of the clips in stenter
  c) Condition of clips in stenter
  d) Proper operation of the steam valves
  e) Condition of the width controllers
  f) Checking whether the overall condition of the machine is good and operates as per the programme set

- **Setting related**
  a) The concentration of caustic lye and the requirement
  b) The temperature of the caustic lye maintained and requirement
  c) The number of mangles selected matching with the type of fabric

    d)   Maintaining of the temperature of immersing zone as scheduled

    e)   The stretch maintained and planned

    f)   Proper working of the width-controlling unit

    g)   Working of the steam recuperation as needed

    h)   The concentration of acetic acid taken for neutralization and the requirement

    i)   The steam pressure set in the drying cylinders and the requirement

- **Performance related**

    a)   Following of the neutralizing sequence as per programme

    b)   Maintaining pH of neutralizing liquor as specified

    c)   Barium activity number of the mercerized material and the norms

    d)   Achievement of the production as anticipated

    e)   Uniformity of the width and its consistency

    f)   The consumption of caustic soda against the planned consumption

- **Documentation related**

    a)   The fabric details and the quantity received

    b)   Quantity to be mercerized and actually mercerized in each variety

    c)   Quantity of caustic soda consumed

- **Work practice related**

    a)   Following safety requirements all the time

    b)   Cleaning of the machine after completing each run

    c)   Proper storing of chemicals

    d)   Housekeeping around the machine

- **Logbook related**

    a)   The quantity (number of metres) mercerized in each design

    b)   Stock of fabric in batches ready to be mercerized and its details

    c)   Quantity of chemicals consumed

    d)   Problems faced in the shift

    e)   Special instructions for the next shift

- **Management information system related**

  a) Design number and details

  b) Quantity of fabric (number of metres) received for mercerizing

  c) Quantity mercerized

  d) Quantity approved as fresh quality

  e) Quantity rejected as seconds

  f) Consumption of caustic soda

  g) Consumption of acid

- **General**

  a) Whether the men employed are adequately trained

  b) Whether the consumption of chemicals, steam and water is as per plan

  c) Quality and quantity of water discharged for effluent treatment plant

4.2.7 Dos and don'ts for chain mercerizing using caustic soda

In any process, it is necessary to understand clearly what are supposed to be done without fail and what should not be done at any cost. Some examples are given below.

Dos

a) Check the design number and lot numbers of the fabric received for mercerizing and verify whether it is as per your plan.

b) Understand the process before deciding the process sequence or cycle.

c) Check the purity of caustic soda and the quality of water before preparing caustic solution.

d) Stick to the process parameters rather than trying to do some hasty things to increase production.

e) Accumulate the fabrics to be mercerized and prepare a big batch (of identical width and GSM) so that machine can run continuously.

f) Check the concentration of caustic solution before feeding it to mercerizing machine.

g) Check the temperature of the immersing chamber.

h) Check the squeezing rollers and ensure uniformity in surface and hardness as needed.

i) Verify the process sheet before setting the process parameters on the machine.

j) Continuously monitor the width of the cloth while the machine is running.

k) Periodically ensure that the condensates are drained from steam pipes and water is drained from compressed air pipes.

l) Check the condition of leader cloth throughout the machine before stitching fresh cloth for operation and while running out any material.

m) Decrease the air pressure in case of stoppage of the machine even for a small time.

n) Clean the filter with water spray at the end of the programme.

o) Clean the machine while the leading cloth is running at slow speed.

Don'ts

a) Do not accept a batch if variations between different fabrics joined are very high.

b) Do not allow any fold to remain in the fabric while feeding for mercerization.

c) Do not stop the machine for correcting fabric but reduce the speed and correct.

d) Do not allow the temperature to increase while mercerizing.

e) Do not try to get fast results by increasing the concentration of alkali or acid.

f) Do not stretch the fabric beyond limits.

g) Do not wash off the drain water excepting for yarn-dyed fabrics; collect them in the storage tank as it can be reused.

h) Do not recirculate the wash liquor below the chain if yarn-dyed materials have worked in the last programme.

4.3   Liquid ammonia (LA) mercerizing of cellulose fibres

'LA finishing' or 'LA mercerizing' refers to the process that revives the cotton through the expansion of LA at an ultra-low temperature inside the fibre. When the cotton fibre is treated at $-33°C$ LA, ammonia at ultra-low temperature will permeate immediately into the crystallographic structure of the fibre. Stress will be released through interior expansion, which makes the fibre cavity round and smooth and rearranges the molecular structure; thus, the crystallographic structure becomes slack and stable. This physical change makes the surface of the entire fabric smooth and bright, with solid and soft feel, so elasticity and wash-and-wear are achieved.

4.3.1   Benefits of LA mercerizing

There are many benefits of LA mercerizing. The superb appearance, feel and brilliancy of dyed shades make the buyer select the ammonia-mercerized garment/fabric rather than the regular caustic-mercerized one. Other benefits of LA mercerizing are low shrinkage post washing, increase in wrinkle resistance, increase in fibre elasticity, softer to touch and brighter, enhanced tensile strength, dimensional stability, resistance to abrasion, dyeing uniformity, dyestuff affinity, colour solidity and good wash-and-wear properties.

The machine has a number of sections such as fabric entrance section, impregnation section, and Palmer section, steaming zone, the washing section, drying section, cooling cylinder and delivery section (Figs. 4.15–4.23). Let us take an example of Kyoto LA mercerizing range.

The Kyoto LA mercerizing range has an entrance section with a fabric impregnation section, stabilization with washing section with 3 wash boxes and drying section with 18 drying cylinders, and at last, there are 2 cooling cylinders.

Figure 4.15   Liquid ammonia (LA) mercerizing continuous processing machine.

Figure 4.16   Mycom screw compressor for ammonia unit.

Figure 4.17   Kyoto LA mercerizing machine.

Figure 4.18   Preheater.

Figure 4.19   Air cooler.

Figure 4.20   Washer—neutralizer—washer.

Figure 4.21   Padding, palmer unit and curing unit.

Figure 4.22 Curing unit.

Figure 4.23 Drying unit.

4.3.2 Standard operating procedure for LA mercerizing

Following steps are general guidelines for operating an LA mercerizing machine:

1. Clean the machine with air and water and wipe with a clean cloth before starting.

2. Take the programme from the shift officer and prepare the batch in the sequence of the designs aligning the production requirements.

3. Drain the water from compressor pipes and steam condensates from steam pipes.

4. Open the compressed air valve.

5. Check the condition of the washer liquor; if it is coloured, replace the water.

6. Check the guide rollers for scratch and wipe with clean water and cloth.

7. Fill the wash water tank manually.

8. Open the steam valve manually of dryer and washer.

9. Check the stock level of LA by means of a metre gauge at the beginning of the shift while starting a new programme and at the end of each programme and inform the supervisor.

10. Keep the chemical tank for ammonia (to be provided separately outside in open ground of 10 000-kg capacity) clean and cool.

11. Start the circulation of LA from the tank to impregnation tank and at end sending the balance LA back to the tank.

12. Monitor continuously the width of the cloth while the machine is running.

13. In case any damaged material is observed, reduce the speed of the machine and correct the defect while the machine is working. The machine should not be stopped fully as it results in an uneven reaction.

14. The speed is to be set at 20 ± 5 m/min for fabrics going for piece dyeing and 40 ± 5 m/min for yarn-dyed fabrics. The maximum speed can be taken at 45 m/min for large batches with dyed yarns.

15. Before starting the programme, run the machine at low speed till the leader cloth is completely out of the machine.

16. Set the air pressure at mangles at 0.3 MPa.

17. Check the tensions in different drives of the machine and ensure that all dancing rollers are free while working.

18. In case of LA mercerizing yarn-dyed fabrics, chill the LA solution to $-13°C$.

19. Set the temperature in the panel to control the temperature of water at the washers.

20. Enter the details of the batch LA mercerized such as the batch number, the design numbers, the quality numbers, total metres of fabric mercerized, the stoppages and quality problems faced, if any, in the register.

4.3.3 Dos and Don'ts for LA mercerizing

Dos

a) Accumulate the fabrics to be LA mercerized and prepare a big batch so that machine can run continuously.

b) Monitor continuously the width of the cloth while the machine is running.

c)   Check the surface of all rubber rollers and ensure that they are in order.

d)   Periodically check the concentration of liquor ammonia and availability.

e)   Periodically ensure that the condensates are drained from steam pipes and water is drained from compressed air pipes.

f)   Check the condition of leader cloth throughout the machine before stitching fresh cloth for operation, and while running out any material.

g)   Decrease the air pressure in case of stoppage of the machine even for a small time.

h)   Clean the filter with water spray at the end of the programme.

i)   Clean the machine while the leading cloth is running at slow speed.

Don'ts

a)   Do not stop the machine for correcting fabric but reduce the speed and correct.

b)   Do not accept a batch if variations between different fabrics joined are very high.

c)   Do not allow any fold to remain in the fabric while feeding the machine.

d)   Do not allow the temperature of the liquor ammonia to increase.

4.3.4   Normal problems in fabric mercerizing

The normal problems in mercerizing a fabric are local mercerization leading to patchy dyeing, low barium activity number and width variations. They have been discussed in detail in a separate chapter.

4.4   Responsibilities and authorities of supervisor in mercerizing

4.4.1   Responsibility

The supervisor in fabric mercerizing section has following responsibilities:

a)   Ensuring that the required designs and lots of fabric are mercerized with least rejections irrespective of various problems being faced in the department

b)   Ensuring that the mercerized fabric is wound on to batch and supplied to the user in time

c)   Monitoring the consumption of chemicals

d)   Recovering the caustic/LA as much as possible for reusing

e)   Ensuring all safety precautions while handling chemicals and running machines

4.4.2   Authorities

The supervisor in the mercerizing section must be authorized as follows:

a)   To reject the batch if the quality difference between fabrics is more in the batch prepared

b)   To reject the caustic lye and the acids if they are found contaminated or weak

c)   To question the operators when the quality or productivity is not as expected

4.5   Mercerizing of knitted fabrics in tubes

The regular fabric mercerizing machines are not suitable for mercerizing knitted fabrics because of the following reasons:

•   Difference in constructional features of the fabric

•   Sensitivity towards tension and stretch applications

•   Unsuitability of woven cloth expanders to knitted fabrics

Recently, machines are developed for mercerizing knitted fabrics in tubular form. Knit Merceriser model TKM-2010 (Fig. 4.24) of Swastik Machineries is an example which is suitable for mercerizing knitted cotton tubular fabrics of up to 1.2-m width.

The important features of a Tubular Knit Merceriser are special entrance scaffolding with guide rollers suitable for knit fabrics, impregnation compartment, ring expanders, dimensional stability section, fabric-straightening section and washing section.

A suitable feeding device with a stretcher at entry, a caustic treatment section with wet fabric stretchers—specially designed to feed fabric in a creaseless manner, followed by a dwell zone. A pneumatic squeezer is provided at the end of dwell zone. Squeeze rollers are made from special rubber to get a better squeeze.

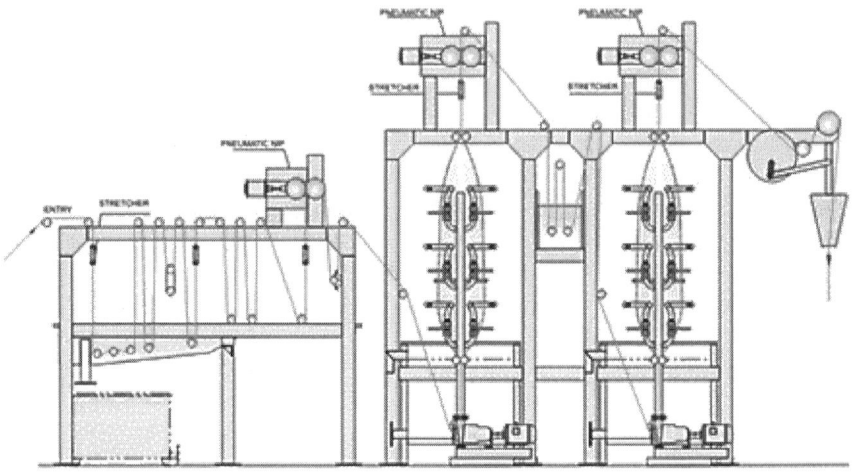

Figure 4.24   Knit Merceriser—TKM 2010.

A highly efficient washing system with tower washers, with rows of water spraying devices, a large water collecting tank, with a filter and pump, is provided for recirculation.

A suitable overflow system is provided to collect the residual caustic from the tank, which is taken to caustic recovery system. The number of washers depends upon the production requirement. A small neutralizing tank is provided before the last tank.

Padding process

5.1 Padding—purpose and systems

Getting a uniform shade when a large quantity of fabric is dyed in a single shade is a challenge. The uniform application of dye liquor to the fabric is most critical part of a dyeing. Directly passing the fabric through dye liquor cannot give uniform shade. Therefore, padding mangle is used which impregnates dye uniformly on the fabric first, which is then taken for treatment later. Satisfactory performance of padding mangle is absolutely essential for the success of such processes.

Padding mangle is a familiar machine used in dyeing, pretreatment or finishing. It applies chemicals or even dyes to an open-width fabric in a uniform manner. Padding mangle is not a complete dyeing machine; other machines are necessary for the fixation of dyes in the padded fabric. Hence, padding mangle is only one part of a process in which chemicals or dyes are transferred to the fabric uniformly. The reaction or dye fixation is carried out separately by a suitable combination of time and temperature in another machine (batching device, steamer, curing chamber, etc.)

This machine is used in continuous and semi-continuous methods of chemical/dye application to fabrics. It is suitable for the application of low substantivity dyes (dyes having low affinity to the fibres in use) or chemicals to fabrics. If the dyes used in the padding mangle have high affinity to the fibres, their concentration in the trough decreases with time, resulting in tailing effect.

The mangle consists of two cylindrical rubber bowls with a stainless steel mandrel. The diameter of bowls is equal with 55–70° shore hardness. The bowls should be perfect cylinders with a smooth surface. The size is generally 170–200 cm in length and 30–40 cm in diameter. Generally, the lower bowl is fixed and is driven by an electric motor. The top bowl is mounted on arms pivoted at the side in such a way that there is a gap (~2 cm) between the two bowls when the machine is not in operation. The upper bowl moves by contact friction generated by the pneumatic pressure. For operation, pneumatic pressure is applied on the top bowl which may be as high as 50 kg/cm (10 t for a 2-m bowl).

The padding operation consists of two steps. The fabric is first immersed in the dye liquor to achieve thorough impregnation, and then the fabric is passed between two rollers to squeeze out air and to force dye liquor inside the fabric. The excess liquor is sent back along the fabric. The first step is known as 'dip', while the contact between the squeeze rollers as well as passing between the rollers is known as 'nip'.

In the dip process, the fabric to be padded is passed through a padding bowl or trough which is situated below and in front of the mangle. The bowl is filled with padding liquor. As the fabric passes through the trough guided by guide rolls, it gets saturated by the liquor. Since the passage time in the liquor is very small, the fabric should be well prepared and be highly absorbent. In the next step, that is, nip process, the fabric passes through the bowls in such a way that the squeezing pressure is applied on the saturated fabric. When the bowls under pressure move in opposite directions, the fabric is transported forward. The squeezing action of the bowls expels the air in the interstices of the fabric, which, in turn, is replaced with padding liquor. The liquor is distributed over the entire fabric surface uniformly and the excess squeezed liquor flows back to the trough.

Padding may be done by using either two bowls or three bowls. The nip rollers (often called bowls) are the key to successful pad dyeing. In general, two-bowl nips are preferred for lightweight or standard fabrics running at moderate speeds, whereas three-bowl arrangements are intended for heavier or more densely woven qualities that may be more difficult to wet out and thus require a double-dip and double-nip treatment.

Since the liquor is continuously taken out of the fabric, it needs to be replenished continuously. The level in the trough is maintained by overflow. Sometimes, due to tight construction of some fabrics, application of liquor may not be uniform in two-bowl padding mangles which operate in one-dip–one-nip mode (Fig. 5.1). Such fabrics may be handled by three-bowl padding mangles by two-dip–two-nip sequences (Fig. 5.2).

For dyeing most fabric at 50 m/min, a single dip and single nip is sufficient, whereas for dyeing heavy fabrics or dyeing at high speeds of 120 m/min, double dip and double nip along with double squeezing using a three-bowl padding mangle is preferred.

The liquor retained in the fabric after padding is expressed by weight as % of weight on the dry fabric. It is referred as % pickup or % expression.

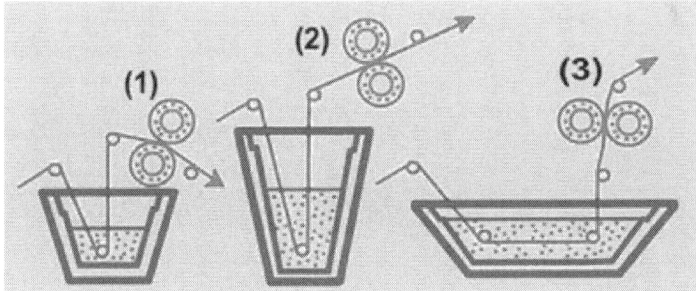

Figure 5.1 Two-bowl padding mangles with varying troughs and bowl arrangements.

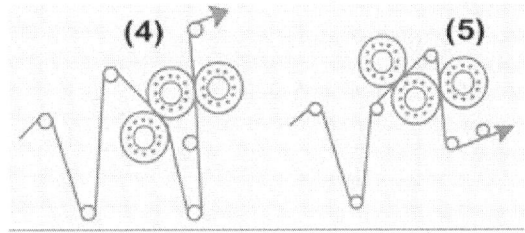

Figure 5.2 Three-bowl padding mangle.

Generally, getting expression less than 60% is difficult in padding mangle. Eighty per cent expression is normal. An increased pressure at the nip results in lower per cent expression, but at the same time, the penetration inside the fabric is better.

Mangle width for textile application may be as high as 4–5 m, but 1.7–2 m is more common. Most of the padding bowl mangles have a steel mandrel covered with about 15-mm deep hard rubber with a comparatively soft rubber surface having 55–70° shore hardness.

In old machines, one roller used to be of rubber covered and the other stainless steel. As it was giving face-to-back difference in the fabric, the present-day machines have same hardness on both the rollers.

The mangles should be equal in diameter, usually 30–40 cm. Larger diameter rollers are used for high-speed machines. For padding of dyes, minimum 1-cm contact distance is desirable.

The dye trough or the pad trough is usually a deep U-shaped vessel with a single roller attached to the base of a displacement block that leaves a narrow passage to accommodate the moving fabric.

5.2   What padding should and should not do?

5.2.1   Should do

a)   Should apply colours and chemicals uniformly on the fabric intended for dyeing.

b)   Should ensure uniform impregnation of colours and chemicals inside the fabric.

5.2.2   Should not do

a)   Padding should not leave colours and chemicals to remain as superfluous on the fabric surface.

b)   Padding should not result in uneven impregnation of colours and chemicals in the fabric.

5.3   Different padding sequences

5.3.1   Pad-steam method for scouring

This technique is mainly used for scouring woven fabric and dyeing in a continuous process. It is particularly suitable for the application of direct, vat, sulphur and reactive dyestuffs. It includes the following steps:

- Impregnation by padding

- Steaming (at about 100°C)

- Additional impregnation of the fabric with developing agents (e.g. reducing agents in vat or sulphur dyeing)

- Washing and rinsing

5.3.2   Pad-dry process

In this continuous dyeing process, the fabric is padded with the dye solution with 60–80°C and then fabric is passed through the drying chamber. Then, the fabric is padded with chemical (NaOH + salt). Then, the fabric is passed through a chamber at 102°C for 30 s. Then, fabric is washed in open-width washer.

5.3.3   Pad-batch process

This is used for semi-continuous dyeing process (Fig. 5.3). The fabric is first impregnated with the dye liquor in a padding machine. Then, it is subjected to batch-wise treatment in a jigger. It could also be stored with a slow rotation for many hours. In the pad batch, this treatment is done at room temperature.

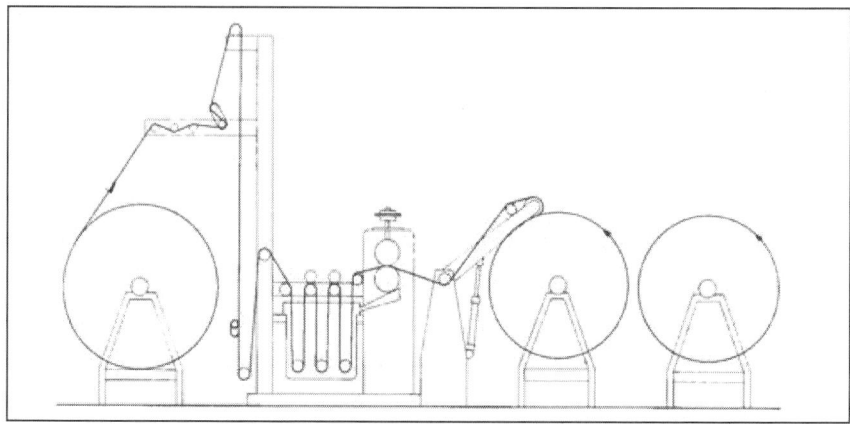

Figure 5.3   Pad-batch process.

5.3.4 Pad-roll process

In the pad batch, the treatment is done at room temperature, while in pad roll, it is done at increased temperature by employing a heating chamber. This helps in fixation of the dyes on to the fibre. After this fixation process, the material in full width is thoroughly cleansed and rinsed in continuous washing machines.

5.3.5 Pad-steam process

Pad steam is a process of continuous dyeing in which the fabric in open width is padded with dyestuff and is then steamed (Fig. 5.4). It is an ideal machine for reactive dyeing. Light, pale and medium shades can be dyed in this machine.

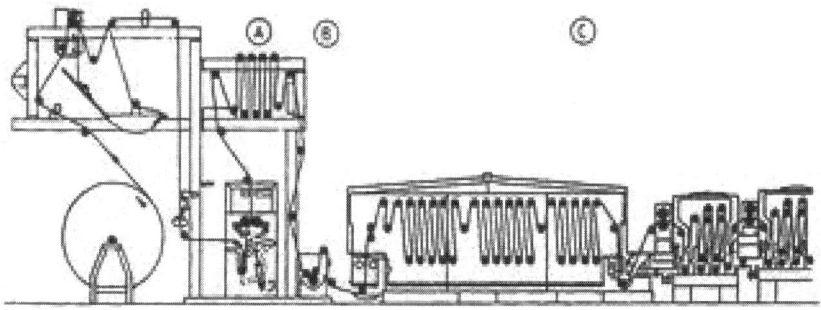

A) Padding of liquor (Bicoflex foulard) and air passage
B) Kiss padding of chemicals
C) Steaming and aftertreatment

Figure 5.4   Pad-steam process.

Continuous roller steamer is used for diffusion of reactive, vat, sulphur and direct dyes into cellulosic fibres in an atmosphere of heat and moisture that is created by saturated steam injected into the steamer.

5.4 General activities in padding

The fabrics to be dyed are first assembled into batch by rolling on a roller with uniform tension. Care is taken to see that the fabrics of similar density and those required to undergo similar treatment are grouped and assembled into a batch.

The fabrics fed are first immersed in the dye liquor of required concentration and then passed through the bowls of padding mangle. The squeezing pressure is adjusted to get the required impregnation and to remove excess liquor. The padded materials are rolled on a batch and kept in rotating station before delivering to next process.

5.5 Precautions to be taken for padding

a) The rubber rollers should be even and have uniform hardness throughout its surface.

b) The pressure applied should be such that the fabric is nipped for minimum 1 cm.

c) The fabric fed is guided by width guides so that the fabric moves exactly in the predetermined path.

d) The dye liquor should be clear without any lumps of dyestuff.

e) The dye liquor should be concentrated enough to prevent early absorption of dyes while padding.

5.6 Control points and checkpoints

It is essential to have clarity on the points to be controlled to achieve the targets and those to be checked to ensure the process is in control. These points need to be reviewed from time to time and modified to suit the requirements of individual companies and their targets. Each mill should prepare its own 'control points and checkpoints' and display them in the work area so that the people refer and follow them. Following are just for reference.

5.6.1 Control points

a) The dye recipe depending on the colour and the depth of shade to be dyed and the GSM of fabric

b) The selection of padding process—two bowls or three bowls

c) The pressure on squeezing rollers

d) The speed

5.6.2 Checkpoints

- **Material related**

  a) The design number and the fabric density matching to the plan

- **Padding machine related**

  a) Condition of the mangle rollers and the trough

  b) The surface hardness of rubber rollers

- **Setting related**

  a) The dyeing recipe and its concentration

  b) The level of dye liquor in the trough

  c) The pressure set on squeezing rollers

  d) The speed set

- **Performance related**

  a) Uniformity of padding observed after completing the dyeing process

  b) Per cent pickup

- **Documentation related**

  a) Design number and lot number padded

  b) Dye recipe used

- **Work Practice related**

  a) Whether the machine was cleaned thoroughly before starting

  b) Whether the surroundings are kept clean all the time

- **Logbook related**

  a) Design and lots completed padding

  b) Design and lots waiting for padding

  c) Problems faced if any during the shift

- **Management information system related**
  a) Design and lots fed
  b) Metres in each design and lot
  c) Colours consumed
  d) Other chemicals consumed
- **General**
  a) Whether the safety precautions are taken as needed
  b) Whether the nip guards are set properly

5.7   Normal problems in padding

a) Fabrics absorbing the dyes fast and the concentration of pad liquor becoming dilute

b) Improper pressure on squeezing rollers resulting in uneven pickup of dyes

c) Low pressure of squeezing leading to dyes remaining superfluous on the fabric

d) Folds and creases in the fabric leading to fold marks and crease marks

5.8   Dos and don'ts for padding

It is necessary to have a clarity on what is supposed to be done without fail and what should not be done at any cost. Some examples are given below.

5.8.1   Dos

a) Understand the process before deciding the process parameters

b) Check the preparation of the recipe before allowing it in the immersion trough

c) Stick to the process parameters rather than trying to do some hasty things to increase production

d) Set the selvedge guides and ensure that the fabric does not go to the edge of the mangle

e) Check the surface of rubber rollers periodically and ensure that there are no cracks or uneven hard surface

f) Maintain the level of liquor in the immersion chamber; do not allow it to vary

g) Periodically check the concentration of liquor in immersion chamber: if it is found light, increase the concentration.

h) Monitor the pressure of the squeezing rollers.

i) Select the number of bowls depending on the density of the fabric.

j) Check the setting of nip guards and ensure that they are proper.

5.8.2 Don'ts

a) Do not stop the machine in between when the padding operation is going on.

b) Do not allow folds and creases in the fabric to enter the padding zone.

c) Do not use dyes with high affinity for padding.

5.9 Responsibilities and authorities of supervisor in padding

5.9.1 Responsibilities

a) Ensuring that the required designs and lots of fabric are padded with least rejections irrespective of various problems being faced in the department.

b) Ensuring that the padded fabric is wound batch-wise and supplied to the user in time.

c) Monitoring the consumption of chemicals.

d) Ensuring all safety precautions are taken while handling chemicals and running the machines.

5.9.2 Authorities

a) Authorized to reject the batch if the quality is different than the programmed.

b) Authorized to reject the dye recipe prepared if found with lumps and undissolved particles.

c) Authorized to question the operators when the quality or productivity is not as expected.

6

Fabric dyeing in open width

6.1 Introduction

Fabrics can be dyed either in open width or in a rope form. Open width dyeing is practiced for woven fabrics those are more stable, whereas delicate fabrics are preferably dyed in rope form. The oldest system of dyeing fabrics in open width, which is still in practice also called as jigger dyeing. Although high-temperature–high-pressure beam dyeing machine can be used for open width dyeing of fabrics, they are preferably being used for dyeing weaving beams.

The open width dyeing can be in batch system, semi-continuous or in continuous systems.

6.2 Batch dyeing

Batch dyeing process, also sometimes referred to as exhaust dyeing, is the most popular and common method used for dyeing of textile materials. In this process, the dye gets slowly transferred from a comparatively large volume dyebath to the substrate or material that is to be dyed. The time taken is also longer. The dye is meant to 'exhaust' from dye bath to the substrate.

In batch processes, textile substrates can be easily dyed at any stage of their assembly into the desired textile product. This includes fibre, yarn, fabric or garment. Some type of batch dyeing machines can function only at temperatures up to 100°C. For example, cotton, rayon, nylon, wool and so forth can be dyed at 100°C or lower temperatures, while polyester and some other synthetic fibres are dyed at 100°C or even higher temperatures.

There are three general types of batch dyeing machines: circulating the fabric; circulating the dye bath and circulating both the bath and material. Examples of batch dyeing process are Beck, Jet, Jigs, Beam Package dyeing machines and so forth.

For a batch dyeing process, the following techniques can prove to be effective for optimum utilization.

1. Use machineries that are fitted with automatic controllers of fill volume, temperature and other dyeing cycle parameters such as indirect system of cooling and heating, hoods and doors that lessens vapour losses

2. Choosing the machinery that is exactly designed for the size of the batch which needs to be processed

3. Ensuring that the machine is operated exactly within the specified range of nominal liquor ratios for which it is designed. It has been seen that machines that are operated with a consistent liquor ratio while being loaded at 60% level of their nominal capacity gives optimum results. With yarn dyeing machines, this level can stretch to even 30% of the nominal capacity

4. Opting machineries that adhere to the following requirements:

    • Liquor ratio that is low or ultra low

    • Complete in process separation of bath from substrate

    • Mechanism that involves smooth internal separation of process liquor from the washing liquor

    • Mechanical liquor extraction that brings the carry over to minimum and improves washing efficiency

5. Reduced cycle duration

6. Replacement of overflow-flood rinsing method with methods such as drain and fill

7. Proper reuse of rinsed water for the next dyeing session

Reuse of the dye bath if technical considerations allow

6.3   Semi-continuous dyeing process

In the process of semi-continuous dyeing that consists of pad-batch, pad-jig and pad-roll, the fabric is first impregnated with the dye liquor using a padding machine. Then, it is subjected to batch-wise treatment in a jigger. It could also be stored with a slow rotation for many hours. In the pad-batch, this treatment is done at room temperature, while in pad-roll, it is done at increased temperature by employing a heating chamber. This helps in fixation of the dyes on to the fibre. After this fixation process, the material in full width is thoroughly cleansed and rinsed in continuous washing machines.

In semi-continuous dyeing, the dye is applied continuously by a padding, whereas the fixation and washing remaining discontinuous. The liquor ratio in semi-continuous dyeing is not of much importance and is not taken as a parameter. One of the widely used techniques for semi-continuous dyeing process is the pad-batch dyeing.

6.4   Continuous dyeing process

A continuous dyeing process consists of dye application, dye fixation with heat or chemicals and finally washing as a continuous process. It has been found to be most suitable for woven fabrics. Mostly, continuous dye ranges are designed for dyeing blends of polyester and cotton. The step of padding plays a key role in the operation of continuous dyeing. Sometimes nylon carpets are also dyed in continuous processes, but the design ranges for them is unlike that for flat fabrics. Warps are also dyed in a continuous process. It is observed that continuous dyeing is a popular dyeing method and accounts for around 60% of the total length of the products that are dyed.

In a continuous dyeing range (CDR), the textile substrates are fed continuously into a dye range; the speeds can vary between 50 and 250 m/min. A continuous dye range has been found useful and economically sustainable for dyeing long runs of a given shade.

The main consideration in the selection of dyeing method such as batch dyeing or continuous dyeing is the tolerance factor for colour variation. The variation in shade is more for continuous dyeing as compared to batch dyeing. This is because of the speed of the process and presence of a large number of process variables which affects dye application in continuous dyeing.

Continuous and, to some extent, semi-continuous dyeing processes are less prone to water consumption than batch dyeing but results in high concentration of residues. If some strict control measures are taken up, it is possible to reduce these losses of concentrated liquor.

6.5   Jig dyeing machine

Jigger is one of the oldest dyeing machines used for cloth dyeing operations in open width (Fig. 6.1). It is suitable for dyeing of woven fabrics, up to boiling temperature without any creasing. They exert considerable lengthwise tension on the fabric and are more suitable for the dyeing of woven than knitted fabrics. Since the fabric is handled in open width, it is very suitable for fabrics which crease when dyed in rope form.

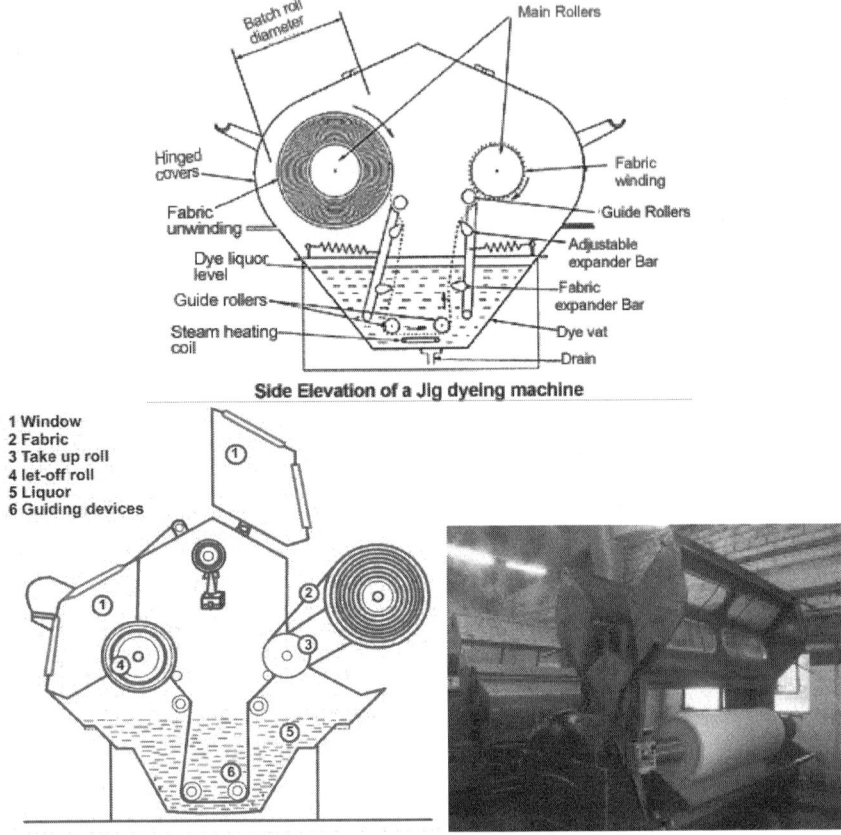

Side Elevation of a Jig dyeing machine

1 Window
2 Fabric
3 Take up roll
4 let-off roll
5 Liquor
6 Guiding devices

Figure 6.1   Jig dyeing machine.

Jigger consists of a V-shaped trough. Two steel or ebonite rollers are fitted above the trough over which the fabric to be treated is wound in roll form. Fabric passes from one roller trough guide rollers in the trough to the other roller. Since most of the fabric in roll form remains out of the trough, a low material to liquor ratio (3:4) can be maintained.

During the process, the fabric in open width form unwinds from one roller, passes through guide rollers in the trough and winds on the other roller. When all the fabric from the first roller is transferred to the second roller, a reversal mechanism is activated and the direction of the fabric movement is changed from the second roller to the first. This process continues till the treatment is over. The complete transfer of the entire fabric length from one roller to the other is known as one end or turn. For a given process, the number of turns depends on the residence time required for the process. Each end may take 10–15 min. In jig dyeing, the duration of the process is measured on the basis of the number of passages or ends of the fabric passing through the dye bath from roller to roller. The end of dyeing phrase is known as the passing of fabric through dye liquor from one roller to another.

The liquor in the trough may be heated by steam running through pipes in the trough and having perforations. Early jiggers were open but nowadays they come equipped with top covers to minimize heat loss through evaporation.

An additional roller is placed on fabric wound roller such that it squeezes out liquor from the fabric into the trough.

Jiggers can be used for desizing, scouring or even bleaching of textile fabrics.

There are two types of jiggers, namely open jiggers and closed jiggers.

1. **Atmospheric jig or open jigger:** Atmospheric jigs operate at atmospheric temperatures and pressures. These machines are applied to natural fibres. Here, the temperature limit is typically 100°C.

2. **High-temperature jig or closed jigger:** A high-temperature jig functions in the same way as an atmospheric jig but comes with the addition of a pressure vessel that is designed to function at 130°C. The pressure vessel also helps in having a close control of the dyeing temperature. Typically it is applied for dyeing synthetic fibres.

The jigger machines have two main rollers which revolve on smooth bearings and are attached to a suitable driving mechanism, which can be reversed as required. The fabric is wound on one of the main rollers and fed from the other. The fabric moves from one roller to the other through the dye liquor trough located at the lower part of the machine. There are various arrangements of guide rollers at the bottom of liquor trough. During each passage, the cloth passes around these guide rollers.

The concentrated dye liquor is usually introduced directly into the dye bath in two equal portions, which are added just before commencing the first and second ends. The liquor is agitated by the movement of the fabric through the dye bath. Several horizontal spray pipes are fitted across the full width of the trough in order to expedite fabric rinsing.

Live steam injected into the bottom of the trough through a perforated pipe across the width of the jig heats the liquor. Some modern jigs also have heat exchangers for indirect heating.

Covering the top of the jig minimizes the heat loss to the atmosphere, keeps the temperature uniform on all parts of the fabric and minimizes exposure of the liquor and the cloth to air. Minimizing exposure to air is important when using sulphur or vat dyes since these dyes can be oxidized by atmospheric oxygen.

A few metres of leading fabric, similar in construction to the cloth under process, is stitched to each end of the cloth batch, to allow the entire length of the fabric to pass through the dye bath during the dyeing process. When jig processing is completed, the fabric is run onto an A-frame through a nip or suction device to remove extraneous water during unloading.

Modern machines such as automatic and jumbo jiggers have full automation in the drive, tension regulation and control, fabric speed and metering, smooth and jerk-less stop and start, counters for number of turns, gradual and noiseless reversal, automatic temperature regulation and control and so forth.

Old jiggers used to have a trough capacity of 200 L and the fabric length processed at a time was 500–1000 m, whereas modern high capacity jiggers have capacities up to 750 L, which can handle around 5000 m of fabric at a time.

6.5.1   What jigger should and should not do?

Should do

- Should dye the fabric in open width uniformly with required depth and with required fastness
- Should wash and remove the excess dye from the surface of the fabric

Should not do

- Jigger should not develop patches of dyeing
- Jigger should not develop shade variation from one end to another

6.5.2   Dyeing process using a jigger

- The fabrics are received in a batch after padding and completing the rotations in rotating station
- The required length of fabric is wound on one of the rollers of the jigger. While dyeing on jigger machines the cloth revolves around two main rollers
- The open width fabric passes from one roller through the dye bath at the bottom of the machine and then onto a driven take-up roller on the other side
- When all the fabric has passed through the bath, the direction is reversed. Each passage is called an end
- Dyeing always involves an even number of ends

The dye bath has one or more guide rollers, around which the cloth travels, and during this immersion achieves the desired contact with the dye liquor. During this passage, the fabric picks up adequate quantity of dye liquor, excess of which is drained out but still a good quantity is held in the fabric

6.5.3 The factors controlling the rate of dye absorption in jigger dyeing

a) The amount of interstitial dye liquor retained in the interstices of the fabric weaves

b) The exhaustion of the interstitial liquor in the dwell period between successive immersions

c) The degree of interchange of liquor during one immersion (interchange factor)

d) During rotation of rollers, the dye penetrates and diffuses into the fabric. The real dyeing takes place not in the dye liquor but when the cloth is on the rollers since only a very small length of fabric is in the dye bath and major part is on the rollers. Therefore, the speed of cloth during immersion in dye liquor has a very little effect on the percentage of shade produced

6.5.4 Typical operating procedure of dyeing on a jigger

The operating procedure depends on the type of jigger, the type of fabric being dyed, the dye group being used and the depth of shade required. Each mill has to prepare its own operating procedure depending on the circumstances. The following is just for an illustration (Fig. 6.2).

a) Clean the machine and its surroundings before starting any work on the machines

b) Open the water valve and start heat exchangers

c) Switch on the machine power

d) Bring the required fabric batch for dyeing

e) Verify the material brought against the programme given to you for that machine. Check both the quality and the length in metres for each batch

f) After verifying and ensuring that the materials are correct as per programme, load the material into the machine by rolling the fabric on the main roller

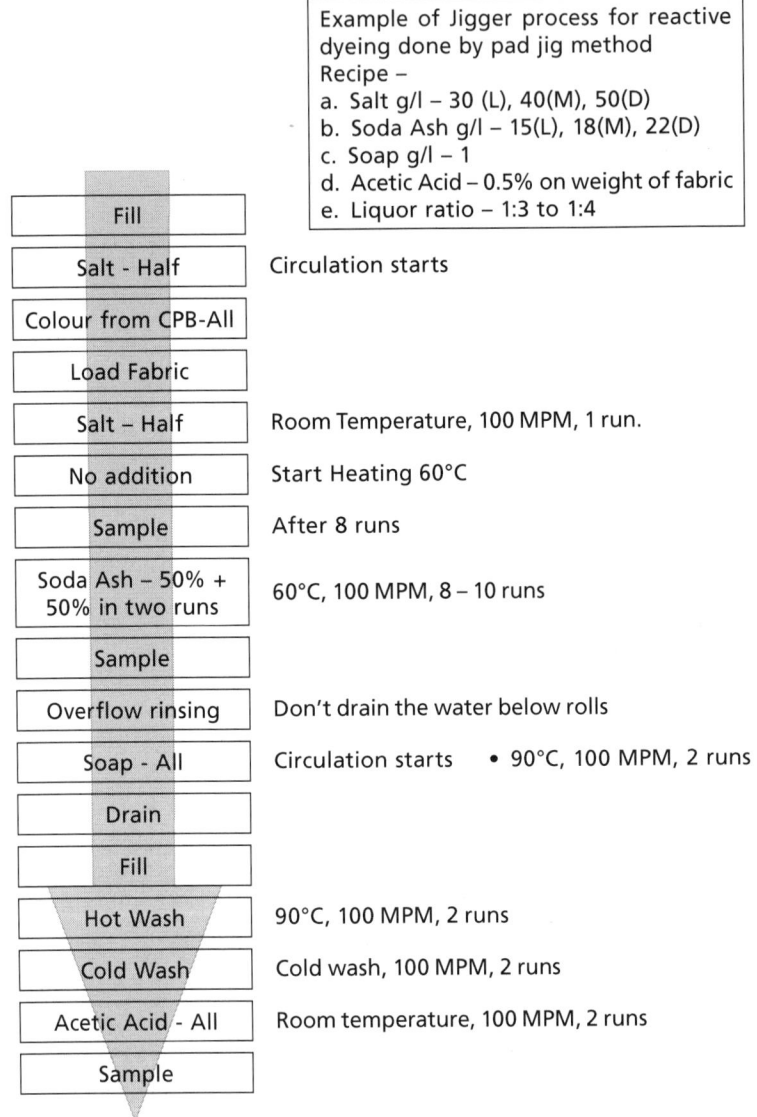

Fill	
Salt - Half	Circulation starts
Colour from CPB-All	
Load Fabric	
Salt – Half	Room Temperature, 100 MPM, 1 run.
No addition	Start Heating 60°C
Sample	After 8 runs
Soda Ash – 50% + 50% in two runs	60°C, 100 MPM, 8 – 10 runs
Sample	
Overflow rinsing	Don't drain the water below rolls
Soap - All	Circulation starts • 90°C, 100 MPM, 2 runs
Drain	
Fill	
Hot Wash	90°C, 100 MPM, 2 runs
Cold Wash	Cold wash, 100 MPM, 2 runs
Acetic Acid - All	Room temperature, 100 MPM, 2 runs
Sample	

Example of Jigger process for reactive dyeing done by pad jig method
Recipe –
a. Salt g/l – 30 (L), 40(M), 50(D)
b. Soda Ash g/l – 15(L), 18(M), 22(D)
c. Soap g/l – 1
d. Acetic Acid – 0.5% on weight of fabric
e. Liquor ratio – 1:3 to 1:4

Figure 6.2   Process flow in jigger.

g)   Take the quantity of water as per the process card

h)   Refer the process card for setting the process parameters and processing the fabrics

i) Bring the colours and chemicals from the colour room after weighing the exact quantity required as per the process sheet

j) Put chemicals and colours as directed in the process sheet for each of the fabric lots, which are specific to lot and design

k) Refer the process sheet relating to the percentage of colours and chemicals, liquor ratio, the timings of different operations, the temperatures to be maintained, the rate at which temperatures are to be raised and brought down, the timing for holding the materials, number of ends (runs) while holding and so forth. In case any information is not clear, refer to your supervisor and get the information, and enter the same in your production report as well as on process card

l) If case samples are to be passed, get the sample passed along with relevant samples

m) If addition is needed to match to shade, then process the material again in the machine to get the correct shade

n) Once the sample is likely to be passed, follow the instructions as in process sheet and discuss with the supervisor in case of changes

o) Check the pH of the cloth after process and washing, and then unload the materials from the machine

p) Keep the material at the designated place as instructed by the supervisor

q) In jigger dyeing continuous running wash is not possible. You need to drain the water and refill after every wash

r) Enter the details of the design number, batch number, lot number, quality number, the process done, the quantity of material processed and the problems faced in the production record

s) If you find any irregularity or discrepancy, inform the supervisor immediately and take directions to correct the situation

t) If you notice any problem in the machine, may be electrical or mechanical, inform the concerned supervisor immediately and get them rectified. Enter complete details in the production report book

6.5.5   Some typical problems that may be encountered in jig machines

Following are some typical problems encountered in jig machines. The problems related to the dye group adapted are discussed in detail in another chapter.

1. Temperature control from side-to-side and end-to-end of the roll, not properly leading to uneven dyeing .

2. Tension control from end-to-end, not proper giving shade variation from end to end

3. Inconsistent speed control from end-to-end resulting in shade variations

4. Prevention of creases, not proper resulting in fabric dyed having creases

5. Prevention of air, especially in vat dyeing or reactive dyeing is very important to prevent oxidation immaturely. Improper prevention leads to the improper development of shades

6. Side-to-centre colour variations, called listing

7. Lengthways colour variations, called ending

6.5.6   Precautions to be taken for jigger dyeing

Following precautions are needed to be taken while dyeing on a jigger:

a) Not winding the fabric slack on the rollers

b) Closing of the hoods in case of dyeing vat colours and reactive colours to prevent it from getting oxidized due to atmospheric oxygen

c) Wearing safety gadgets such as gloves, aprons and gum shoes while handling chemicals

d) Wearing a mask when the chemicals are active reacting with cellulose

e) Not storing alkali and acid side by side

f) Not to keep the chemical containers open to air

6.5.7   Control points and checkpoints

It is essential to have clarity on the points to be controlled to achieve the targets and those to be checked to ensure the process of jigger dyeing is in control. These points need to be reviewed from time to time and modified to suit the requirements of individual companies and their targets. Each mill should prepare its own 'control points and checkpoints' and display them in the work area so that the people refer and follow. Following is an example:

Control Points

a) Deciding and selection of process parameters, namely:

   • Machine setting

   • Speed

- Running time for each end
- Temperature
- Dye recipe
- Washing and fixing sequences

b) Deciding the acceptance criteria for shade matching and colour fixation

c) Deciding on chemicals and their quality to be procured

d) Employing trained and qualified employees

e) Evolving production norms

f) Evolving norms for consumption of chemicals, water and steam

Checkpoints

a) **Material related**
   - The fabrics received against the plan received for dyeing
   - The quantity of fabric received for loading in jigger—metres and kilograms

b) **Jigger machine related**
   - The condition of the machine
   - The working of various valves and controls
   - Condition of the spray nozzles

c) **Setting related**
   - The speed set
   - Number of ends (runs) set
   - The recipe prepared

d) **Performance related**
   - Uniformity of shade throughout the fabric
   - The shade obtained against the requirement
   - Fastness properties of the dyed materials

e) **Documentation related**
   - The design number and the shade dyed
   - Quantity dyed

f) **Work practice related**

- The cleaning of the jigger thoroughly before loading the material
- Proper alignment of the batch while loading on to jigger
- Proper closing of the door while dyeing process is on in case of vat and reactive dyeing
- Proper cleaning of spray nozzles
- Removing of the water completely and refilling with fresh water in each wash

g) **Logbook related**

- Jiggers working
- Starting time of the running lots
- Activities done and to be done

h) **Management information system related**

- Jigger number
- Batch number
- Number of metres
- Weight in kilograms
- Colour used
- Chemicals used

i) **General**

- Use of safety gadgets such as gloves and gum shoes while handling chemicals
- Removing the completed batch and delivering to the next process

6.5.8 Limitations of jigger

Following are the normal limitations of a jigger:

a) Jigs exert considerable lengthwise tension on the fabric and hence not suitable for the dyeing of knitted fabrics

b) The swelling and dissolution of size makes the fabric slippery and unstable in roll form

c) The low liquor ratio makes washing-off difficult

d) There is little mechanical action in a jig machine and it is less suitable where vigorous scouring is required before dyeing

e) Moiré effects or water marks may arise on some acetate and nylon fabrics because of pressure flattening the structure of the rolled fabric

6.5.9 Dos and Don'ts for jigger dyeing

It is necessary to understand clearly what is supposed to be done without fail and what should not be done at any cost. Some examples are given below.

Dos

a) Understand the process before deciding the process sequence or cycle

b) Study the process sheet in detail before programming the machine

c) Check the fabric weight (calculate by considering the GSM and the length and width) and then work out the requirement of dyes and chemicals

d) Clean the jigger thoroughly after completing dyeing of a batch

e) Check the purity of dyes and chemicals and the quality of water before preparing dye solution

f) Stick to the process parameters rather than trying to do some hasty things to increase production

g) While batching the fabric, ensure for alignment of the selvedge

h) Check the pH of cloth before unloading from the machine

Don'ts

a) Do not allow the running speed to go above or below the specified limit

b) Do not unload the material before getting the test results of the sample sent for testing

c) Do not process a fabric if it has creases or folds

6.5.10 Responsibilities and authorities of supervisor in jigger dyeing

Responsibilities

a) Ensuring that the required designs and lots of fabric are dyed with least rejections irrespective of various problems being faced in the department

b) Getting the fabric tested for shade and dye fixation before unloading it from the jigger

c) Programming the machine as per the process sheet

d) Ensuring that the dyed fabric is wound on to batch and supplied to the user in time

e) Monitoring the consumption of water, dyes and chemicals

f) Ensuring all safety precautions are taken while handling chemicals and running the machines

Authorities

a) Authorized to reject the batch and send for re-batching if the folds and creases are found in the fabric

b) Authorized to reject the dye liquor if lumps are found in it

c) Authorized to question the operators when the quality or productivity is not as expected

6.6 Continuous dyeing system

Continuous dyeing is usually defined as a dyeing method where a relatively concentrated dye solution is applied evenly across the entire width of the fabric passing through it in a continuous manner. The application of colorant solution is usually accomplished by padding but also can be done by other means. Padding is followed by subsequent fixation of the dye by chemical or thermal means. Continuous dyeing is predominantly used for woven fabrics. However, machinery is also available for both open width and tubular knits. When processing knits, the fabric must be subjected to low and uniform tension for maintaining the desired aesthetics. Padding techniques must be altered to properly handle tubular knit goods because edge lines can occur if good dye penetration is not obtained or if the hardness of the pad rolls is not correct.

In the pad-batch method, the fabric ready-for-dyeing is impregnated with dye liquor, excess liquor is squeezed out on a mangle, the fabric is batched onto rolls or held in boxes for 2–12 h, and then covered with plastic film to prevent adsorption of carbon dioxide from air or evaporation of water. Subsequently, the fabric is washed off in any of the conventional ways, depending upon the available equipment.

The CDR is excellent machine for bulk order. Consistency in shade is obtained because of less material handling. All bulk order such as defence uniforms, police uniforms, school uniform, civil uniform, exports bulk order and so forth, can be successfully executed in CDR machine.

Normal process sequence:

1. Pad-dry cures. For example—pigment dye and so forth.

2. Pad-dry-chemical padding-steaming-washing-drying. For example—
    vat dye, reactive dye (vinyl sulphone), sulphur black and so forth

3. Pad-dry-thermosol. For example—disperse/reactive dye and disperse
    dye (selected dye otherwise staining chance will be there)

4. Pad-wet steaming-washing-dry. For example—selected reactive dye
    shades

Above all processes, sequence no. 2 process is running smoothly for maximum
shade. Main important feature in CDR machine can be divided into following
sections:

a) **Dye padder** with continuously variable crown roller profile. This
    consists of two S-Rolls padding machine for minimizing centre side
    variation in fabric

b) **Infrared drier** The fabric having high moisture content is passed
    through this zone to reduce the moisture content on the fabric ultimately
    to avoid migration of dye stuff

c) **Hot flue drier** The fabric is subjected to heat treatment gradually to
    dry or dye fixation depending on the machine configuration

d) **Chemical padding** section For reactive (vinyl sulphone) dye in
    chemical padding section

Generally continuous dyeing process can grouped into three main topics,
namely pad-bake, pad-steam and pad-batch. These three continuous dyeing
processes are applied in various types. Let us discuss on pad-steam system
of continuous dyeing.

The pad steam processes can be further grouped as pad-dry-pad-steam
process, pad-(pad)-steam process, pad-dry-steam process and pad-dry
e-control process (Fig. 6.3).

Pad steam is a process of continuous dyeing in which the fabric in open
width is padded with dye stuff and is then steamed. Pad steam method is ideal
for reactive dyeing. Light, pale and medium shades can be dyed in this machine.
Continuous roller steamer is used for diffusion of reactive, vat, sulphur and
direct dyes into cellulosic fibres in an atmosphere of heat and moisture that
is created by saturated steam injected in to the steamer.

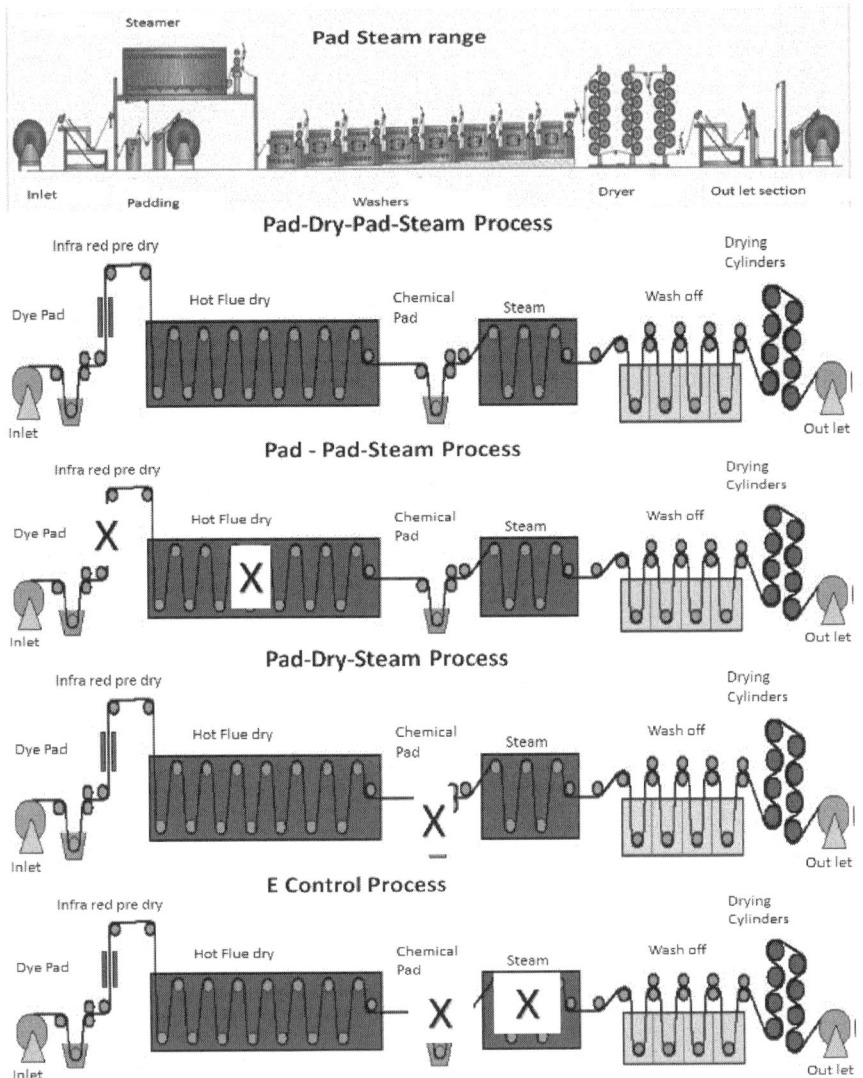

Figure 6.3   Different combination of Kuster pad-steam system.

Pad steam can be used as follows:

- As a pad batch for reactive dyeing in which batch is left for12–18 h for the completion of the reaction. For time saving the fabric passes through the steamer for 1 min and the reaction is completed

- Reduction clearance in which polyester cotton fabric is treated with caustic and sodium hydrosulphide to remove the disperse dye from cotton

- Stripping of the fabric, that is, colour can be removed completely by adding higher amount of caustic and sodium hydrosulphide
- For the development of vat dyes
- Washing of dyed fabric

Pad steam is generally suggested for long metres of fabric. It is suitable for fabric more than 500 m as it has a good stability of bath. Long metres heavy mercerize weaved fabric is dyed by the method of pad-steam with distance drying. The pad-dry-pad-steam process is unproductive for short time lots.

Pad steam is appropriate for bright colours. It has more control parameters. It is necessary to obtain suitable condition.

Main sections of the machine are inlet section, padding section, steamer, washer, dryers and outlet section.

E-control process is improved by the cooperation of Montfort's company. It is economical in using of chemicals as it is rapid. The washing is done immediately after dyeing. In this process, steaming does not exist. With the 25% humidity, at the 110–130°C cellulose dyeing is done and polyester dyeing is done at 210°C. E-control is usable with hot air drying and fixation continue dyeing process as it includes less procedures and uses less chemicals. Tailing problem is encountered in e-control process like the other process.

Padders are used for padding colours and chemicals. In a separate chapter, the padding mangle is discussed in detail. The pressure of the padders is 1.5–2 bar. Two types of pressure are used in Kuster padders, that is, hydraulic and pneumatic. The central pressure is hydraulic and side's pressure is pneumatic. One can adjust the pressure of the padders as required. Liquor is first picked in the fabric; afterwards, the excessive liquor is squeezed out by means of padders. The pressure determines the pickup per cent.

In steamer, temperature required for the fixation of colour to the fabric is given. This temperature is achieved by saturated steam. The saturated steam is used to avoid drying of the chemicals on the surface of fabric preventing fabric from stains. The roof is heated to avoid water dropping which can lead to spotty dyeing. Here, water is not provided at the entry of steamer to prevent reaction of chemicals that just applied before entering into steamer. Water lock is provided at the end of steamer.

The functions of water lock/seal are:

1. To stabilize steam coming out from the steamer
2. To avoid thermal shocks

3. To avoid dyes to be diluted by keeping its temperature at about 40–45°C. Keeping temperature at 40–45°C serves the purposes of cooling the fabric, removing excess dye from surface and to condense the steam that comes along the fabric

Washing is carried out in order to remove unfixed dyes. After steamer, fabric flows from seven to nine washers. Most commonly, first to four washers are used for washing of salt or chemicals which are being applied in trough of pad steamer. In fifth and sixth washer, oxidation is done if required. If oxidation is not required then soaping is done in fifth and sixth washers. Neutralization is done in seventh washer by using acetic acid. In counter flow system, eight chambers are used.

There are different types of washes as mentioned below:

1. **Normal wash**: This wash is for printed fabrics the temperature is kept at 50–60°C.

2. **Mercerized wash**: In this only hot washing is done. The washing temperature in first chamber is 50°C and two to eight is 98°C and last one has 70°C.

3. **Dyed wash**: This is used for dyed fabrics and in this soap is added for washing. For light shades, only soap is added, whereas for dark shades soda ash is also added with soap which decreases the depth of colour and improves rubbing fastness. The temperature in first and second tank is at room temperature and from third to eight is 98°C, whereas in the last one, it is 50°C. The temperature in washing tanks (Fig. 6.4) is controlled by a temperature sensor.

At the end of the pad steam machine, there are three groups of drying cylinder for drying the fabric. Each group has 12 hot cylinders but last one has 10 hot and 2 cool cylinders. All the cylinders are Teflon coated. Their purpose is to remove water molecules from fabric.

Figure 6.4   Washing tanks.

The fabric from drying cylinders (Fig. 6.5) passes through some tension rolls and also from antistatic rods which absorbs the charge from the fabric. This rod has 5 KV of voltage and 1800–18 000 A of current. After passing through these, this fabric is wound on the batcher (Fig. 6.6).

Figure 6.5   Drying cylinders.

Figure 6.6   Batcher.

6.6.1 Problems in continuous dyeing

Typical problems encountered in pad dyeing are lengthwise shade variation (also called tailing or ending) and widthways shade variation (also known as listing or side-centre-side shade variation). The dyeing problems occurring in a CDR can be attributable to the dye padder, pre-drying, the thermosol unit, the chemical padder, the steamer and the wash boxes.

Fabric dyeing in rope form

7.1 Introduction

Dyeing of fabric in rope form is in practice from the oldest days of hand dyeing of fabrics. As the speed of dyeing and handling more fabric at a time was demanded, machines such as winch and jet dyeing were developed. Dyeing in rope form can be done for woven fabrics as well as knitted fabrics. For delicate and light fabrics, rope dyeing is preferred.

7.2 Winch dyeing

The winch is an old equipment used for processing fabrics in rope form (Fig. 7.1). It is suitable for fabrics which can withstand creasing during rope form processing. Winch applies a low amount of tension on fabrics during processing. During processing, most of the fabric remains submersed in a deep trough which increases material to liquor ratio (1:20 or 25) as compared to a jigger.

In winch, the fabric is processed in the form of an endless loop. The fabric is piled over the sloping bottom of a deep trough. At the deeper end of the trough, the fabric rises and passes over a 15-cm diameter reel known as jockey or flyer and then over a much larger elliptical winch reel. The rotation of the winch reel provides the motion to the fabric. There is no positive grip between the fabric and the reel. The fabric movement takes place due to the friction and the weight of wet fabric.

At a time, one or more ropes can be dyed. When dyeing multiple ropes, separation is maintained by a peg rail. The peg rail runs through the width of machine and is located at about 20 cm above liquor level.

The main chamber of the winch is divided into two parts by a perforated stainless steel sheet. The smaller chamber situated towards the front of the

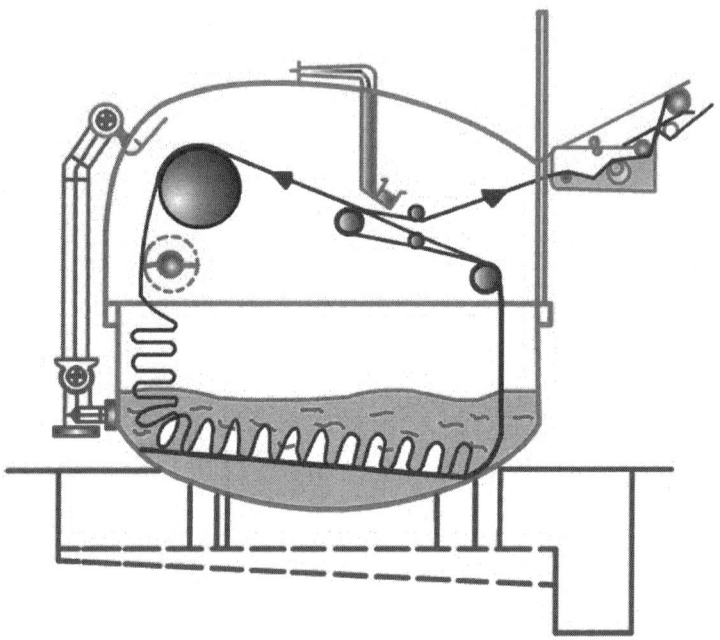

Figure 7.1   Schematic diagram of a winch dyeing machine.

machine is used for adding dyes and auxiliaries as well as for housing the steam pipes.

Winch provides little sideways movement of the fabric, as it can result in entanglement. Most of the fabric remains in the form of a slowly moving pile at the machine bottom. This can result in uneven dyeing. The high material to liquor ratio also results in poor dye exhaustion of low substantivity dyes.

Winches can be used for desizing, scouring, bleaching or even washing of textile fabrics.

7.3   Jet Dyeing

Jet dyeing machine dyes fabrics in rope form, where the fabric gets transported through the machine in a loosely collapsed form that looks like a rope (Fig. 7.2). For transporting the fabric through the tube, a jet of dye liquor is supplied through a Venturi. The jet creates turbulence. This helps in dye penetration along with preventing the fabric from touching the walls of the tube. This is more suited for light and delicate materials and knit goods.

1. Dyestuff and auxiliaries feeding container
2. Heat exchanger
3. Pump for the circulation of liquor
4. High pressure pipe work
5. Dyeing drum
6. Winch

Figure 7.2 Jet dyeing machine.

As the fabric is often exposed to comparatively higher concentrations of liquor within the transport tube, very little dye bath is needed in the bottom of the vessel. This is just enough for the smooth movement from rear to front. Aqueous jet dyeing machines generally employ a driven winch reel along with a jet nozzle. The following diagram explains the functioning of a jet dyeing machine:

The process can be used for batch dyeing operations such as dyeing, bleaching, washing and rinsing. In this process, dyeing is accomplished in a closed tubular system, basically composed of an impeller pump and a shallow dye bath. The fabric to be dyed is loosely collapsed in a form of a rope and tied into a loop. The impeller pump supplies a jet of the dye solution, propelled by water and/or air, which transport the fabric within the dyeing system, surrounded by dye liquor under optimum conditions. Turbulence created by the jet aids in dye penetration and prevents the fabric from touching the walls of the tube, thus minimizing the mechanical impact on the fabric.

In this machine, the dye tank contains a dye such as disperse dye, dispersing agent, levelling agent and acetic acid. The solution is filled in the dye tank and it reaches the centrifugal pump and then to filter chamber. The solution will be filtered which reaches the heat exchanger and then to the tubular chamber.

The material to be dyed will be loaded and the winch is rotated so that the material also gets rotated. Again the dye liquor reaches the heat exchanger and the operation is repeated for approximately 20–30 min at 135°C. Then the dye bath is cooled down after the material is recovered. The holding time and temperatures depend on the actual dyestuff, the material being dyed and the depth of shade selected.

The fabric in a rope form is fed into the machine without any tension. The fabric is lifted positively by a driven reel and then carried to the far end of the machine by the jet of the dye liquor. Dye additions are done through a secondary pressure pump when operating at room temperature. As the fabric is lifted with a driven reel, the lengthwise tension and consequent creasing on it is minimized.

The jet dyeing machine offers the following striking advantages that make them suitable for fabrics such as polyesters.

a)   Low consumption of water

b)   Short dyeing time

c)   Can be easily operated at high temperatures and pressure

d)   Comparatively low liquor ratios typically range between 1:4 and 1:20

e)   Fabrics are handled carefully and gently

Rope dyeing has the tendency for abrasion of fabrics and results in permanent cracks, creases and streaks.

7.4   Types of jet dyeing machines

In deciding the type of dyeing machine the features generally taken into consideration are:

a)   Shape of the area where the fabric is stored, that is, long-shaped machine or J-box compact machine

b)   Type of the nozzle along with its specific positioning, that is, above or below the bath level

Depending more or less on these criteria modern jet machines are developed as developments on the conventional jet dyeing machine.

a)   Overflow dyeing machine

b)   Soft flow dyeing machine

c)   Airflow dyeing machine

7.4.1   Overflow dyeing machine

Overflow dyeing machines are designed for delicate knitted and woven fabrics that are made up of natural as well as synthetic fibres (Fig. 7.3). They are also extensively used in the production of carpets. The main difference between jet and overflows machines is that in jet machines the fabric gets transported by

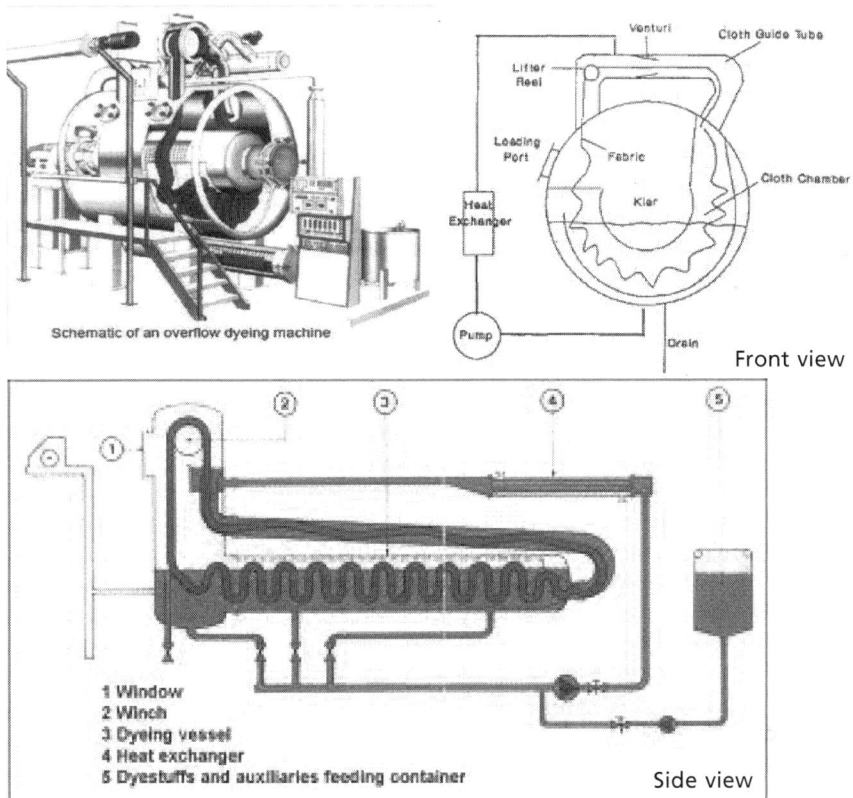

Figure 7.3   Overflow dyeing machine.

a bath that flows at high speed through the nozzle, while in overflow dyeing machine, it is the gravitational force of the liquor overflow that is responsible for fabric transportation.

7.4.1.1 Functioning of an overflow dyeing machine

A winch that is not motor driven is usually located on the topside of the machine where the fabric is hanged. A longer length of fabric is made to hang from the exit side of the winch as compared to the inlet side. By applying the gravitation force, the longer length of textile is pulled downward more strongly than the shorter one. Consequently, the fabric is soaked in the bath without any sort of tension.

Different designs of overflow dyeing machines are available in the market and some of them can operate under pressure and consequently at higher temperatures. Typical liquor ratios for overflows range between 1:12 and 1:20.

7.4.1.2   *Advantages of overflow dyeing machine*

- **No evaporative losses**—As the dyeing vessel is closed, there are no evaporative losses stemming from the dye bath. Further, depending on the situation the temperature may be raised to more than 100°C.

- **No build-up of steam condensate in the dye bath**—The latest technology implies that the dye bath gets heated by a heat transducer which is steam driven. This technology apart from being very efficient ensures that there is no build-up of steam condensate in the dye bath.

- **Low liquor ratios**—Dyeing is conducted at relatively low liquor ratios, for example, 10:1 and may be lesser resulting in substantial savings in water and energy.

- **Excellent dye liquor contact**—Excellent dye liquor contact with the fabric rope results in better and more improved level dyeing.

- **Computer control**—The machines are operated by computer and hence, operator error is eliminated.

7.4.2   Soft flow dyeing machine

The 'soft-flow' machines use the transport tube principle same as overflow machines where the fabric is transported in a stream of dye liquor (Fig. 7.4). In overflow machines, the reel is not motor driven, whereas in soft flow equipment, the reel and the jet work in constant harmony to remove the fabric from the forepart of the storage area, expose it briefly to a high concentration of liquor within the transport tube, then return it to the rear of the vessel. The fabric rope is kept circulating during the whole processing cycle right from loading to unloading. The liquor or fabric circulation is not stopped for usual drain and fill steps. The soft flow machines are gentler on the fabric than conventional jet overflow machines.

Figure 7.4   Soft flow dyeing machine.

7.4.2.1 Working principle of soft flow

In soft flow, there is a system for fresh water to enter the vessel through a heat exchanger to a special interchange zone; at the same time, the contaminated liquor is allowed to channel out through a drain without any sort of contact with the fabric, or for that matter, the new bath in the machine.

The fabric in a rope form is fed into the machine without any tension that is lifted positively by a driven reel and then carried to the far end of the machine by the jet of the dye liquor. The rate of flow is lesser in soft flow compared to jet dyeing machines, and hence the name soft flow. Dye additions are done through a secondary pressure pump at room temperature. As the fabric is lifted with a driven reel, the lengthwise tension and consequent creasing on it is minimized.

It is not possible to have running wash in soft flow machines. It has to be washed, and the water is drained and again washed with fresh water.

7.4.2.2 Key Features of soft flow dyeing machine

- Significant savings in processing time

- Savings in water (around 50%)

- Excellent separation of different streams resulting in optimum heat recovery and a distinct possibility of further use or a dedicated treatment

7.4.2.3 Types of soft flow dyeing machine

There are different types of soft flow dyeing machines. A few of the commercially popular machines along with their particular technical specifications are discussed here. The categories are not exhaustive as such.

7.4.2.3.1 Multi-nozzle soft flow dyeing machine

Multi-nozzle soft flow dyeing machines offer high productivity with low liquor ratio in fabric dyeing compared to the conventional dyeing machinery (Fig. 7.5). Multi-nozzle low/high temperature dyeing technology works as soft flow as well as jet dyeing machine. The main advantage of this development is to increase the machine versatility to process all types of fabrics. The pilling effect is not the specialty of this process.

In this technical development, the soft nozzles have large diameter and fix nozzle has an individual flow control valve. The main nozzle is adjustable. By increasing and decreasing the gap of the nozzle, flow and pressure increases and decreases as per fabric quality demand. To run this machine as a soft flow machine, increase the gap that decreases the pressure and increases water flow. At that time, all three nozzles work as soft flow and water flow gets control individually by control valve as per fabric quality demand. To run the machine as a jet dyeing machine, decrease the gap of the nozzle that increases the pressure and decreases the water flow. At that time close the soft nozzle control valve.

Technical features of multi-nozzle soft flow machines are:

- Very low liquor ratio—around 1:1 or 1:2 (wet fabric)
- Can reach high temperatures up to 140°C
- Easily dye 30–450 GSM fabrics (woven and knitted fabrics)
- Number of very soft flow nozzles
- Counterflow as well as reverse flow
- Reduce pollution plant size
- Wide capacity—10–300 kg
- Work as a jet dyeing machine for polyester and cotton blend fabrics
- Work as a soft flow machine for woven and knitted cotton fabrics

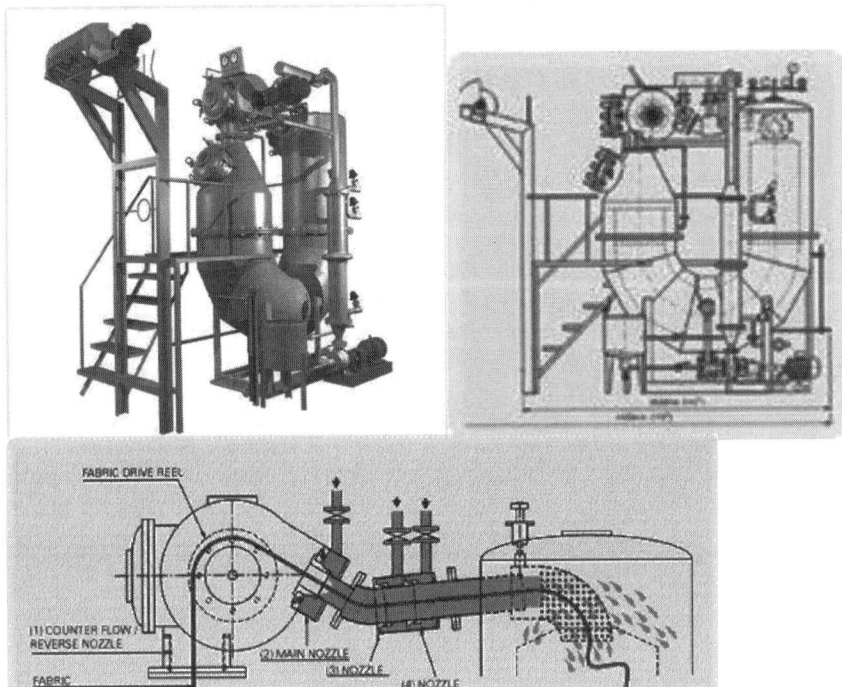

Figure 7.5　Multi-nozzle soft flow dyeing machine.

7.4.2.3.2   High temperature–high pressure soft flow dyeing machine
These machines use a jet having lower velocity than that used on conventional jet dyeing machines (Fig. 7.6). The vigorous agitation of fabric and dye formulation in the cloth increases the dyeing rate and uniformity. It minimizes creasing as the fabric is not held in any one configuration for very long. The lower liquor ratio allows shorter dye cycles and saves chemicals and energy.

Technical features of high temperature–high pressure (HTHP) soft flow machine are:

- Compact body made of stainless steel

- High-efficiency heat exchanger for quick heating/cooling

- Heating rate—around 4°C/min up to 900°C and around 3°C/min up to 135°C, at steam pressure of 6 bar

- Cooling rate—around 4°C/min at water pressure of 4 bar and 15°C

- Maximum working temperature is 135°C

- Maximum working pressure of 3.2 bar

- Manual as well as automatic controls

- Heavy duty stainless steel pump

Figure 7.6   High temperature–high pressure soft flow machine

7.4.3 Airflow dyeing machine

The main difference between the airflow machine and jet dyeing machine is that the airflow machine utilizes an air jet instead of the water jet for keeping the fabric in circulation (Fig. 7.7). Typically, the fabric is allowed to pass into the storage area that has a very small amount of free liquor. This results in a reduction in consumption of water, energy and chemicals.

Here, the fabric does not remain in touch with the liquor (the bath used is below the basket that holds the fabric in circulation). This invariably means that the bath conditions can be altered without having any impact on the process phase of the substrate.

7.4.3.1 Working principle of airflow dyeing machine

In this process, dyeing is accomplished in a closed tubular system, basically composed of an impeller pump and a shallow dye bath. The fabric to be dyed is loosely collapsed in a form of a rope and tied into a loop. The impeller pump supplies a jet of dye solution, propelled by water and air, which transport the fabric within the dyeing system, surrounded by dye liquor under optimum conditions. Turbulence created by the jet aids in dye penetration and prevents the fabric from touching the walls of the tube, thus minimizing the mechanical

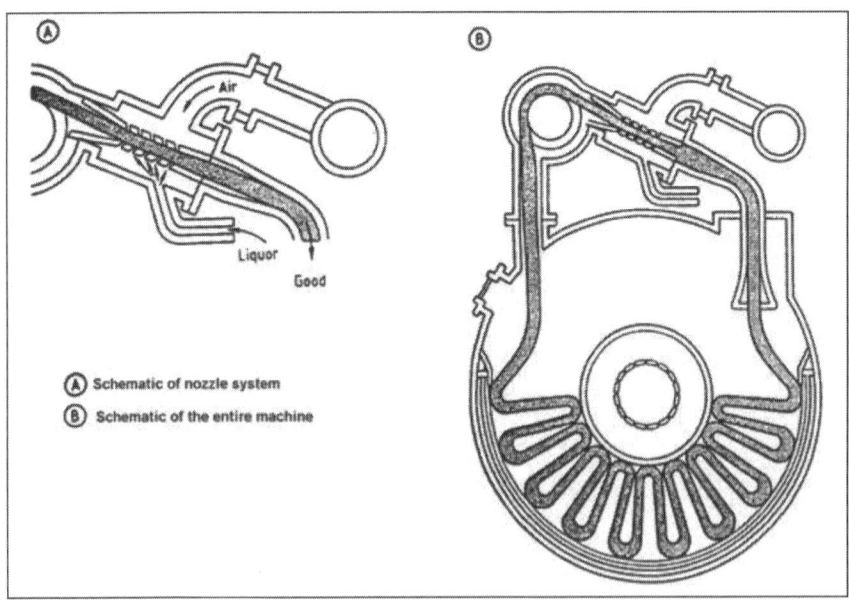

Figure 7.7 Airflow dyeing machine.

impact on the fabric. The process can be used for batch dyeing operations such as dyeing, bleaching, washing and rinsing.

In this machine, the dye tank contains a dye such as disperse dye, dispersing agent, levelling agent and acetic acid. The solution is filled in the dye tank. The solution will be filtered which reaches the tubular chamber. The material to be dyed will be loaded and the winch is rotated so that the material also gets rotated. Again the dye liquor reaches the heat exchanger and the operation is repeated for approximately 20–30 min at 130–135°C. Then the dye bath is cooled down after the material is taken out. The holding time and temperatures depend on the actual dyestuff, the material being dyed and the depth of shade selected.

The fabric in a rope form is fed into the machine without any tension. The fabric is lifted positively by a driven reel and then carried to the far end of the machine by the jet of the air. Dye additions are done through a secondary pressure pump at room temperature. As the fabric is lifted with a driven reel, the lengthwise tension and consequent creasing on it is minimized.

7.4.3.2 Advantages of airflow machine

The advantages claimed by airflow process are as follows:

a) Completely separated circuit for liquor circulation without getting in touch with the fabric

b) Bathless dyeing operation

c) Rinsing process offers all the added benefits of continuous processing as it is no longer a batch operation

d) Extremely low liquor ratio

e) Virtually nonstop process

f) Comparatively lesser energy requirement due to faster heating/cooling and optimum heat recovery from the hot exhausted dye liquors

g) Reduction in consumption of the chemicals (e.g. salt) dosage of which is based on the amount of dye bath

h) Lesser water consumption (savings up to 50% from the conventional jet dyeing machines)

i) Sensitivity towards ecology

j) Economical operation

k) More safety while dyeing

7.4.3.2   Unique Water Saving Capacity of Airflow Machine

Airflow dyeing machine can operate at a liquor ratio that is even below 1:5 while a conventional hydraulic dyeing system generally operates with a liquor ratio of about 1:10. It is worthwhile to know that exchanging the liquor ratio of 1:10 with a single 300-kg dye lot to ratio of 1:5 can result in water savings corresponding to an average monthly water consumption of one person.

7.5   Process steps in jet dyeing

The following is just an example. The real process parameters are to be derived depending on the colour being used, the GSM of the fabric and the actual lap-dip shade matching procedure.

Step no.	Process for 200-kg fabric batch	Parameter	Min	Water (L)
1	Fill Water	1100 ± 100 L	5	1100
2	Load the fabric		8	
3	Stitch the ends and make continuous rope		2	
4	Dose dyes and chemicals	As per slip	3	
5	Increase temperature by direct heating	Up to 60°C	10	
6	Hold at 60°C		5	
7	Increase in temperature from 60 to 90°C	At 2°C/min	15	
8	Hold at 90°C		5	
9	Increase in temperature from 90 to 110°C	At 1.5°C/min	14	
10	Hold at 110°C		5	
11	Increase in temperature from 110 to 130°C	At 1.5°C/min	14	
12	Hold at 130°C		30	
13	Bring down from 130 to 90°C	At 4°C/min	10	
14	Drain half bath		3	
15	Fill water		3	600
16	Drain half bath		3	
17	Fill water		3	600
18	Drain fully		5	
19	Fill water		5	1100
20	Unload		5	

7.6 Control points and checkpoints

The control points and checkpoints are to be decided by the type of the machine, the type of fabric processed, normal customer complaints and the problems faced earlier in the process. The following are just for illustration.

Control points

a) Selection of suitable jet dyeing machine for the material to be processed.

b) Deciding and selection of process parameters, namely:

- Process parameters
- Speed
- Running time for each operation
- Temperature
- Pressure
- Dye recipe
- Liquor ratio
- Washing and fixing sequences

c) Deciding acceptance criteria for shade matching and colour fixation

d) Deciding on chemicals and their quality to be procured

e) Employing trained and qualified employees

f) Evolving production norms

g) Evolving norms for consumption of chemicals, water and steam

Checkpoints

- **Material related**

  a) The fabrics received for dyeing against the plan received for dyeing

  b) The quantity of fabric received for loading in jet dyeing—metres and kilograms

- **Machine related**

  a) The condition of the machine

  b) The working of various valves and controllers

  c) Whether the spray nozzles are clean

- **Setting related**
  a) The force of water/air jet
  b) The recipe prepared
  c) The speed of reel

- **Performance related**
  a) Uniformity of shade throughout the fabric
  b) The shade obtained against the requirement
  c) Fastness properties of the dyed material

- **Documentation related**
  a) The design number and the shade dyed
  b) Quantity dyed

- **Work practice related**
  a) Whether the jet dyeing machine was thoroughly cleaned before loading the material
  b) Whether the ends of the fabric was stitched properly after feeding into the machine
  c) Whether the pressure and temperature maintained are as per norms in HTHP jet dyeing machine

- **Logbook related**
  a) Machines working
  b) Starting time of the running lots
  c) Activities done and to be done

- **Management information system related**
  a) Dyeing machine number
  b) Design number
  c) Number of metres
  d) Weight in kilograms
  e) Colour used
  f) Chemicals used

- **General**

  a) Use of safety gadgets such as gloves and gum shoes while handling chemicals

  b) Removing the completed batch and delivering to the next process

7.7   Normal Problems in Jet Dyeing

a) Cracks, creases and streaks in the fabrics dyed due to abrasion

b) Materials not being carried by air/water jet in case of higher GSM fabrics

c) Improper dye additions through a secondary pressure pump leading to improper shade

7.8   Dos and Don'ts

It is necessary to understand clearly what is supposed to be done without fail and what should not be done at any cost. Some examples are given below.

Dos

a) Study the process slip and follow instruction of supervisor in detail before programming the machine

b) Check the pH of cloth before unloading from the machine

Don'ts

a) Do not allow the reel speed to go above 200 m in normal jet dyeing machines

b) Do not allow the reel speed to go above 600 m in case of airflow machines

c) Do not allow the reel speed to go above 120 m in case of soft flow machines

d) Do not unload the material before getting the test results of the sample sent for testing

Airflow processing of fabrics

8.1 Concept

In airflow processing, air currents at high speed are used to give softening effect on fabrics, drying of fabrics and also for a number of other purposes in fabric finishing. By using air currents in place of conventional water and chemicals, the problem of pollution is avoided while the fabric is given different finishes.

In 1985, Biancalani introduced the AIRO concept, which revolutionized the thinking of 'wet processing'. AIRO improves the hand and look of the textile product with considerable effects on its quality, allowing a number of finishing combinations, ranging from a basic drying and softening treatment to sophisticated wet processes with added chemicals. It is based on the combination of an air-only fabric transport system and the high-speed crashing of the material against a special grid.

Some commercially available machines using air as conveyor are:

a) AIRO Fabrics Softening Washing and Drying Machine Model AIRO DUE by Biancalani S. P. A

b) AIRO Fabrics Softening Washing and Drying Machine Model AIRO Quattro by Biancalani S. P. A

c) AIRO Fabrics Drying Machine Model AIRO*SL by Biancalani S. P. A (Fig. 8.1)

d) Air-Flow Softening Finishing Machine for Washing, Drying and Softening (LD-4) Zhangjiagang Polygee Environmental Protection Technology Co., Ltd.

Figure 8.1   AIRO Fabrics Softening Washing and Drying Machine model AIRO.

Figure 8.2   Fabric guiding rings.

8.2 Principle of operation

a)   The fabric, in rope form, is driven by an intense airflow and accelerated without any tension inside the processing tubes, and then ejected against an impact grid positioned in the rear of the machine, where it discharges all the accumulated kinetic energy.

b)   After falling on to a Teflon-coated slide, the fabric flows to the front of the treatment vat, ready for being guided by a roller (Fig. 8.2) to the beginning of the new cycle.

c)   The exclusive air transport system keeps the fabric safe from any form of mechanical crease or abrasion and delivers a strong and delicate action at the same time without any friction and with a complete absence of defects during washing or squeezing process as well (Fig. 8.3).

Figure 8.3 Fabric crashing the grid.

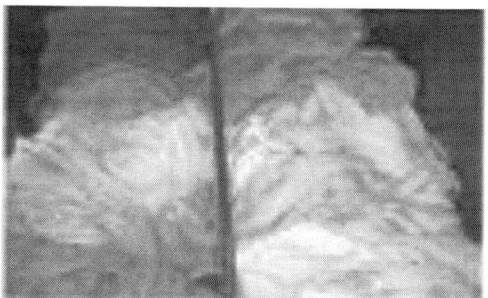

Figure 8.4 Fabric getting massage inside by air.

d) The fabric is continuously massaged by air inside the tubes and immediately opened at the exit (Fig. 8.4).

e) The air enveloping the ropes forces the water to deeply penetrate the fabric and also saturates in the inner parts of the fibres, preparing them for the strong mechanical action that will determine their optimum swelling and largely improving the efficiency of washing process.

f) The fabric is completely opened by the airflow as soon as it is ejected from the tubes, preventing the formation of any kind of marks, and undergoes a strong and controlled crash against the impact grid. This contributes to release and expel the water and provides a mechanical and permanent effect of softening and natural compaction of the fibres delivering a special hand, known as AIRO Hand (Fig. 8.5).

g) Sophisticated hydraulic system with introduction and recirculation pumps, bath filtering, direct and indirect heating, chemical tank with

Figure 8.5   Fabric in action at AIRO.

mixer and pH meter, in combination with air action, enables different types of wet treatments.

h)   Multipoint steam injection system is provided for intensive and uniform steaming. The first set of nozzles injects steam at the entrance of the treatment tubes, where it is pushed by the air to penetrate deeply through the fibres creating an intense messaging effect. A second set of nozzles acts upon the fabric at the point where it opens and helps to maintain optimal steam concentration inside the machine.

i)   The equipment is provided with cyclone-based system with cascade filters (Fig. 8.6) and water recycling, which provides an automatic self-cleaning system that rapidly cleanses the whole interior of the machine.

j)   All parameters are managed by a programmable logic controller and controlled through a touchscreen. Programming of parameters, such as temperature, air capacity, speed, water inlet, chemicals inlet and steam supply, is done using the touchscreen.

k)   The machines are provided with a filtering system to guarantee zero-level emissions and minimum maintenance.

l)   The model AIRO*SL, which is only for drying, is provided with an automatic self-cleaning disc filter with lint compactor while the versions with washing capability are provided with a cyclone-based system with cascade filter and water recycling.

m) Defibrillation can be achieved by the exclusive mechanical action of beating, which removes the pilling completely.

Figure 8.6   Cyclone-based system with cascade filters.

n)   By adapting particular air transporting system and the design of ejector tubes, it is possible to achieve either fibrillation or defibrillation on light- and heavyweight fabrics, either woven or knitted, of lyocell and Tencel.

o)   Multiple steam injection can be used to ensure an intensive and uniform steaming, allowing to achieve excellent dimensional stability of woven and knitted fabrics to obtain perfect results of lambskin and polar fleece. A first set of nozzles injects the steam at the entrance of the treatment tubes, where it is pushed by the air to penetrate deeply through the fibres, creating an intense massaging effect. A second set of nozzles acts upon the fabric at the point where it opens, that is, at the exit of the tubes and right before the contact with the grid, and helps to maintain the optimal steam concentration inside the machine.

8.3   Various activities that can be done using airflow machines

The airflow machines, although looks very simple, can perform a number of activities. Following are some examples.

1.   Activities those can be done on all versions are as follows:

   a)   Mechanical, chemical-free softening of dry fabric, with or without steam application

   b)   Drying and mechanical softening of wet fabric

   c)   Intensive steaming treatments

   d)   Polymerization of resin-treated fabrics

   e)   Compaction and relax.

2.   Activities those can be done additionally on machines fitted with washing arrangement are as follows:

   a)   Application of chemicals (i.e. softener) by exhaustion

   b)   Washing

   c)   Enzyme treatments

   d)   Fibrillation and defibrillation

   e)   Biopolishing

   f)   Desizing

   g)   Bleaching

   h)   Milling

   i)   Rapid scouring

3.   Various results that can be obtained using airflow processing are as follows.

   a)   Softness, volume and drape

   b)   Lamb skin and polar fleece

   c)   Inhibition of pilling formation

   d)   Opening of the chenille: three dimensionality on jacquards

   e)   Swelling and volume of velvet, corduroy and pile fabric in general

   f)   Compaction and dimensional stability on knitted and stretch fabrics

   g)   Delave and colour reduction

   h)   Tencel and lyocell fibrillation and defibrillation

   i)   Resilience and crease resistance (easy care, no iron)

   j)   Enzyme washing of silk fabric

   k)   Sense effect on cupro and blends

   l)   AIRO Hand on particularly rigid fabric or coupled laminated material

m) Wrinkled effect and casual look

n) Elasticity and sewability

o) Softness and uniformity of the surface of printed cloths

p) Peachy hand on emerized and brushed surfaces

q) Grain-enhancement and natural/aged look on synthetic leather

r) Opalescence

s) Paper hand

t) Gummy hand

u) Frosting effect

8.4 Typical work procedure for softening cotton fabrics

The typical operating procedure (standard operating procedure) for softening cotton fabrics using AIRO DUE machine has following basic steps.

a) Clean the machine thoroughly and check both the tubes for clean and smooth inner surface.

b) Check the squeezing mangle and ensure that the surface is smooth without any cracks.

c) Clean the squeezing mangle before starting the machine.

d) Check the air pressure and ensure it as required.

e) Check the steam pressure (if required) and air pressure ratio and ensure it as required.

f) Check the water spray nozzle and ensure it to be clean.

g) Check the temperature if required.

h) Check the cleanliness of the stitching machine.

i) Take programme from a supervisor.

j) Load the fabric of specified length as per process sheet.

k) Verify the batch number, lot number and trolley number with the job card and process sheet as applicable.

l) Stitch the ends of the fabric and make it a continuous rope.

m) Inspect the fabric after the process of AIRO. Check for damage such as stain, distortion and torn fabric.

n) Cover the trolley with polyethylene paper at the delivery end and keep at the proper place.

o) Keep proper tag on the batch.

Normal process parameters to soften cotton fabrics

Parameters	
Speed	500 rpm
Airflow	99
Softening temp.	90°C
Drying temp.	100°C
Process sequence—steps	**Time taken (min)**
Load	20
Washing	10
Drain	2
New water fill-up	5
Softening	40
Squeezing	5
Drying	60
Cooling	10
Unloading	20

Terminating the production

a) Allow the water spray to work on both sides.

b) Close steam valve and air pressure valve.

c) Clean the inside of the machine with water.

d) Cool the machine with air after draining the water.

8.5 Control points and checkpoints

It is essential to have clarity on the points to be controlled to achieve the targets and those to be checked to ensure the process of control. These points need to be reviewed from time to time and modified to suit the requirements of individual companies and their targets. Each mill should prepare its own 'control points and checkpoints' and display them in the work area so that the people refer and follow. The control points and checkpoints should consider the model of airflow processing machine and the processes being conducted. The following is a general and basic minimum guideline.

8.5.1 Control points

a) The process parameters to be adapted considering the type of fabric and the actions required

b) The speed

c) The length of the fabric to be processed in each run

8.5.2 Checkpoints

- **Material related**

  a) The design number of the fabric as per the work order

  b) The fabric density matching to the plan

- **Machine related**

  a) Condition of the nozzles and the pipes

  b) Condition of the fabric crashing grid

- **Setting related**

  a) The temperature set

  b) The speed set

- **Performance related**

  a) Uniformity of process observed after completing the process

  b) Check for the process problems such as stains, cloth tearing out and distortion

- **Documentation related**

  a) Design number and lot number processed

  b) Combination of processes used

- **Work practice related**

  a) Whether the machine was cleaned thoroughly before starting

  b) Whether the surroundings are kept clean all the time

- **Logbook related**

  a) Design and lots completed

  b) Design and lots waiting for the process

  c) Problems faced if any during the shift

- **Management information system related**
  a) Design and lots fed
  b) Metres in each design and lot
- **General**
  a) Whether the safety precautions are taken as needed

8.6 Dos and don'ts

8.6.1 Dos

- Take the process parameter from the supervisor as AIRO 1000 can be used for different processes.
- Take a correct length of fabric as required for the process.

8.6.2 Don'ts

- Do not assume same parameters for all batches.
- Do not take longer lengths or shorter lengths as it changes the number of hits on impact grid and; hence, the finish obtained.

9.1 Drying

Drying is a process of removing moisture content from any material. It can be done by various methods listed as follows:

a) Heating

b) Blowing dry air on the materials

c) Centrifuging and throwing the moisture out

d) Using radio frequency (RF) waves

e) Using vacuum

f) Exposing to open sunlight and air and so on

In textile manufacturing and while using textiles, drying is necessary to eliminate or reduce the water content of the fibres, yarns and fabrics after wet processes. Drying, in particular by water evaporation, is a high-energy-consuming step and hence attention is needed to reduce overall energy consumption by reuse/recycling options.

9.2 What drying should and should not do?

9.2.1 Should do

Excess moisture present in a fabric might hinder the further processes, and hence it is necessary to remove that excess moisture. Drying should remove the excess moisture from the textile material.

Drying should ensure maintenance of minimum moisture required for the material to remain stable while processing.

9.2.2 Should not do

a) Drying should not burn or damage the textile materials.

b) Drying should not result in lower moisture content than that is required for the textile material. It means the material should not be over dried.

c) Drying should not result in the change of chemical composition of any materials added on to textiles.

d) Drying should not reduce the strength of textile materials.

e) Drying should not lead to fading or dulling of the colour of textile materials.

9.3 Areas where drying is needed in textiles

a) Drying can be applied to textiles in different forms such as loose fibres, yarn in hank form, yarn on cheeses, yarn sheet in sizing, fabrics in rope form in processing, fabrics in open width in processing, garments after washing and clothes after domestic washing and so on. Sometimes, the clothes worn might get wet due to coming in contact with water such as rains, slipping into a water body, and so forth, and drying is needed in such cases also.

b) Fibre drying is normally carried out after bleaching of cotton fibres used for surgical purposes, after dyeing of fibres to get special fibre dyed effects. In some rare cases, when fibres get wet due to exposure to rainwater or water leakage from humidifiers, they are dried by spreading them in the open air.

c) Yarns are dried in hank form after wet processing.

d) Yarns are dried in cheese form after wet processing in high temperatures–high pressures dyeing machines.

e) Yarn sheets are dried after applying size in a sizing machine.

f) Fabrics undergo various drying operations after every process in wet processing such as desizing, mercerizing, scouring, bleaching, dyeing, printing and finishing operations.

9.4 Different drying techniques adapted in textiles

Drying techniques may be classified as mechanical or thermal. In general, mechanical processes are used to remove the water which is mechanically bound to the fibre. This is aimed at improving the efficiency of the following step. Thermal processes consist of heating the water and converting it into steam. Heat can be transferred by means of convection, infrared radiation and direct contact and by RF. Normal drying systems are as follows.

a) The water content of the fibre is initially reduced by either centrifugal extraction or by mangling before evaporative drying. The evaporative drying is done by using dry and/or hot moving air.

b) The water content in hanks is initially reduced by using centrifugal extraction followed by evaporative drying.

c) The drying of cheeses after wet processing is normally carried out by centrifugal extraction followed by radio frequency drying (RFD).

d) The drying of yarn sheet after drying is done normally by using drying cylinders. In some cases, it is coupled with hot air drying or infrared drying.

e) The fabric drying in rope form is done by using mangle squeezing followed by blowing hot dry air.

f) The fabric in open width is dried by using a chain of drying cylinders or by using hot air drying after initial squeezing by mangles.

9.5 Centrifugal hydroextraction

Textile centrifugal extractors (hydroextractors) are a robust spin dryer and normally operate on a batch principal (Fig. 9.1). When using conventional batch hydroextractors, fibre is unloaded from the dyeing machine into specially designed fabric bags which allow direct crane loading of the centrifuge. When the perforated drum is revolved, the water is thrown out because of the centrifugal force. An extraction cycle of 3–5 min reduces the residual moisture content to approximately 1.0 L/kg dry fibre (in the case of wool). In case of hanks, they are unloaded directly into the rotating perforated cage. Hydroextraction reduces the moisture content approximately 0.4 L/kg dry weight. Hydroextraction is used for drying the washed clothes in domestic washing machines.

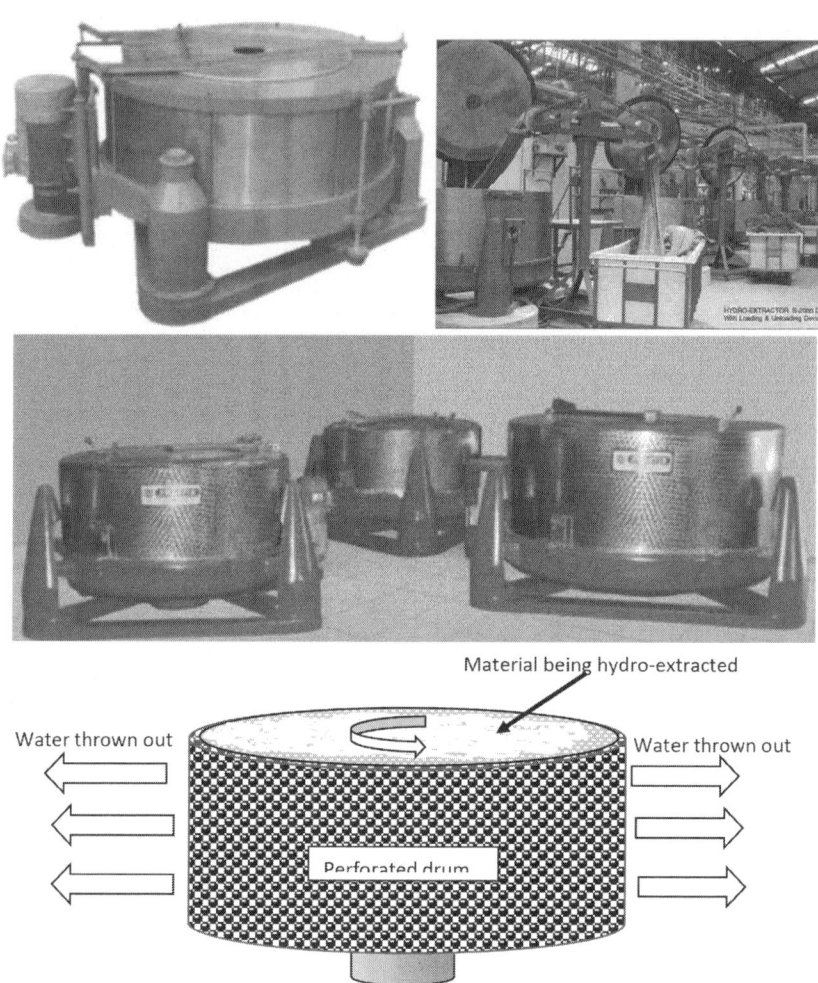

Figure 9.1   Hydroextractor.

9.6 Mangling

Pneumatically loaded mangles (heavy solid rollers) can be used to reduce the water content of textile materials to be dried such as dyed loose fibres, dyed hanks, sized yarns or treated fabrics (Fig. 9.2). The materials pass in between two heavily loaded solid rollers pressing each other, which compresses the textile materials and removes the water which is superfluous.

In case of fibre drying, a mangle is often associated with a fibre opening hopper which is designed to break up the dye pack and present the fibre to a

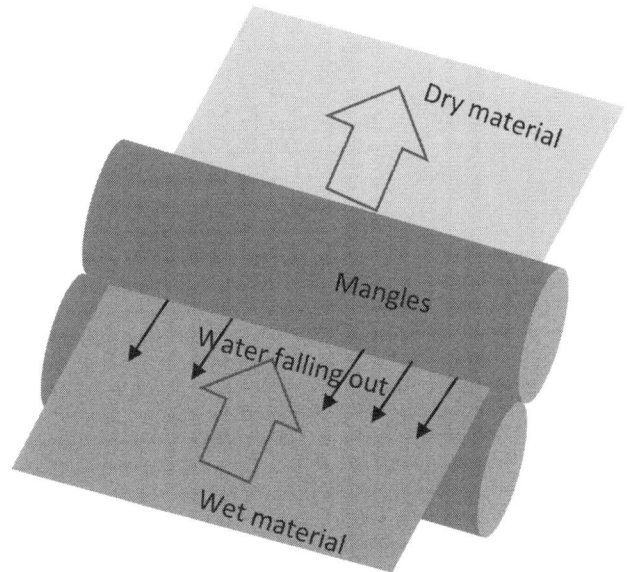

Figure 9.2   Mangling.

continuous dryer as an even mat. Mangling is invariably less efficient than centrifugal extraction. In case of hank drying, the mangling may be done prior to the hanging of hanks in open air for drying. In case of yarn sizing, the drying cylinders are used to dry the yarn coming out of mangles. In all open-width fabric processing systems, the fabrics can be dried using multiple drying cylinders or hot air drying chambers after preliminary mangling operation. Normally, squeezing by mangles is not suggested for delicate fabrics.

9.7 Drying Cylinders

Drying cylinder is a steam heated cylinder used for drying sized yarn or fabrics in the processing area. There may be many of heated revolving cylinders for drying fabric or yarn. They are arranged either vertically (Fig. 9.3) or horizontally (Fig. 9.4) in sets, with the number varying according to the material to be dried. They are often internally heated with steam. Entry cylinders are normally coated with a nonstick, chemical and heat-resistant polymer such as Teflon to prevent sticking.

Cylinder drying is an efficient and cheapest system for fabric drying, with energy saving, since it is direct contact drying. Cylinder dryers are designed for single or double end fabric, running side by side for getting higher production.

Figure 9.3   Vertical drying range.

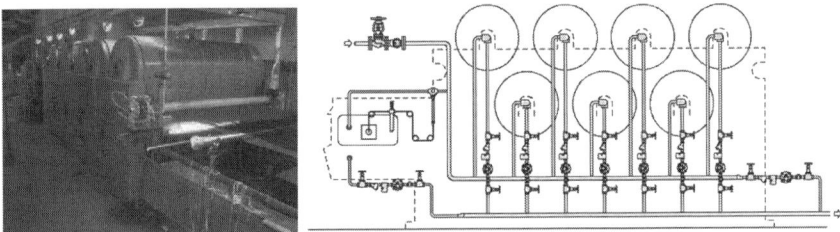

Figure 9.4   Horizontal drying range.

They are used in continuous operation with wet processing, finishing lines and also as a separate drying machine.

Fabric entry to the first cylinder can be from top or bottom, as per customer's requirement. Normally, the first two cylinders shall be Teflon coated to avoid sticking to the cylinder. The last cylinder shall be a cool cylinder.

High efficient drying capacity is ensured at higher temperature working. The evaporated water particles should be removed from the drying surface in order to get the best efficiency of drying. Hence, exhaust hood with the fan is provided at the top. The steam gets condensed inside the cylinders, and hence needs to be removed periodically. Condense water are removed by separate pipe for the initial start of the machine.

The temperature of the drying cylinders is controlled by controlling the steam pressure. Maintenance and calibration of steam pressure gauges are very important.

The speed of cylinders is also an important parameter to ensure correct and uniform drying of the material.

Heavy-duty ball bearing with pedestal is provided for longer life to the drying cylinders.

9.8 Hot air drying

In this machine, the drying unit is a closed chamber containing a number of guide rollers through warp yarn. If moisture remains inside the chamber it may condense and again fall on the yarn. Hence, continuous removal of moist air is required.

In hot air drying system, air is heated either by closed steam coils or by gas burners (Fig. 9.5). There shall be number of chambers such as two, three, four or five depending on the type of fabric to be dried. Different range of temperature can be maintained in each chamber depending on the process and the fabric.

Hot air (Fig. 9.6) is blown into the chamber causing the evaporation of moisture in the yarn or fabric. Exhaustion should be used to remove the moisture. If moisture remains inside the chamber it may condense and again fall on the materials such as yarn or fabric being dried. Hot air should be continuously passed through the chamber, so the process becomes somewhat expensive.

In a hot air drier for drying garment panels and fabrics, two endless mesh conveyors are placed lengthwise into the chamber, namely conveyor net and filter net. Each chamber contains a burner, which supplies hot air. This hot air is guided through the ducting line by suction fan. There are nozzles placed in between filter net and conveyor net. When the fabric passes on the conveyor

Figure 9.5   Hot air drying of garment panels.

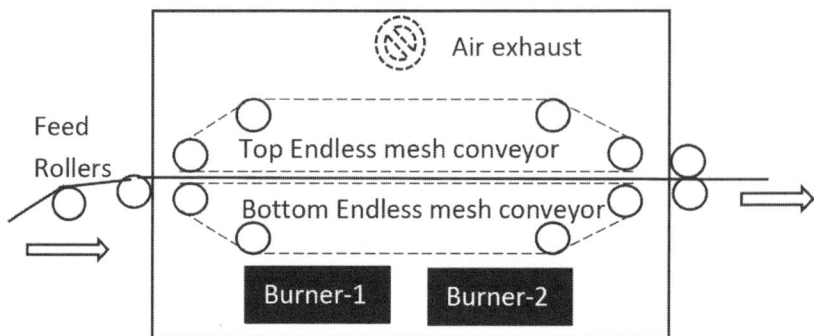

Figure   9.6 Hot air chamber.

net, hot air is supplied to the wet fabric. There are exhaust fans which suck the wet air and deliver to the atmosphere through the ducting line.

The speed of the dryer depends on the temperature of the machine and the density [grams per square metre (GSM)] of the fabric. If the machine temperature is high then machine speed is also kept high and when the machine temperature is low then machine speed is kept low. The vibration speed of the machine for heavy fabric is around 730 m/min and the normal fabric is 480 m/min.

The drying is done at low temperature for light shade fabrics, whereas for dark shades, higher temperature is needed. Following is an example of two chamber drying range.

Shade	Chamber 1 (°C)	Chamber 2 (°C)
Light	120	130
Medium	135	140
Deep	150	170

1. The temperature is set between 120 and 130°C for white and 150 and 170°C for colour fabric (GSM temperature or moisture content temperature).

2. The overfeed is set up to 10~20% or as required to get finish GSM.

3. The speed is as per GSM (6~20 m/min).

Depending on the fabric construction, the temperature and speed set shall differ. The following table gives an example of knitted fabric drying using Santex 2 chamber drying machine.

For cotton fabric

Fabric type	Overfeed (%)	Temperature (°C)		Speed (m/min)		Folder speed (m/min)
		Light colour	Deep colour	High grams per square metre (GSM)	Low GSM	
Single jersey	− 15 to −20	145	165	6.5~7	8~9	2~3
Single lacoste	−20 to −25	145	165	6.0~7	8~9	2~3
Polo pique	−20 to −25	145	165	6.0~7	8~9	2~3
Interlock	−20 to −25	155	170	5~6.5	7~8	2~3
Rib	−5	145	165	4~4.5	5~5.5	3~4
Grey mélange	−20	150	165	4~4.5	5~5.5	3~4

For polyester fabric

Fabric type	Overfeed (%)	Temperature (°C)		Speed (m/min)		Folder speed (m/min)
		Light colour	Deep colour	High GSM	Low GSM	
Single jersey	−5	135	115	10~12	8~10	4~6
Single lacoste	−5	135	115	10~12	8~10	5~6
Polo pique	−5	140	115	10~12	8~10	5~6
Interlock	−5	135	110	6~8	8~9	4~5
Rib	−5	145	115	4~4.5	4.5~6.0	5~6
Grey mélange	−5	130	115	6~8	5~5.5	5~6

• If the fabric is redder than the standard one, then reduce the temperature.

• If the fabric is yellower than the standard one, then increase the temperature.

• If the fabric is bluer than the standard one, then increase the temperature.

9.8.1   Drying of hanks using hot air chamber

Evaporative dryers consist of a number of heated chambers with fan-assisted air circulation, through which the hanks pass suspended on hangers or poles or supported on a conveyor. The hank sizes employed in carpet yarn processing require a slow passage through the dryer to ensure even final moisture content, and a residence time of up to 4 h is common. Air temperature is maintained below 120°C to prevent yellowing (wool yellows above the boiling temperature).

All designs are capable of continuous operation. Thermal input is normally provided by a steam heated exchanger and many designs incorporate air-to-air heat exchangers on the dryer exhaust to recover heat. In rare cases, hanks may be dried by employing a dehumidifying chamber. Moisture is recovered by condensation, using conventional dehumidification equipment. In comparison to evaporative dryers, yarn residence time tends to be longer, but energy consumption is lower.

Hot air drying has the advantages of regular drying, no shinning effect, nonsticky property and high-speed drying. The disadvantages are expensive process, for the closed chamber, required more time, less suitable for fine yarn and difficult to maintain temperature.

9.9   Yarn package drying

The moisture of dyed packages is initially reduced by centrifugal extraction. Specially designed centrifuges, compatible with the design of the dyeing vessel and yarn carriers, are employed.

Traditionally, packages were oven dried, very long residence times being required to ensure adequate drying of the yarn on the inside of the package. Two methods are currently used: rapid (forced) air drying and RFD; the latter sometimes being combined with initial vacuum extraction. Forced air dryers are generally operated by circulating hot air from inside of vacuum extraction. Forced air dryers generally operate by circulating hot air from the inside of the package to the outside at a temperature of 100°C followed by conditioning, in which remaining residual moisture is redistributed in a stream of air passing from the outside to the inside of the package. RF dryers are operated on the conveyor principle and are perhaps more flexible than the types mentioned above. Lower temperatures can be used and energy efficiency is also high.

9.10   Radio Frequency Drying (RFD)

A wet product submitted to a RF field absorbs the electromagnetic energy so that its internal temperature increases. If a sufficient amount of energy is supplied, the water is converted into steam, which leaves the product; that is to say, the wet product gets dried.

In RFD system, the alternating electric field between two electrodes is created by RF generator (Fig. 9.7). The material to be dried is put crossed between the electrodes, where the alternating energy causes polar molecules in the water to continuously reorient them to face opposite poles – in the same way as the

Figure 9.7   Radio frequency textile yarn dryer.

magnets move in an alternating magnetic field. This movement causes the friction and due to this water in the material rapidly evaporates throughout the material.

To determine the amount of heat generated in the product, the frequency, the dimensions of the product, the square of the applied voltage and the dielectric 'loss factor' of the material is required. The dielectric loss factor is essential to measure the ease with which the material can be heated by this method. The main reason for this is water, which is far more amenable than other materials. This reduction in loss factor or receptivity to RF energy provides valuable protection against overheating of the dried material. Where uniformity of product dryness is essential, RFD is the ideal drying process.

The heating/evaporation processes carried out by means of RF are:

a) **Endogenous**: The thermal energy is not supplied to the wet product by an external heat source, but rather it is generated directly throughout its mass.

b) **Selective**: The electromagnetic energy is absorbed mainly by the moisture and not by the product itself.

c) **Controlled**: The heating/evaporation rate is directly proportional to the amount of electromagnetic energy supplied, and this energy can be controlled precisely.

9.10.1   Why RFD?

a) Water is more receptive than any other dielectric material. Therefore, in process of RFD, RF power will be absorbed in higher amount from wetter areas, resulting in uniform moisture distribution.

b) In the conventional drying process, many times, uneven shrinking consequences into surface cracking. In RFD, due to uniform moisture distribution, surface cracking is reduced.

c) In RFD, RF is direct form of applying heat, so no wastage of heat shall be there in this process.

d) RF dryer required one-fifth or one-eighth space required over the conventional dryer.

e) RFD is 2–20 times faster than conventional drying methods.

f) Here, heating begins directly in the product so the dwell time is far less than in a conventional dryer.

g) This process is eco-friendly process.

h) Maintenance cost is low as compared to other drying processes.

RFD has the following advantages:

a) **High energetic efficiency**: The electromagnetic energy is transferred directly to the whole of the wet product instantaneously, without losses to the surroundings, and is entirely exploited for the drying process. The energy transfer is not negatively affected by variable parameters such as the dimensions or the density of the product.

b) **Outstanding quality of the dried product**: All problems normally caused by the heat transmission phenomena are totally eliminated, so that the product dries quickly and uniformly down to the desired residual moisture level, with a beneficial steaming and bulking effect of the fibres which greatly improves the physical properties such as elasticity, softness, hand and colour effect.

We shall discuss the operations of RFD dryer in detail in a separate chapter.

9.11 Hydroextraction by suction

The fabric is transported flat over a 'suction drum' which is linked to a pump. The external air is sucked through the fabric and thereby removes the excess water. The resulting residual humidity is still about 90%.

9.12 Stenter

a) This machine is used for complete drying of the fabric. The fabric is conveyed through the machine in open width. A hot current of air is blown across the fabric thereby producing evaporation of the water.

b) The fabric is sustained and moved by two parallel endless chains. The fabric is hooked undulating and not taut to allow its shrinking during

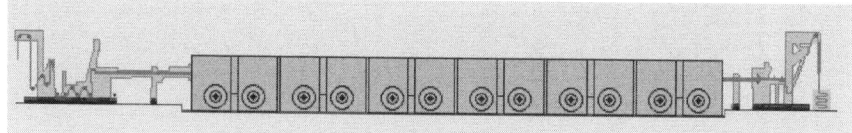

Figure 9.8   Schematic diagram of a stenter.

drying. Here, the fabric is straightened weftwise and warpwise by the tension provided by the tensioners and then the fabric is heated in wet form. There are different numbers of heating chambers depending on the maker of the machine. In these chambers, the fabric in wet form is treated with air at high temperature in order to stabilize it in required dimensions. The temperature of every successive chamber is higher than the previous one.

c)   Most common stenter designs are horizontal and multilayer, but many new designs exist (Fig. 9.8). In the horizontal stenter machine, the fabric enters wet from one side and exits dried from the other. In the multilayer type, it enters and exits from the same side. In the first one, the fabric moves horizontally without direction changes, while in the second, it is derivate of many timers, which makes this equipment unsuitable for delicate fabrics. On the other hand, horizontal stenter frames occupy more space and are less efficient (in terms of energy consumption).

d)   Hot flue dryer: This machine is composed of a large metallic box in which many rolls derivate the fabric (in full width) so that it runs a long distance (about 250 m) inside the machine. The internal air is heated by means of heat exchangers and ventilated.

e)   Contact dryer (heated cylinder): In this type of machinery, the fabric is dried by direct contact with a hot surface. The fabric is longitudinally stretched on the surface of a set of metallic cylinders. The cylinders are heated internally by means of steam or direct flame.

f)   Conveyor fabric dryer: The fabric transported within two blankets through a set of drying modules. Inside each module, the fabric is dried by means of a hot air flow.

g)   This equipment is normally used for combined finishing operations on knitted and woven fabrics when, along with drying, a shrinking effect is also required in order to give the fabric a soft hand and good dimensional stability.

We shall discuss the operations of a stenter in a separate chapter in detail.

9.13   Relax dryers

Relax drying means drying the wet processed fabric in a tensionless condition. The purpose of using relax dryer is to dry and/or shrink dyed or bleached delicate woven fabrics, retaining the fabric properties.

Relax dryers are hot air drying machines constructed in a way so they can suit the fabric which has to be passed in a relaxed condition (Fig. 9.9). The fabrics are carried by conveyors (Fig. 9.10) and hence there shall be no tension. The conveyors are made of special high heat-resistant mesh.

The dryer is integrated with motorized centring and tensioning device for individual conveyors. These dryers are ideal for knits, terry towels and delicate fabrics. All hassle free drying can be accomplished with relax dryer.

The normal standard operating procedure for operating a relax dryer is as follows:

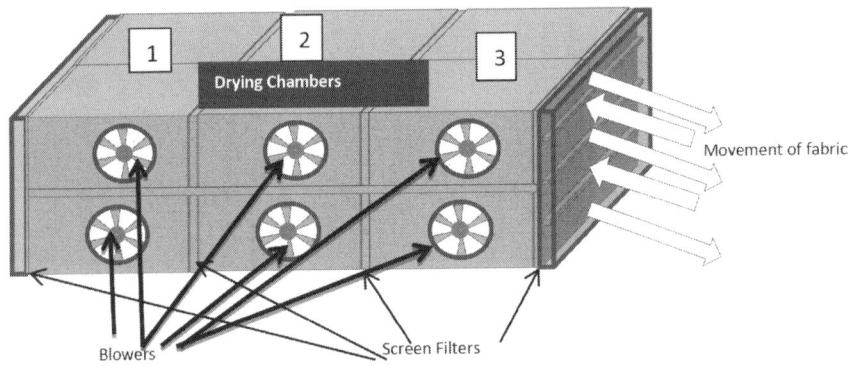

Figure 9.9   Drying unit of relax.

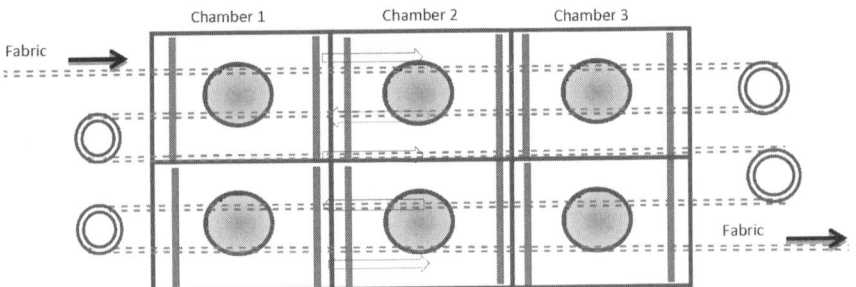

Figure 9.10   Passage of fabric in drying unit at relax dryer.

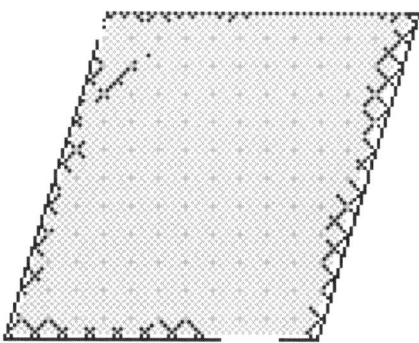

Figure 9.11 Filter mesh filled with dust and ash.

i. Clean the machine with compressed air and then wipe with a clean cloth before starting the work. Clean all sensors and photocell units with clean dry cotton cloth.

ii. Drain condensate water from the pipe lines and then open the steam valve.

iii. Switch on the machine. Start all blowers and material conveyor belts.

iv. Clean all the filter mesh in the drying chambers (Fig. 9.11). (Note: one chamber has eight filter frames, and there are three chambers.)

v. Start heating the steaming chamber. After the machine reaches full temperature, feed the material.

vi. While the heating process is on, run the conveyor belts. If heating process is done by keeping the feeder stationary, the conveyor belt is likely to lose its life.

vii. Set the machine parameters such as speed and temperature for the material to be processed by discussing with supervisor.

viii. The machine operates between 110 and 130°C. The normal working temperature is between 125 and 130°C. Do not operate the machine if the temperature is less than 110°C.

ix. The machine runs at the required speed which is decided depending on the quality.

x. Write the complete details of the material being processed in the production book.

xi. Stitch the ends and ensure continuous feed. The machine will not stop for any operations such as feeding or taking material out.

Figure 9.12 Conveyor sheets of a relax dryer.

xii. Put the fabric to be dried between two conveyor sheets (Fig. 9.12).

xiii. In case of observing any abnormality, inform the concerned supervisor immediately and get it rectified.

xiv. In case of any electrical or mechanical problem in the machine, get it rectified by the concerned engineering person. Write the details in the production book.

xv. Cover the processed material with polythene sheet and keep at the designated place.

xvi. Use only polytetrafluoroethylene-coated fibreglass and silicone adhesive tapes for joining the conveyors as they have temperature resistance up to +260°C.

We shall discuss the operations of a relax dryer in a separate chapter in detail.

Hot-air stenters

10.1 Purpose and functions

Stentering means stretching. Stenters have an important role to play in fabric dyeing and finishing works (Fig. 10.1). The work of drying, heat setting and curing of fabric is done at stenters which also have an effect on the finished length, width and properties of the fabric. The purpose of the stenter machine, therefore, is to bring the length and width to predetermine dimensions. It has other purposes such as heat setting, applying finishing chemicals and adjusting the shade variation. The main function of the stenter is to stretch the fabric widthwise and to recover the uniform width.

The following are the normal functions of a stenter:

1. Heat setting for thermoplastic synthetic and blended fabrics and Lycra fabric

2. Controlling the width of the fabric

3. Applying finishing chemicals and fixing them on fabric

4. Controlling the loop of the knit fabric

5. Controlling the moisture level in the fabric

6. Controlling spirality in the fabric

7. Controlling grams per square metre of the fabric

8. Drying the fabric after finishing the process

9. Controlling shrinkage property of the fabric

10. Curing treatment for resin, water-repellent fabric.

In a stenter, the fabric can be processed at different speeds varying from 10 to 100 m/min and at temperatures up to and in excess of 200°C depending on the fabric and the treatment being given.

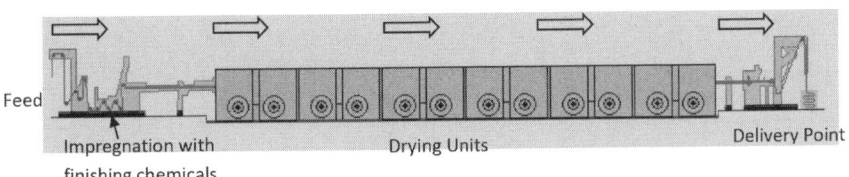

Feed

Impregnation with   Drying Units   Delivery Point
finishing chemicals

Figure 10.1   Schematic diagram of a stenter.

Sophisticated feed and transport mechanisms are provided in stenter to present the fabric to the oven in a way to ensure that the finished product meets customer requirements.

To offer better control, stenters are split up into a number of compartments, usually between two and eight 3-m sections each fitted with a temperature probe, burner/heat exchanger, fans, exhaust and damper.

The fabrics are finally delivered either in a batch or in a trolley as per the requirement.

The capacity of a process house is normally referred by the capacity of the stenters installed.

10.2   What stenters should and should not do?

10.2.1   Should do

a)   Giving various finishes while controlling the width of the fabric.

b)   Heat setting for polyester blends and also for other materials as needed.

c)   Maintain the moisture level in the fabric delivered.

10.2.2   Should not do

a)   Should not overdry the fabrics. Overdrying reduces the strength of the fabrics.

b)   Should not overstretch the materials. Overstretching not only reduces the strength of the fabric but also results in uneven shrinkage after washing.

c)   Should not deform the fabric by pulling it to one side.

10.3   Working principle

10.3.1   Feeding

The fabric enters the machine from a batcher or a trolley and is passed on to the solution tank through the guide rollers and the tensioners. The solution tank contains different finishing auxiliaries. The fabric is padded in the solution tank and then taken out, passed through the squeezing rollers that squeeze the fabric according to required percentage. Then, the fabric is passed through the weft straightener.

In the feeding unit, the fabric is collected from the batcher or trolley to the scray and then it is passed through the padders where the finishes are applied and sometimes shade variation is corrected (Figs. 10.2 and 10.3). The machines are supplied with either a J-scray or an accumulator for continuous operation.

Modern stenters (e.g. Swastik OPTIMA 2510 Stenters) are provided uncurlers and infrared (IR) edge sensors, which actuate the infeed device provided with

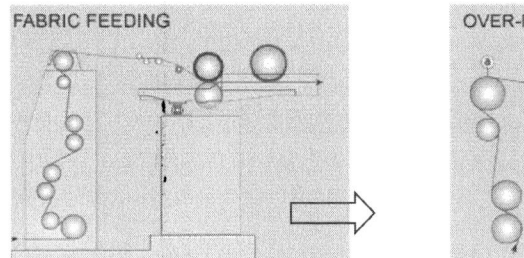 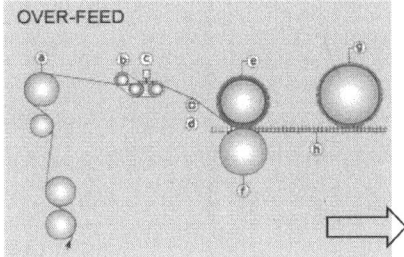

Figure 10.2   Feeding unit.

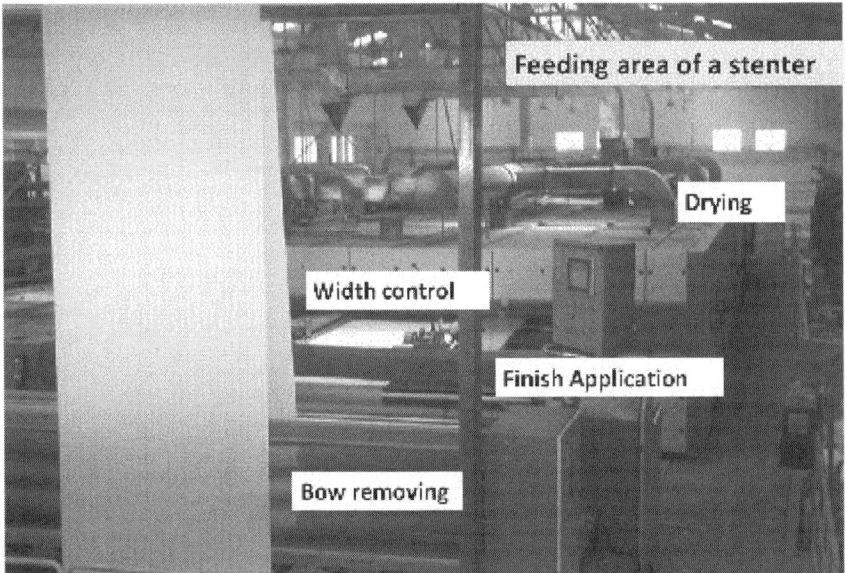

Figure 10.3   Feeding area.

rack-and-pinion system, ensuring perfect fabric holding in clips or pins even at high speeds exceeding 100 m/min. Overfeed systems are provided using variable-frequency drives (VFDs) with AC geared motors. They ensure perfect and precise overfeed controls to the fabric feed as required by the fabric, ranging from − 10 to + 50% by control of motor speed. Separate drives are provided to the fabric tension rollers and selvedge tension rollers. A closed-loop control system with encoder is provided for precise control of the drives.

Big batch units are normally provided with pneumatic cylinder for easy movement in modern stenters. Fabric supporters are provided at feeding end.

10.3.2　Padding of the finishing chemicals

Padding of chemicals is done in solution tank (Fig. 10.4), which is similar to a padding mangle. The padding rollers are pressured, by which the per cent impregnation of chemicals is controlled.

Modern stenters are usually equipped with a heavy-duty two or three bowl mangles designed to impregnate various chemicals and finishes such as starch, finishing chemicals, tinting agents and squeeze all types of fabrics, including delicate textile fabrics made from natural fibres, man-made fibres and their blends. Pressure rollers are usually made from heavy-duty pipes with heavy shaft and covered with synthetic or natural rubber. Generally, the driving roller is hard rubber or ebonite covered. Troughs are of stainless steel with guide rollers suspended from the side walls. Trough can be easily lifted or

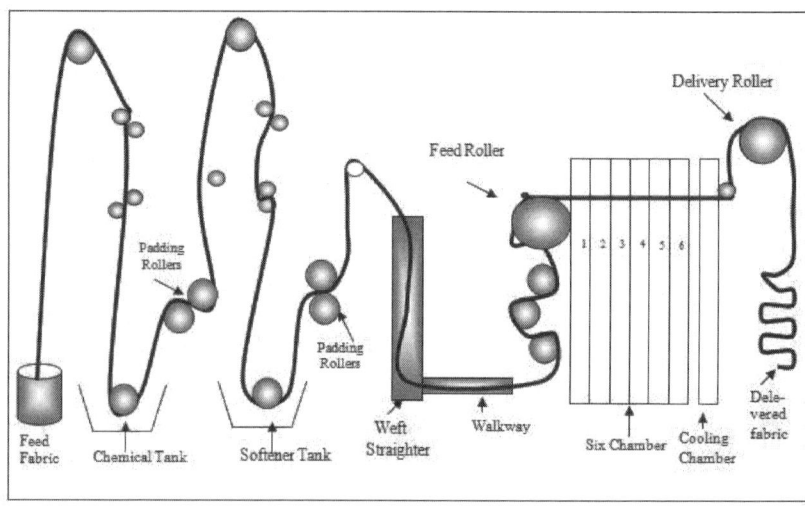

Figure 10.4　Schematic representation of solution tank in a stenter.

Figure 10.5   Weft straightener.

lowered for easy cleaning. Chemical dosing arrangement is also provided. To avoid centre–selvedge variations, the mangles are supplied with the top roller in 'anti-deflection' construction.

10.3.3   Weft straightener

Weft straightener contains a number of rollers through which the fabric is passed (Fig. 10.5) and there are sensors that sense the fault in weft and removes the fault. The function of the weft straightener is to set the bow and the weave of the fabric by gripping properly either by the clips or pins.

The fabric next comes on the head of the machine and is passed through a number of guide rollers and its selvedge is gripped by the chains or pins. Here, the fabric is straightened weft-wise and warp-wise it is straightened by the tension provided by the tensioners and then the fabric is heated in wet form. The pins have a disadvantage; it makes a hole in the fabric at selvedge but the stretching is much better compared to clips.

10.3.4   Drying unit

Stenters can be heated in different ways such as direct gas firing, indirect gas firing, thermal oil heating, steam heating and electric heating (Fig. 10.6).

The most common means of heating nowadays is by direct gas firing, with the burnt gas fumes being fed into the stenter oven. Indirect gas-fired stenters are also available but their efficiencies are poor when compared to direct-fired systems. Gas-fired stenters are highly controllable over a wide range of process temperatures.

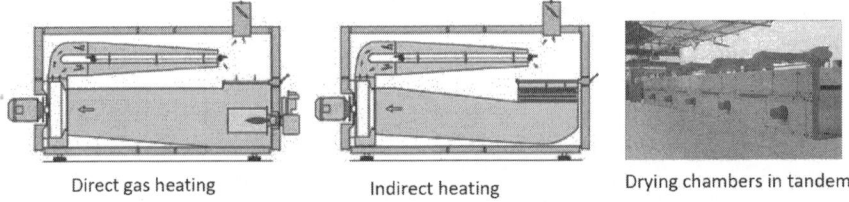

Direct gas heating   Indirect heating   Drying chambers in tandem

Figure 10.6   Drying chambers in a stenter.

Thermal oil heating requires a small thermal oil boiler (usually gas fired) and all its associated distribution pipework. It is less efficient than direct gas firing with higher capital and running costs. This can be used over a wide range of process temperatures. The oil itself can be used as a means of heating stenters. Because of the problems of incomplete combustion, this can only be done indirectly through a heat exchanger. This is relatively inefficient and a very few stenters nowadays use this mode of heating.

There are a number of steam-heated stenters. However, because of temperature limitations (usually a maximum of up to 160°C), they can only be used for drying and not for heat setting or thermo-fixation. The air is heated and forced against the fabric and then recirculated. A fraction of this air is exhausted and mixed with fresh air.

There are a number of heating chambers depending on the maker of the machine. In these chambers, the fabric in wet form is treated with high air temperature in order to stabilize it in required dimensions by heat setting. The temperature of every successive chamber is higher than the previous one.

There are different models of dryers available for stenters. Following are some examples.

Standard machines are provided with a suitable number of exhausts to efficiently remove the moist and contaminated air. The exhaust blowers can be supplied with frequency controllers and moisture sensors to maintain uniform moisture and excellent drying results.

Modern dryers are modular in construction which can be assembled as required (e.g. Dryer Model RDB-3000 of Swastik Processing Machines). The low-volume design Dryer RDB-3000 is equipped with a specially designed duct with the heat exchanger/burner at one end. The blower is fixed at the other end. The interconnecting duct does not permit any infiltration of air into the circuit without heating. Jet nozzles are designed to obtain maximum evaporation from the fabric and are specially designed for the specific fabric requirements. This is a compact, sturdy, simple and efficient dryer equipped with a closed-circuit air circulation system, using radial blowers directly coupled with AC motors which can be controlled by the invertors. This also occupies less space and has lower energy consumption. All drive elements are fixed outside the heated area and are accessible from outside. The dryer is equipped with insulating panels filled with high-density insulating material for minimum radiation losses. Specially designed jet nozzles at the entry and delivery end ensure minimum heat loss through escaping hot air.

The blowers are mounted directly on the motors mounted outside the machine and are easily accessible. The blower speed is regulated through VFDs. The high-volume jet nozzle elements provide even airflow to the fabric. They also provide better fabric support, gentle heat transfer and reduced energy intake. Different nozzle shapes for different applications ensure higher productivity for different quality and types of fabrics, distortion-free for delicate and knitted fabrics and are suitable even for pile fabrics. Air distribution to the fabric from the top and bottom nozzles is regulated through dampers fixed in the circuit.

The dryers can be supplied with different heating media. The machines are equipped with multipass finned tube type heat exchangers with a large heat exchange surface for quick heating with steam or hot oil circulation or large firing area for direct-fired and gas-fired burners.

In new-generation stenter model 'Split Air', each half chamber of the dryer is equipped with a twin blower system incorporating individual inverter-driven blowers for top and bottom set of jet nozzles. This results in infinitely adjustable air balance on each face of the fabric which is very useful in processing a wide variety of fabrics such as knits and coated fabrics of pile fabrics.

Variable flow air circulation system with separate blowers for the top and bottom nozzles is provided in modern stenters, which impinge hot air throughout the chamber in an absolutely uniform manner from centre to selvedge and selvedge to selvedge. Air ratio between the top and bottom is controlled electronically as each blower motor is driven by VFD. There are no mechanical dampers and therefore the airflow is very smooth. These high-speed blowers are designed to increase the airflow which has increased productivity. Even one-side flow is possible with increased or decreased air pressure depending upon fabric construction and air permeability. These are utilized especially for heat setting Lycra-based fabrics and also for coating applications. Filtering of circulated air and insulations is very effective to reduce fuel consumption and so forth.

Modern stenters such as Optima 2510 are equipped with suitable heat recovery systems to preheat atmospheric air with the help of exhaust air, to feed to stenter at high temperature or to heat water to be used in processing machines or in the boiler as feed water.

10.3.5　Clips and pins

The clips are made of several individuals, pressure die-cast, easily replaceable components. These components are made from corrosion-resistant aluminium alloy and are distortion-proof even at the high heat setting temperatures. These are normally designed considering the lowest individual details and to ensure long trouble-free performance with all types of fabrics.

The steel shoe holding the clips is fitted with sintered metal liners and connected with steel links. The roller chain glides smoothly on special cast iron rails. The sliding properties of sintered metal considerably reduce lubrication requirements even at the high heat setting temperatures, thus ensuring safe operation and long life. Some machines are also equipped with special polymer liners for lubrication-free working. Optima 2610, for example, is equipped with a vertical return pin chain.

In modern stenters for width control, clip-type, pin-type or both pin- and clip-type inlet desks are provided (Fig. 10.7). This ensures versatility of stenters.

A suitable chain cleaning device with steam, consisting of a set of hard-chrome plated nozzles with flexible steam connections, for easy and fast cleaning of pin–clip chain is normally provided in modern stenters. Stenter chain rail guides are made of properly seasoned cast iron chain, which supports the moving pin–clip chain. It is strong and distortion-free even when operating at elevated operating temperatures and at speeds in excess of 150 m/min. Various combinations of entry-and-exit chain rails are supplied for different applications. Extended entry rail tracks up to 6.3 m are provided which can

Figure 10.7　Clip and pin guides.

support the fabric feeding device, steaming device, selvedge gumming and predrying unit and so forth. Extended exit rail tracks up to 5.8 m length are provided to support the conventional air cooling zone, selvedge trimmer, chain cleaning device and so forth.

Modern stenters are provided with a simple motorized, quick-acting two-roller bow and four-roller skew weft corrector for the correction of weft distortions. Alternatively, some machines are equipped with automatic weft corrector with optical sensors and digital control. Additional sensors can be fixed on the machine exit to ensure perfect weft straitening.

The fabric is first gripped from selvedge through chains and then again the fabric is passed through weft straightener. If some faults are left in the first straightener, then these are removed in the second one. Then, the fabric is wound on a batcher.

Conicity of rails is obtained by acting upon the individual width-controlling screws of each compartment, de-clutching the individual unit from the main width adjustment control system and adjusting the individual rail to precise requirements, by acting upon each gearbox provided on individual adjusting screw with the help of a handwheel. Suitable indications are normally provided on each gearbox for precise control. It can also be automatically regulated as preprogrammed by acting on individual motors, with programmable logic controller (PLC) control.

Stop mark preventing damper system is provided with electronic bow and weft straightening device.

The modern stenters are equipped with requisite capacity of reduction gearbox with direct coupled motors avoiding chain and belt drive. The system helps in increasing the efficiency of individual drive console besides reducing power consumption and maintenance cost.

Tensions between padder and stenter draw roll and overfeed roller are controlled by unique drive system with load cell feedback and PC controls, making it easy to set the machine tension to suit the fabric.

10.3.6 After drying

After drying, cooling cylinders are provided for additional cooling of fabric and to reduce the sudden shrinkage. This is coupled with moisture indicator cum controller. Electronic control over the airflow is provided for easy handling of lightest to heaviest fabric with the same efficiency.

Filtering of circulated air and insulations is provided which is very effective to reduce fuel consumption.

Figure 10.8   Lint filter.

A suitable filter is provided just above the heat exchanger or in case of direct fired natural gas burners in the duct (Fig. 10.8). The filter is easily accessible and can be cleaned standing outside the chamber while the machine is in operation. A second security filter is also provided.

Some of the modern machines are generally equipped with a centring-cum-opening device; one or two padders and a weft corrector at the infeed. Driven scroll rollers, mechanical decurlers, pneumatically actuated pin-on device, selvedge gumming device, steaming device with IR predryer are normally fixed in the entry zone. The exit zone is generally equipped with an efficient cooling arrangement, edge trimmer, a fabric take-off roller, plaiter with conveyor and a batching device.

For example, Swastik Stenter Optima 2610 and 2620 are the stenters suitable to process open-width knit fabrics with a vertical return chain (Model 2610) and a vertical return chain with conveyor belt (Model 2620). Machines can be fully automatic with PLC and touchscreen to control all the processing parameters at the fingertip and have repeatability in the process. Accessories such as centralizing and stretching device, fully automatic bow and weft straightener, fabric temperature indicator and dwell time controller, exhaust humidity controller, auto overfeed control system are also available to enhance the quality and repeatability of the fabric finished.

For standard machines, the control system is with contactors and relays. The centralized operation system is provided on the operator console fixed on the entry box. The fabric speed is indicated on digital indicators placed at a strategic location. Two width-indicating scales, indicating fabric width, are also placed on the entry and exit sides. In case of a machine equipped with weft shrinking device, individual width indicators are provided at each rail joint.

Temperature-indicating controllers for each chamber are placed on the control desk along with the digital speed indicator at the entry of machine. Device for temperature recording, metre counter, control valves for oil, steam and gas lines are also normally supplied as optional units according to customer's requirement. The required indicators for voltage, current, air pressure and so forth are normally fixed on the operator console provided on the machine.

In case of a fully automated machine set-up, a PLC and a touchscreen human machine interface are provided to control all machine operating parameters. The PLC is generally equipped with recipe management software to ensure reproducible results.

10.4   Typical operating procedure for operating a modern hot-air stenter

The work procedure depends on the machine installed, the finishing to be done and the fabric in use and so forth. Following are the general indications of the general procedures. Each mill has to establish procedures suiting their actual requirements.

1. Get the details of the activity to be done; the lots to be run through stenters; the process details along with the speed, temperatures, the chemicals to be used and critical checks to be made before starting the work.

2. Clean the inside portions of all dryers and the filter meshes before starting the machine.

3. Give start request by pressing the start button. The machine will give a warning for 5 s by sounding horn and signal light flashes so that no one should remain in the danger zones marked. The machine shall start after 2 s after the warning signal.

4. Start the transport chain to move at creep speed.

5. Set the temperature and speed parameters on the control panel.

6. First, start the exhaust fans and then the air circulation fans. Connect heating unit and set temperatures. Heating unit shall start after the flushing period.

7. Cut off entry and exit and disconnect selvedge guards.

8. Adjust the machine data such as exhaust air fans, circulation air fans, fan controller, heating and temperatures, feed section and exit, expander roller, feed, selvedge uncurler, selvedge guard and fabric width.

9. Introduce the fabric using 'introduction' mode and start the winding on to the batcher using the 'winding start' mode. During this time, the maximum speed is 15 m/min.

10. Prepare the recipe as per instructions. Fill the trough up to 50 mm below its upper edge.

11. Fill the heating chamber of the double-walled trough of the padder with water. Release transportation screws so that the contact roller can rest fully on the material.

12. The spring-loaded condensate relief valve (rubber ball) opens when the steam pressure exceeds the maximum of 0.7 bar. Adjust the steam throughput so that the condensate can escape through the rubber valve with a slight overpressure.

13. Insert material according to cloth run plan and fill the trough with liquor. Heat trough up, switch on all drive units and check correct processing of the machine.

14. Open the hand valve for compressed air and set pressure according to the pressure diagram at the control desk.

15. The run of cloth is saturated with liquor in the padder. After this, the liquor is squeezed out of the cloth to specified residual moisture content. Uniform squeezing by means of the padder rollers is to be adjusted through contact pressure at the pressure roller. The pressure adjustment depends on the length of the roller. The squeezing pressure is dependent on the fabric width.

16. When there is a power failure or an emergency stop, the rollers are raised automatically to save the rollers from being damaged.

17. Press the relevant button on the control desk to swing the trough downwards for draining the liquor.

18. Clean the padder after use. Hose down the parts coming in contact with the liquor. The rollers should run slowly under contact pressure during cleaning. Hot soapy water with little hydrosulphite should be used for cleaning.

19. Ensure that hydraulic chain tension unit is filled with recommended hydraulic oil. Check oil level after every 200 working hours.

20. Check the chain tension and ensure that there are no slack links.

21. Set width as per requirement on each spindle.

22. Monitor the bending angle of the entry chain.

23. Ensure the gas pressure where gas burners are provided for heating and the thermal oil pressure in case of thermal oil systems.

24. Clean completely the inside of all gas ducts before connecting the gas control system and the gas burners. Blow out all gas ducts thoroughly with pressurized air.

25. Monitor the tension and overfeed for correcting the weft distortion while bow straightening at the take-up.

26. Deliver the treated materials either on to large batches or on to trolleys by plaiting as per the requirement.

27. Switch off the plant in sequence so that the parts of the machine are not damaged.

28. Remove the fabric and allow the dryer to continue without it. Then, switch off the heating system. Allow the chain to continue running at creep speed. Allow the air circulation fans and exhaust fans to continue and adjust the speed to minimum. Wait till the temperature reduces below 100°C. Stop the chain and then switch off the air circulation fans and exhaust fans. Switch off the controller voltage through the key-operated switch and then switch off the mains.

10.5   Control points and checkpoints

It is essential to have clarity on the points to be controlled to achieve the targets and those to be checked to ensure the process is in control. These points need to be reviewed from time to time and modified to suit the requirements of individual companies and their targets. Each mill should prepare its own 'control points and checkpoints' and display them in the work area so that the employees refer and follow. Following are some examples of the control points and checkpoints in a stenter operation.

10.5.1   Control points

a) Deciding and selection of process parameters considering the fabric and the finish, namely:

   • Machine setting

- Speed
- Temperature at different chambers
- Stretch
- The finishing recipe

b) Deciding acceptance criteria for the finish given

c) Deciding on chemicals and their quality to be procured

d) Employing trained and qualified employees

e) Evolving production norms

f) Evolving norms for consumption of chemicals, water and steam

10.5.2   Checkpoints

- **Material related**

    a) The fabrics received against the design number and lot numbers specified in the programme

    b) Checking the chemicals and auxiliaries and ensuring that they meet the acceptance criteria

- **Machine related**

    a) Condition of the mangles and trough

    b) Condition and alignment of the clips/pins in stenter

    c) Proper operations of the steam valves

    d) Condition of the width controllers

    e) The overall condition of the machine

    f) Machine operating as per the programme set

- **Setting related**

    a) The recipe prepared for finishing as required

    b) Maintaining the temperature maintained as needed

    c) The number of mangles and the pressures selected matching with the type of fabric being treated and the finish

    d) The stretch maintained and planned

    e) Working of the width-controlling working as set

- **Performance related**
  a) The production achieved and the norms
  b) The width uniformity
  c) The consumption of chemicals and the planned
- **Documentation related**
  a) The fabric details and the quantity received
  b) Quantity to be stentered and actually stentered in each variety
  c) Quantity of chemicals consumed
- **Work practice related**
  a) The people following safety requirements all the time
  b) Thorough cleaning of the machine after completing each run
  c) Proper storing of the chemicals
  d) The housekeeping around the machine
- **Logbook related**
  a) The quantity (number of metres) stentered in each design
  b) Stock of fabric in batches ready to be stentered and its details
  c) Quantity of chemicals consumed
  d) Problems faced in the shift
  e) Special instructions for the next shift
- **Management information system related**
  a) Design number and details
  b) Quantity of fabric (number of metres) received for stentering
  c) Quantity stentered
  d) Quantity approved as fresh quality
  e) Quantity rejected as seconds
  f) Consumption of chemicals
- **General**
  a) Whether the men employed are adequately trained
  b) Whether the consumption of chemicals, steam, gas and water is as per plan

10.6   Dos and don'ts

It is necessary to understand clearly what one is supposed to do without fail and what should not be done at any cost. Some examples are given below.

10.6.1   Dos

a) Keep running the transport chains while heating and cooling the drying chambers.

b) Check the marked danger area before starting the machine and see that all the people leave the danger area within the warning time.

c) Wait after giving start request for 5 s. After 5 s, if the machine does not start in next 2 s, give the start request again.

d) Before starting the production, heat up the complete dryer to the desired operating temperature.

e) Ensure nip guard setting of minimum 4 mm towards squeezing rollers.

f) Check that all safety devices are in position before starting.

g) Do the width setting when the machine is warmed up and the chain is stopped. Smaller fine settings of 20–30 mm can be done with running cloth transport chain.

h) Cover the rubber rollers at take-up as hard fabric selvedges abrading the take-up roller damage the rubber covering when bow straightening.

i) While stopping the machine, stop it in a sequence.

j) Clean the padder immediately after each use by water, soap and hydrosulphite.

10.6.2   Don'ts

a) Do not cut off transport chain during heating or cooling. Discontinuous heating may damage the transport chain.

b) Do not allow any person to be near the danger zone while starting the machine.

c) Do not cover, short circuit, remove or set out of function the warning signals.

d) Do not feed the fabric before the dryers have reached the required temperature.

e) Do not allow the steam pressure to cross maximum 0.7 bar in the padding trough as it can damage the trough.

f) Do not operate the padding rollers in dry condition when pressure is applied.

g) Do not run the padder without any liquor otherwise the slip bearings in the trough will run dry and be destroyed.

h) The revolutions per minute of upper and lower air fans should not differ by more than 15%.

i) Do not turn on the main switch immediately after turning it off. Wait for 2 min and then start.

j) Do not leave the trough uncleaned as it starts rotting.

10.7 Tips for saving energy while stenter processing

1. Use less energy-intensive methods first:

    a) To reduce energy consumption, it is important to remove water by using less energy-intensive methods first such as the mangle, centrifuge, suction slot, air knife or drying cylinders. Even though drying cylinders are about five times more energy intensive than a suction slot, they are still about 1.5–2 times less energy intensive than a stenter. Drying the fabric down to about 25–30% regain before passing it through the stenter still makes it possible to adjust the fabric width to the customer's requirements.

    b) Other techniques used to reduce drying costs include IR and radio frequency (RF) drying. Gas-fired IR drying has been used for the predrying of textiles prior to stentering. This can lead to increasing drying speeds by up to 50%, thereby relieving production bottlenecks which tend to be around stenters. Typically, you could expect the IR drying energy requirement to decrease by as much as 50–70% when compared to conventional stenter drying.

    c) If an efficient means of pulling the fabric out to width could be devised for a short hot zone length, then IR could be used to do all the drying. RF drying is used extensively for the drying and dye fixation of loose stock, packages, tops and hanks of wool and sewing cotton. The energy requirement for RF drying when compared to conventional drying in a steam-heated dryer can be as much as 70%. It is, however, limited to loose stock and packages and cannot be modified to accommodate knitted or woven fabric since the traditional stenter transport mechanism, pins and clips would interfere with the RF drying field causing discharge.

2. Do not overdry: As with the contact drying of textiles, it is important not to overdry, more so on stenters since it is a more energy -intensive drying technique. There are automatic IR, radioactive or conductivity-based systems which can be linked to the stenter speed control to achieve as close as possible the fabric regain.

3. Turn off exhausts during idling: Commission dyers and finishers tend to operate with relatively small batch sizes, and so in some extreme cases, the operatives may be required to change to different fabric qualities every hour. It is common practice to leave the exhausts on during these changeovers, which may take 10–15 min or more. With the large air heating requirement, it is important to isolate the exhausts, or at least partially close them down, wherever possible during periods of idling.

4. Dry at higher temperatures: If the fabric allows the drying at a higher temperature, it means that radiation and convection losses become relatively smaller compared to evaporation energy.

5. Shut and seal side panels: On older machines, the side panels may be damaged, thereby upsetting the delicate air balance within the machine. All faulty panels should be repaired or replaced to provide an effective seal around the oven. Improving insulation is usually not practicable. Although on some older machines, it may be cost-effective to insulate the roof panels.

6. Insulation: Insulate the steam pipes and the machine parts and prevent the heat from getting lost due to radiation.

7. Optimize exhaust humidity:

    a) When drying, there is an optimum exhaust rate which should be adhered to. Since a significant number of stenters still rely on manual control of exhausts, which basically means 'fully open all the time', the potential for energy-saving is considerable. Manual control of exhausts is generally very difficult since the expected airflow patterns and the ones found in practice vary considerably, hence the tendency to leave them fully open.

    b) Optimization of exhausts can be achieved by controlling the exhaust humidity between 0.1 and 0.15 kg water/kg dry air. This is called the Wadsworth criterion. It is not unusual to come across stenters where the exhaust humidity is 0.05 kg water/kg dry air, which means a considerable wastage of energy. Instruments are available which automatically control the dampers to maintain exhaust humidity within this specified range, thereby cutting air losses without significantly affecting fabric throughput. These vary from wet/dry bulb temperature systems to fluidic oscillators measuring the variation in sound through a special filter head.

c) When drying of solvent-based work is required, then the high air losses may not be avoidable for safety reasons. Many solvent-based systems have now been replaced by aqueous systems because of the Environmental Protection Act (EPA).

8. Heat recovery:

a) Exhaust heat recovery can be achieved using air-to-air systems such as the plate heat exchanger, glass tube heat exchanger or heat wheel. Efficiencies are generally about 50–60% but there can be problems with air bypass, fouling and corrosion.

b) If other measures are applied first, such as fabric moisture control and exhaust humidity control, then there is usually no or little economic loss for such systems.

c) Air-to-water systems like spray recuperator avoid fouling and clean the exhaust but there may be problems of corrosion. There is also the need for secondary water/water heat exchange and of course the problem of coinciding utilizations.

d) When stenters are exhausting prohibitive amounts of volatile organics or formaldehyde, then a form of scrubber, electrostatic precipitator or even an incinerator may be required to comply with the statutory limits set under the EPA process guidance notes. In these cases, it makes sense to incorporate heat recovery so that at least the installation costs can be recovered.

9. Direct gas firing:

a) Compared to other stenter heating systems, direct gas firing is both clean and cheap. When it was first introduced, there were fears that oxides of nitrogen, formed to some extent by exposure of air to combustion chamber temperatures, would either cause fabric yellowing or partial bleaching of dyes. This has since been shown to be unjustified.

b) Unlike steam and thermal oil systems, there are no distribution losses to worry about. Heating up times are shorter and thermal capacities less in direct gas firing leading to lower idling losses.

Radio frequency dryers

11.1 What is radio frequency (RF) drying

In conventional dryers, such as steam, pressure, hot air and so forth, the heat is generated separately and then transferred to the material by conduction, convection and/or radiation. These are all indirect heating. In radio frequency dryers (RFDs), heat is generated within the material itself by radio frequency (RF) (Fig. 11.1). This method is called as direct heating or dielectric heating.

In RF drying system, alternating electric field between two electrodes is created by RF generator. The material to be dried is placed in a crosswise manner between the electrodes where the alternating energy causes polar molecules in the water to continuously reorient themselves to face opposite poles, which is similar to the way magnets move in an alternating magnetic field. This movement causes the friction and water in the material rapidly evaporates.

In the RFD, the product passes through an electrode system, which is subject to a high voltage at a frequency of 27.12 MHz (i.e. 27 120 000 times per second). Since water has the highest dielectric constant, it easily gets heated when subjected to alternating electric field of frequency 27.12 MHz.

Water molecule contains two atoms of hydrogen (positive charge) and one atom of oxygen (negative charge), closely held together in chain form. The water molecule gains enough energy to break up the closely held hydrogen and oxygen in the molecule and gets separated. In other words, the molecules of a bipolar substance (water) realign themselves with the above RF field and generate frictional heat within and evaporate water rapidly. Nonpolar substances are not affected by this process.

Water content from textile fibres, yarns and fabrics can be removed through RF. A wet product submitted to an RF field absorbs the electromagnetic energy so that its internal temperature increases. If a sufficient amount of energy is

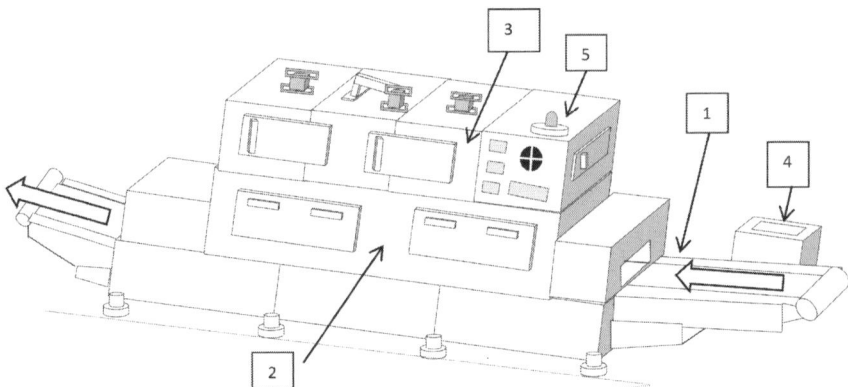

Figure 11.1   Schematic diagram of a radio frequency dryer (RFD)
(1: conveyor assembly, 2: oven assembly, 3: generator assembly,
4: control console and 5: control panel).

supplied, the water is converted into steam, which escapes the product and
the wet product gets dried. Apart from textile industry, this process is used in
paper industry, foundries, chemicals, pharmaceuticals and ceramic industry.

To determine the amount of heat generated in the product, the frequency, the
dimensions of the product, the square of the applied voltage and the dielectric
'loss factor' of the material are required. The dielectric loss factor is essential
to measure the ease with which the material can be heated by this method.
The main reason for this is water, which is far more amenable than other
materials. This reduction in loss factor or receptivity to RF energy provides
valuable protection against overheating of the material. When uniformity of
product dryness is essential, RF drying is the ideal drying process.

In the RF dryer, the heating process is self-regulating and is directly
proportional to the amount of water content in the product. If the input product
moisture of the RFD is uniformly balanced, then the heating is uniform and
the output product drying quality is uniform as well. It is, therefore, suggested
to remove excess moisture by hydro-extracting process before putting the
materials for drying in RFD.

The heating/evaporation processes carried out by means of RF are as
follows:

a) *Endogenous*: The thermal energy is not supplied to the wet product by
    an external heat source but it is generated directly throughout its mass.

b) *Selective*: The electromagnetic energy is absorbed mainly by the
    moisture and not by the product itself.

c) *Controlled*: The heating/evaporation rate is directly proportional to the amount of electromagnetic energy supplied, and this energy can be controlled precisely.

The industries that require quick and uniform drying have turned to RF drying as it is the most effective method for drying and value addition to their products. The textile industry, especially, relies on RF drying for its product enhancement; the steaming and bulking effect improves physical properties such as elasticity, softness, colour and hand.

11.2 Advantages of RF drying

Following are some of the advantages of using RFDs:

a) *Uniform moisture distribution*: Water is more receptive than any other dielectric material. Therefore, in the process of RF drying, RF power will be absorbed in higher amount from wetter areas resulting in uniform moisture distribution.

b) *Reduction of surface cracking*: In the conventional drying process, many times, uneven shrinking leads to surface cracking. In RF drying, due to uniform moisture distribution, surface cracking is reduced.

c) *Improved physical properties*: The steaming and bulking effect improves physical properties such as elasticity, softness, hand and colour effect.

d) *No wastage of heat*: In this, RF is direct form of applying heat so there is no wastage of heat.

e) *Low space*: RFD requires one-fifth or one-eighth of the space required to the conventional dryer.

f) *Faster*: RF drying is 2–20 times faster than conventional drying methods.

g) *Low dwell time*: Here, heating begins directly to the product so the dwell time is far less than in a conventional dryer.

h) *High energy efficiency*: The electromagnetic energy is transferred directly to the whole of the wet product instantaneously without losses to the surroundings and is entirely exploited for the drying process. Furthermore, the energy transfer is not negatively affected by variable parameters such as the dimensions or the density of the product.

   The efficiency of conventional dryer decreases significantly as the lower moisture level is reached and the dried product surface becomes a greater thermal insulator. With more moisture to be removed, RF drying provides an energy-efficient means of achieving the desired moisture objectives because the moisture alone is treated and energy is not wasted on heating the product, the environment as well as the water.

i)   *Outstanding quality of the dried product*: All problems normally caused by the heat transmission phenomena are totally eliminated so that the product dries quickly and uniformly down to the desired residual moisture level, with a beneficial steaming and bulking effect of the fibres which greatly improves the physical properties such as elasticity, softness, hand and colour effect.

j)   *Flexibility*: The same machine can dry tops, packages, hanks, loose stock and bagged products. In operation, the electrode assembly can be lowered or elevated by push-button control; the conveyor belt speed can be similarly varied and so can be the input voltage too. These three variables enable many different product configurations to be handled. The latest machines are built with preset controls, allowing the operator to press one button and alter all settings simultaneously to cater for different loads. Changeover times accordingly are now matters of seconds only.

k)   *Yarn improvement*: RF does offer substantial (and quantifiable) advantages for textile manufacturers. Accuracy of moisture control to 1% is reliable and continuous. By relying on 'direct' heating, it has been repeatedly found that dye migration is reduced; therefore, producing whites are 'whiter'—particularly on cottons. Fibres have better bulk and handle. 'Overdrying' and 'wet spots' are eliminated.

l)   *Productivity*: RF drying equipment offers great possibilities for increasing productivity. The units are immediately available after switching on with no 'warm-up' time being needed. Different packages with any number of different colours can be dried simultaneously. Large or small production batches can be similarly accepted. It is possible to monitor costs, in a way that is frequently difficult with traditional systems because electrical energy is the only input and it is consumed fully by the product to be dried.

m) *Easy to maintain*: RFDs can also be moved easily to different locations and also lend themselves to automated loading and unloading. Maintenance can be performed adequately by normal plant electrical staff.

n)   *Easy to operate*: Operators generally learn quickly to operate RFDs. There is no environmental heat hazard as no radiated heat is released. The units are quiet in operation. The controls are easy to comprehend.

o) *Eco-friendly*: This is an eco-friendly process.

p) *Low maintenance cost*: Maintenance cost is low as compared to other drying processes.

In general, the advantages of radio frequency dryers (RFDs) can be listed as follows:

1. Reduction of energy consumption
2. Ease of operation
3. No steam requirements
4. No dye migration
5. Common unit to dry range of product
6. Pollution free
7. Improvement of product quality
8. Rapid payback
9. Immediate heat—no warm-up time
10. No contamination
11. Accurate final moisture control
12. No radiated heat loss
13. Improved workplace environment

11.3 RF calculations

By international agreement, the following RFs are reserved for industrial use: 13.56 MHz±0.05%, 27.12 MHz±0.6% and 40.68 MHz±0.05%. For microwave heating, the following frequencies are reserved: 2450 and 896–915 MHz (industrial heating).

The heat generated within a given product is expressed by the formula:

$\Rightarrow$ Energy input = voltage × frequency × loss factor (polar molecules).

This formula also explains why certain manufacturers use 27.12 MHz frequency instead of 13.56 MHz, in spite of the design parameters being more stringent. There is a wider frequency tolerance and also a theoretical reduction of 41% in electrode voltage. This can be of importance as it avoids incidence of electrical breakdown within materials being processed and it also permits the drying of textile packages on steel springs. It has been found desirable that heat input density and electrode voltage should be low for all materials where possible.

As a 'rule of thumb', 1-kWh energy is found to be useful to remove 1 kg water. The precise formula is:

$\Rightarrow$ Temperature rise energy (kWh) = (kg × temperature rise × specific heat)·863

However, when calculating requirements, it is best to work from 'worst-case' data, that is, wintertime temperatures and maximum likely moisture content of input materials. Allowances of 10–20% should also be made for unavoidable radiated and convected heat losses which depend on the materials' ratio of surface area to volume.

Setting up speed:

$$V(m/h) = \frac{P \times 1.2}{W \times N}$$

where

V = Conveyor speed in m/h

P = Power available for use in kW

W = Weight of water to be removed from each package (kg)

N = Number of packages placed on 1-m length of belt

11.4   Application of RFD

11.4.1   Packages and muffs

Following the removal of free water from packages by hydro-extraction or predrying with a rapid dryer, these are simply loaded onto a conveyor for passage through the RF unit. As much as 1 h production can be loaded.

Upon entering the RF zone, heat energy is generated throughout each package in proportion to its moisture content with the result that very even heating and drying take place. There is no mechanical damage to yarns and usually the dried materials have a much better feel and appearance. Whites are usually much whiter than from indirect drying.

11.4.2   Tops

After hydro-extracting, tops are an ideal subject for RF drying. The small surface area to volume makes conventional drying techniques very inefficient by comparison. Moisture situated or held in the centres is instantly heated on arrival in the heating zone and is levelled until the required regain is reached. The zone containing higher moisture gives out more vapour till the time the moisture percentage in the entire mass becomes same.

11.4.3 Hanks

Traditionally, hanks have specific problems during the drying cycle. The suspension by mechanical means can result in a permanent wet spot with resultant discolouration. Metal band conveyorized dryers tend to produce wide moisture variations across the bandwidth. Neither system can give automatic moisture control, which is always available with RF ovens.

RF dryers dry hanks in a relaxed state usually on a polyester perforated band. This allows air to be directed through the band and across the surfaces of the hanks to assist steam removal.

11.4.4 Loose stock

With suitable hopper and distribution equipment, loose fibres may be conveniently transferred from continuous or batch hydro-extraction to the RF conveyor. Open-weave polyester bands are not prone to colour contamination, and fibre retention is overcome by brush and air jet under the return pass of the band. Laborious cleaning down operations after each change of colour, associated with other types of dryer, are eliminated.

11.4.5 Hosiery and made-up garments

An interesting application for RF techniques is the drying of hosiery in bags and garments after hydro-extracting. Particularly, in the case of hosiery made in synthetics, the amount of energy required is low and all parts of the articles are reduced to the same level of moisture, including the joints in tights. It is difficult to see how the conventional dryers can do this work efficiently.

11.4.6 Thread and warp drying

RFDs are now being used extensively to dry sized warp threads. Heat is directly generated within the water-based size so that almost instant drying can take place without affecting the quality of the product. Yarn drying is also now being done by RF after the felting process for carpet and knitting yarn with much success.

11.4.7 Bales

In the case of wool, it is usual to transport raw wool in bales using high compression. Sometimes, this material is also stored at low temperatures. Opening up the compressed fibre has been a traditional cause of difficulties. RF heating of the bale before opening resolves the difficulties.

As very high compression bales are being used nowadays, RF can be used to eliminate another problem related to moisture content. With these bales, it is of paramount importance that there are no wet spots prior to baling since these can cause severe damage due to bacteriological effects. Drying with RF before baling ensures that there are no wet spots since these are always preferentially heated with this technique.

11.4.8 Electronic dye fixation

Several RF units have been installed for dye fixation. The process has the advantage of being continuous and easily controllable. Basic research is now probably necessary on the related physical chemistry if this application is to become a widespread and major dye house process. From an electronics viewpoint, the process is simple and reliable.

11.5 Typical RFD

There are a number of manufacturers of RFDs such as Stalam and Monga Strayfield. Salient features of Stalam RFD (as claimed by the manufacturer) is explained here as an example.

a) The RF generators are of the 'lumped components' type, having high efficiency (Q factor) and outstanding reliability; all components and circuits are easily and quickly accessible.

The cooling system of the triodes is made up of a double water circuit; it is designed to allow the longest possible life of the triodes and does not require periodic maintenance operations. Upon request, generators fitted with a forced-air cooling system are supplied.

b) The RF power adjustment is accomplished by means of a semiautomatic circuit which controls the power supplied to the product being dried through a variable capacitor, located in the generator. No continuous raising or lowering of the upper electrode for RF power adjustment is required: the electrode is fixed or automatically positioned at preset heights.

c) The power density is as low as 5–15 kW (RF)/m2 of electrode surface. Therefore, the heating/evaporation process is extremely gentle and the quality of the dried product is greatly enhanced.

d) The construction is modular, in the sense that high-capacity dryers are made up of two or more drying modules, connected in series; this construction principle allows dryers to work with reduced RF power densities in all circumstances. Further, modules can be added to the first at any time so that considerable investment and floor space savings are possible.

e) The construction materials are mainly stainless steel and aluminium, to ensure a long life of the dryers even in difficult working environments, and all electrical and electronic components used are of the best international makes, other than amply oversized, to guarantee the highest level of reliability.

f) Dryers are also equipped with electromagnetic field leakage preventing and RF interference suppressing systems in order to comply with all safety and electromagnetic compatibility regulations in force in the different countries.

11.6 Typical working procedure

a) Before starting RFD, clean the air filter (Fig. 11.2).

b) Start cooling tower first, and then the blower and steam. Wait for 5 min. Then, start the dryer.

c) Feed only hydro-extracted cheeses to the RFD.

d) Keep the material to be dried on the slowly moving conveyor of the drier. Set the speed as prescribed by the dyeing in charge in case of different count or else run with the set speed in metres per hour as follows:

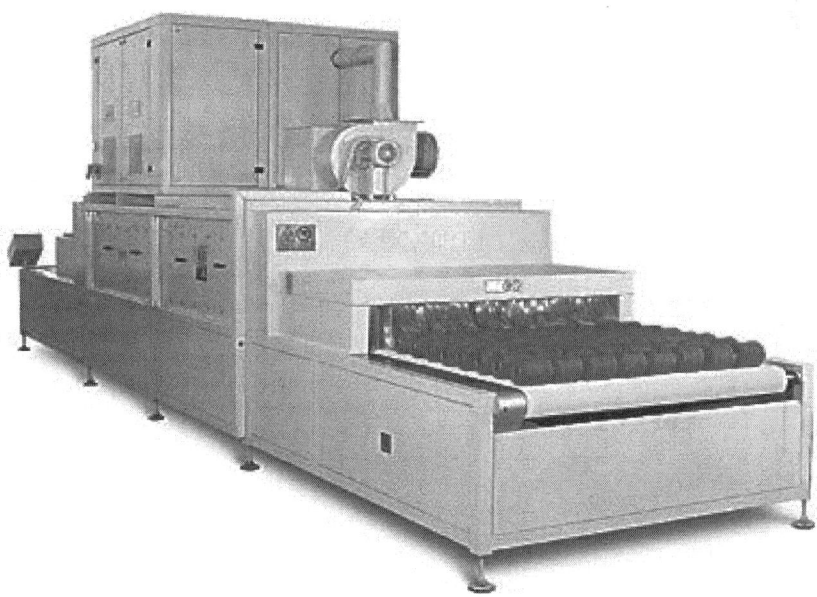

Figure 11.2 Radio frequency dryer.

Figure. 11.3   Peg trolley.

- 85 kW—RF—6.3–6.8
- 100 kW—RF—7.2–7.6.

e)   The material is placed in the dryer and it comes out after drying from the other end. Remove the material after it comes out. Before stopping the RFD, stop the dryer first and then steam and blower. After 5 min, stop the cooling tower.

f)   Remove the dried materials. Put them on pegged trolley (Fig. 11.3), insert the process sheet and send to yarn winding section through yarn godown after getting approval from Quality Assurance Department.

g)   Do not feed full wet material to RFD.

11.7   Dos and Don'ts

It is necessary to understand clearly what is supposed to be done without fail and what should not be done at any cost. Some examples are given below.

11.7.1   Dos

a)   Understand the basic material to be dried using RFD. For example, the easiest of all materials to dry is polyester, while the most difficult appears to be the acrylics.

b)   Ensure that the materials fed to RFD have undergone some preliminary drying process like hydro-extraction and the input moisture content is uniform.

c) Please refer microprocessor controller manual for machine operation, schedule maintenance, product parameters, fault diagnosis and conveyor speed calculation.

11.7.2 Basic product loading

a) Packages with metallic centres or bobbins must be loaded only horizontally.

b) Packages with nonmetallic centres can be loaded horizontally or vertically.

c) Hanks must be loaded as evenly as possible, ensuring there are no concentrations of material and with a uniform height.

11.7.3 Don'ts

a) Do not dry materials like acrylic in RFDs. Some acrylics are highly polar and will continue to heat once dry and will eventually catch fire.

b) Where certain types of whiteners are used on acrylics, excess energy density can cause slight yellowing.

c) Avoid drying some of the polyamides as they are also polar.

d) Do not place complete wet material on conveyors of RFD.

e) Do not feed materials of different polarity together.

11.7.4 Safety notices

a) Operation of electronic equipment involves the use of dangerous voltage. Safety regulations must be observed all the time.

b) Do not change valves or make adjustments inside the equipment with the power supply 'ON'.

c) Do not depend on interlock switches for protection.

d) Always cut off the power supply before removing covers or other protective devices and also ensure that power supply cannot be switched on accidentally.

e) Under certain conditions, dangerous voltages may exist in circuits with power controls in the OFF position due to charges retained by capacitors.

f) Always disconnect power and ground the circuits before touching them.

g) Wherever feasible, when testing circuits, check for continuity and resistance rather than directly checking voltage at various points.

h) Under no circumstance should any person reach within the unit for the purpose of servicing or adjusting the equipment without the immediate presence or assistance of another person capable of rendering aid.

i) Do not tamper with interlocks.

j) RFDs are normally provided with a recycling unit that restarts HF/HT automatically after a generator trip-out. Therefore, switch off before testing.

Relax dryers

12.1 Principle and purpose

The basic purpose of relax drying is to dry and/or shrink dyed or bleached delicate fabrics while retaining the fabric properties. They are constructed to suit the fabric which has to be passed in a relaxed condition while drying. The dryer is normally integrated with motorized centring and tensioning device for individual conveyors. These dryers are ideal for knits, terry towels and delicate fabrics. All hassle-free drying can be accomplished with relax dryer.

Relax dryers have conveyors for the transporting of materials and not clip, pin or drying cylinders as in case of other dryers. Up and down conveyor belt designs facilitate the fabric tumbling in waveform, which is more relaxing and tension-free.

On a tensionless conveyor dryer (Fig. 12.1), the knitted and woven fabrics are treated under total relax form, with tumbling movements, for achieving the dimensional stability by means of overfeeding to obtain length shrinkage, better handle and bulking to the fabric (Figs. 12.2 and 12.3).

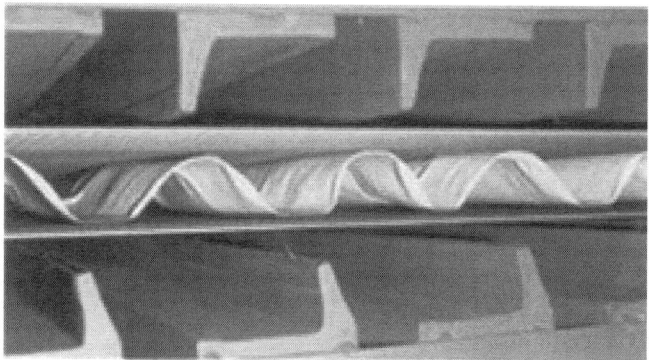

Figure 12.1 Tensionless conveyor.

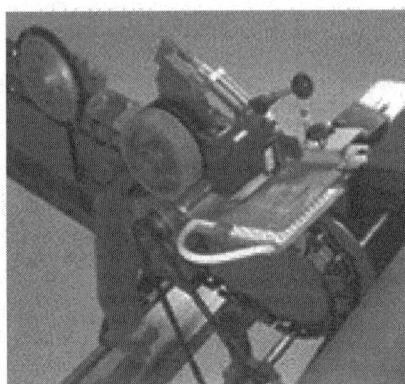

Figure 12.2   Fabric guides and overfeed.

12.2   Different types of relax dryers

There are different types of relax dryers, which can be classified either as rope drying machines with air jets or open-width drying machines either by hot air or steam.

Within open-width relax dryers, there are different varieties such as single-pass (Fig. 12.3), double-pass (Fig. 12.4) and three-pass machines (Fig. 12.5). In single pass, the fabric moves only once in the drying chamber, whereas in double pass, it passes twice in the same chamber and in three pass, it passes three times.

In relax air drying machines, a large volume of air is circulated with the help of specially positioned well-designed jets. The machine is provided with removable nozzles that enable drying of fabrics in between Teflon–glass conveyors in the shortest possible time and with an effect, almost similar to tumbling effect. The staggered high-velocity jets placed in opposite positions provide near-pulsating movements, ensuring optimum residual shrinkage with softer, fuller, bulkier handle of fabrics and improved production with excellent shape stability.

Universal Relax Dryers suitable for both open-width and tubular fabrics are normally equipped with the special in-feed device, with positively driven scroll roller, heavy-duty padder for wetting out followed by the second padder for chemical finishes, a well-designed pin frame entry with vertical return pin-chain with special edge guider and positively driven scroll rolls. A pneumatic pinning device is also incorporated to obtain uniform grams per square metre (GSM) control.

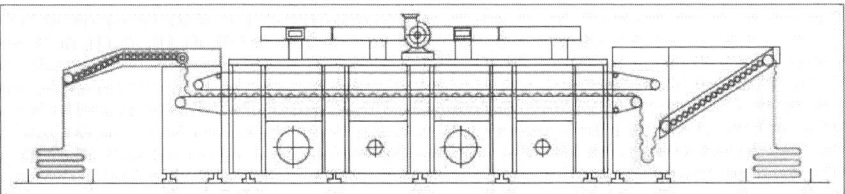

Figure 12.3 Single-pass conveyor relax dryer.

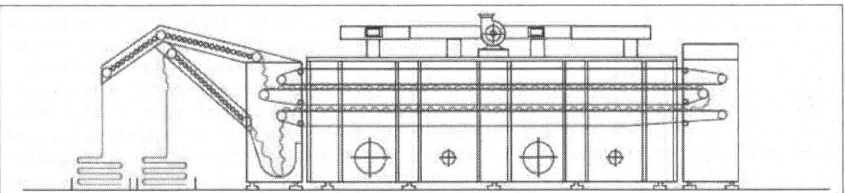

Figure 12.4 Double-pass conveyor relax dryer.

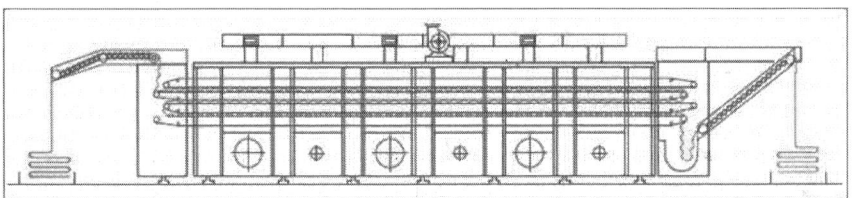

Figure 12.5 Three-pass conveyor relax dryer.

The relax dryers are provided with a possibility of overfeeding up to 40%, which permits good tumbling effect to allow good shrinkage of fabrics. These machines are suitable for production from 4 to 20 t/day, depending on the type of fabrics and number of passes. Some commercial models of relax dryers are discussed herewith.

The RotoSwing Tumble and Relax Dryer developed by Fleissner produces tumbler-resistant goods with residual shrinkage values that usually lie between 0 and less than 5%, depending on the fabric construction and the respective pretreatment (Fig. 12.6). Thus, it became possible to wash and dry knitwear in household washing machines and electric dryers to maintain their dimensional stability and avoid shrinkage of the garments.

The intake section includes spreading devices with threaded expander rollers and selvedge uncurler for open-width goods and a felt roller for both open-width and tubular goods for slip-free conveyance and overfeeding of the fabric onto the dryer belt. The overfeeding ratio can be set individually for

Figure 12.6   RotoSwing Tumble and Relax Dryer.

the respective fabric. The discharge section comprises a discharge conveyor with a plaiter or a winder for woven goods.

In between the top and bottom conveyors, the nozzle boxes with 16 nozzles each for top and bottom air are incorporated. In conventional dryers, the top and bottom nozzles are arranged in staggered order and guide the fabric fed to the dryer with a certain overfeed in a sinusoid movement between the transport conveyors through the machine. Such a nozzle arrangement is not likely to offer optimum conditions for achieving an intense tumbling effect. To achieve the maximum possible shrinkage and a good fabric development in continuous operation, a tumbling movement must be created which considerably increases the fabric movement and comes very close to discontinuous tumbling. In the RotoSwing, this is achieved by moving the complete bottom nozzle system with its 16 nozzles per compartment (mounted in a common frame) to and for staggered by 1 nozzle each.

In a dryer with several compartments, these frames are connected to each other so that all bottom nozzles are moved together. This movement is carried out with an offset of one nozzle, that is, each bottom nozzle moves forward past the opposing top nozzle in material flow direction and back again with the rotational speed of 60 rpm of the crank mechanism. For this continuous movement, three positions or phases of the top and bottom nozzles in relation to each other are of special importance.

In the first phase, the nozzles are offset against each other. The fabric moves in wave-like form between the conveyors through the machine. In this nozzle position, the air flows through the fabric in the same way as in a drum dryer; this is the phase of the maximum drying rate. With each rotation of the traversing drive, this position is reached twice, that is, for 60 rpm, drive speed is 120 times per minute for each top/bottom nozzle pair. For a compartment with 16 nozzle pairs, this results in 1920 times per minute. The transition between the individual phases is fluid.

In the second phase, which occurs four times per rotation of the nozzle drive, the nozzles are arranged opposite to each other in an optimum tumbler position with a minimum offset. Top and bottom air in this position cause the fabric to be 'beaten up and down' at a very high frequency. The frequency of 240 up/down beats per minute between 1 pair of top/bottom nozzles or 3840 beats a minute per compartment allow the fabric to shrink and relax continuously, almost as in the very long fulling cycles of a discontinuous tumbler.

In the third phase, which again reaches twice per rotation, top and bottom nozzles are opposite to each other. A vibrating air cushion is formed keeping the fabric floating. At the same time, air is blown onto and into the fabric from both sides, which improves volume, touch and bulkiness.

The Texshrink Relax Dryer developed by Thakore Exports for ladder-proof fabrics is based on the well-known principle of air percussion drying system (Fig. 12.7). The air movement produces undulations in the fabric and at the same time, it expands the knitted fabric causing dimensional shrinkage and selvedge opening. The fabric is transported through the drying chamber on a conveyor belt being usually overfed to obtain length shrinkage.

Each section is equipped with air recirculation device consisting of alternatingcurrent motor, with fan blower on special air blades. Blower can

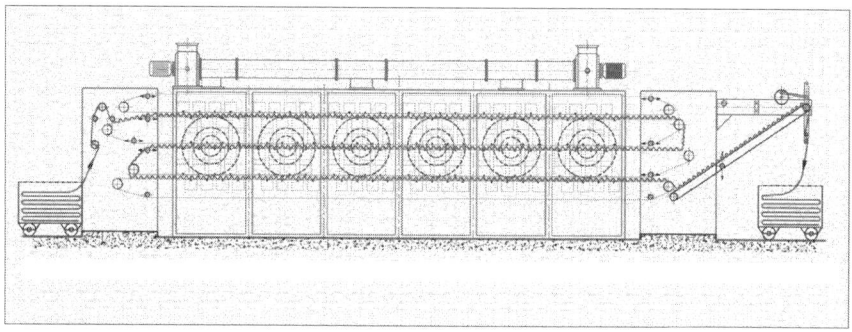

Figure 12.7 Schematic diagram of Texshrink Relax Dryer.

easily be dismantled for maintenance and cleaning. An exhaust control is installed for optimum thermal efficiency. Each drying section has its own temperature control system. Air heating is performed by steam or thermal oil radiators or alternatively equipped with direct gas burners.

Swastik's high-efficiency jet drying relaxing and shrinking machine Relaxair is designed for drying and/or shrinking dyed or bleached knitwear and delicate woven fabrics, in tubular or open-width form, and made of 100% cotton, synthetics or wool and its blends.

A large volume of air, circulated with the help of specially positioned well-designed jets with removable nozzles, enables drying of fabrics in between Teflon–glass conveyors in the shortest possible time and with an effect, almost similar to tumbling effect. Relaxair, with staggered high-velocity jets placed in opposite positions, provides near-pulsating movements and ensures optimum residual shrinkage with softer, fuller, bulkier handle of fabrics and improved production results with excellent shape stability.

The machine is of strong and sturdy modular construction and is designed to achieve reproducible overfeed up to 50% and is provided with a pair of Teflon–glass conveyors, lower fabric transporting conveyor and upper restraining conveyor. Conveyors are equipped with automatic centring device and mechanical/pneumatic tensioning devices.

It is also equipped with a moisture controller acting on the speed of the machine as well as exhaust air moisture control device. Blowers with variable-speed inverters are also available to achieve desired results. It can also be equipped with microprocessor-based controller to control overfeeds, temperature, moisture and speed during the process (Fig. 12.8).

Figure 12.8 Relax dryer.

The Swastik Relaxair is also offered with double-pass and three-pass tensionless drying and shrinking system with three and four Teflon–glass conveyors, respectively. It is also provided with a conveyor, feeding fabrics or with a fabric stretching device with draw roller for multiple tubular webs/stands, as per the requirement. Fabric delivery through conveyor is provided with a plaiting mechanism. The hot fabric is cooled by means of cooling device. These machines are offered with steam/hot thermal oil/natural gas/liquefied petroleum gas/electrical heating system and dry saturated steam at 7 bar as per the requirement.

12.3 A typical procedure for drying with relaxsteam drying machine

The operating procedure depends on the type of relax dryer used, the fabric being treated and the finish. Each mill has to prepare its standard operating procedures. The following steps are the basic guidelines:

a) Clean the machine with compressed air and then wipe with a clean cloth before starting the work. Clean all sensors and photocell units with a clean dry cotton cloth.

b) Drain condensate water from the pipelines and then open the steam valve.

c) Switch ON the machine. Start all blowers and material conveyor belts.

d) Clean all the filter mesh in the drying chambers (one chamber has eight filter frames, and there are three chambers; Fig. 12.9).

e) Start heating the steaming chamber. After the machine reaches full temperature, feed the material.

f) While heating process is on, run the conveyor belts. Do not keep them stationary. If heating process is done by keeping the feeder stationary, the conveyor belt is likely to lose its life.

g) Set the machine parameters such as speed and temperature for the material to be processed by discussing with a supervisor.

h) The machine operates between 110 and 130°C. The normal working temperature is between 125 and 130°C.

i) The machine runs at the required speed depending on quality.

j) Write the complete details of the material being processed in the production book.

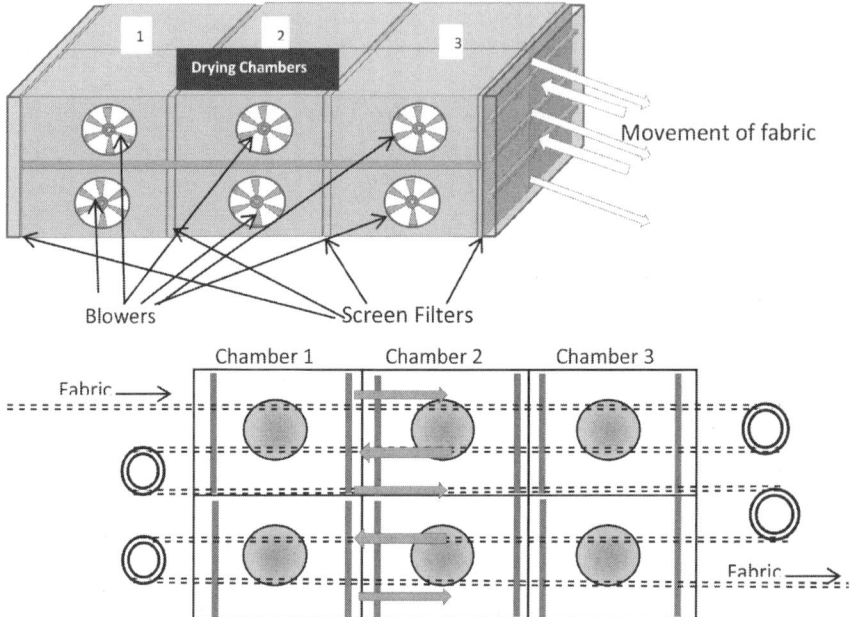

Figure 12.9 Drying unit.

Figure 12.10 Conveyor sheets.

k) Stitch the ends and ensure continuous feed. The machine will not stop for any operations such as feeding or taking material out.

l) Put the fabric to be dried between two conveyor sheets (Fig. 12.10).

m) In case any abnormality is observed, inform the concerned supervisor immediately and get it rectified.

n) In case of any electrical or mechanical problem in the machine, get it rectified by the concerned engineering person. Write the details in the production book.

o) Cover the processed material with a polyethylene sheet and keep at the designated place.

12.4   Dos and Don'ts

It is essential to understand clearly what is supposed to be done without fail and what should not be done at any cost. Some examples are given below.

12.4.1   Dos

a) Put the fabric to be dried between two conveyor sheets.

b) Clean the machine thoroughly with compressed air before starting.

c) Maintain temperature at around 125°C.

d) If conveyor stops or goes to the side, the machine should be stopped.

e) Use only polytetrafluoroethylene-coated fibreglass and silicone adhesive tapes for joining the conveyors as they have temperature resistance up to +260°C.

12.4.2   Don'ts

a) Do not operate the machine if filter mesh is filled with dust and ash. Clean the filter regularly.

b) Do not heat the chamber keeping the conveyor stationary.

c) Do not operate the machine if the temperature is less than 110°C.

Scouring and milling of woolfabrics

13.1 Definitions and introduction

Scouring: Cleaning raw wool and removing impurities such as dirt, sweat and grease by washing is termed as wool scouring. Fabric is also scoured to remove impurities from weaving, singing and carbonizing. We are discussing the operation to remove the sizing and tint used on the warp yarn in weaving and, in general, to clean the fabric prior to dyeing.

Milling: The application of mechanical action to cause the required amount of fibre migration in wool fabrics through the ability of fibres to felt is termed as milling in the textile industry. The process involves beating the wet woollen fabrics number of times by a beater board and making it dense to get a soft feel of the fabric.

Fulling or **tucking** or **walking** ('waulking' in Scotland) is a step in woollen cloth making which involves the cleansing of cloth (particularly wool) to eliminate oils, dirt and other impurities, and making it thicker. Fulling involves two processes: scouring and milling (thickening). Thickening of cloth is done by matting the fibres together that gives it strength and increase waterproofing (felting). This is vital in the case of woollens, made from short-staple wool, but not for worsted materials made from long-staple wool.

13.2 Earlier systems

In Scottish Gaelic tradition, the process of fulling was accompanied by waulking songs, which women sang to set the pace. However, from the medieval period, fulling was often carried out in a water mill. These processes are followed by stretching the cloth on great frames known as tenters. The cloth is held onto tenter frames by tenterhooks. The area where the tenters were erected was known as a tenterground.

In Roman times, fulling was conducted by slaves working the cloth while ankle deep in tubs of human urine. Urine was so important to the fulling business that it was taxed. Stale urine, known as 'wash', was a source of ammonium salts and assisted in cleansing and whitening the cloth.

By the medieval period, fuller's earth had been introduced for use in the process. This is a soft clay-like material occurring in nature as an impure hydrous aluminium silicate. It was used in conjunction with the wash. More recently, soap has been used.

The second function of fulling was to thicken cloth by matting the fibres together to give it strength and increase waterproofing (felting). After this stage, water was used to rinse out the foul-smelling liquor used during cleansing.

From the medieval period, the fulling of cloth often was undertaken in a water mill, known as a fulling mill, a walk mill or a tuck mill. In Wales, a fulling mill is called a pandy and in Scotland, a waulk mill. In these, the cloth was beaten with wooden hammers, known as fulling stocks or fulling hammers. Fulling stocks were of two kinds, falling stocks (operating vertically) that were used only for scouring, and driving or hanging stocks. In both cases, the machinery was operated by cams on the shaft of a waterwheel or on a tappet wheel, which lifted the hammer.

Driving stocks were pivoted so that the foot (the head of the hammer) struck the cloth almost horizontally. The stock had a tub holding the liquor and cloth. This was somewhat rounded on the side away from the hammer, so that the cloth gradually turned, ensuring that all parts of it were milled evenly. However, the cloth was taken out about every 2 h to undo plaits and wrinkles. The 'foot' was approximately triangular in shape, with notches to assist the turning of the cloth.

13.3 Process of scouring and milling

Scouring of woollen fabric is normally carried out in water with the addition of detergent. Synthetic detergents, mainly nonionic and anionic, are widely used, have good stability to hard water and are easy to remove during rinsing. Rope processing is normally carried out at pH 8.0–8.5 and 40°C or slightly higher temperature for open-width processing. In batch processing, the ends of the fabric are sewn together to form an endless loop, which is fed through squeeze rollers for a predetermined time, normally 30–60 min, followed by rinsing. As the fabric passes through the nip rollers, scouring liquor is squeezed from the fabric promoting cleaning and liquor interchange and also 'working' the wool fibre, to develop handle. The fabric is then allowed to 'relax' without tension as it collects in the bottom of the machine before it is lifted to pass again through the nip rollers.

Modern batch scouring machines operate at high speeds (200–250 m/min) to minimize processing time and also to ensure good opening and movement of fabric to reduce the propensity for creasing. Water fans or jets, usually situated before the nip, help to open out the fabric for uniformity of processing. In modern multiple rope machines, individual channel control ensures even finish piece to piece, allowing for variation in fabric lengths. Rapid rope scouring machines operate at higher speed, impacting the fabric against a baffle plate or device, to generate handle and are particularly suitable for blends.

For lightweight fabrics, which may crease if scoured in rope batch machines, open-width scouring processing is used. This may be a batch or continuous process. During scouring, the fabrics are normally sprayed with scouring liquor, followed by nip rollers or vacuum slots to remove liquor but little effect on handle is achieved and a cool, clear finish is retained. New developments in solvent scouring allow continuous cleaning with negligible emissions and almost total recycling of solvent. This provides a good preparation, particularly for blends. The procedure is applicable to elastane-containing fabrics.

Milling normally follows rope scouring and is used to further develop fabric cover and bulk, particularly for flannels and also woollen fabrics. During milling, the liquor is much less compared to that used for scouring and detergent washing. The fabric is passed through a milling box or trough, where it is compressed by rollers. As the fabric is passed through the milling box, the lid or sides of the milling box, which are tapered, restrict the passage of the fabric. This action promotes length shrinkage (felting) of the fabric, generating more bulk, density and cover than possible by scouring alone. In woollen fabrics, the initial base weave is almost entirely obscured during milling. The process also adds strength and density to the fabric by the interfibre felting, providing a suitable base for woollen fabrics, which are to be subsequently raised.

Piece dyeing would follow scouring or milling for pieced dye fabrics.

13.4 Modern scouring and milling machine

There are a number of machines used for scouring and milling of woollen fabrics and blends of wool (Fig. 13.1). Examples are CombiSoft and Supervelox machines. These are discontinuous cloth rope finishing machine which improves the handle and quality as well as softness of finished textiles.

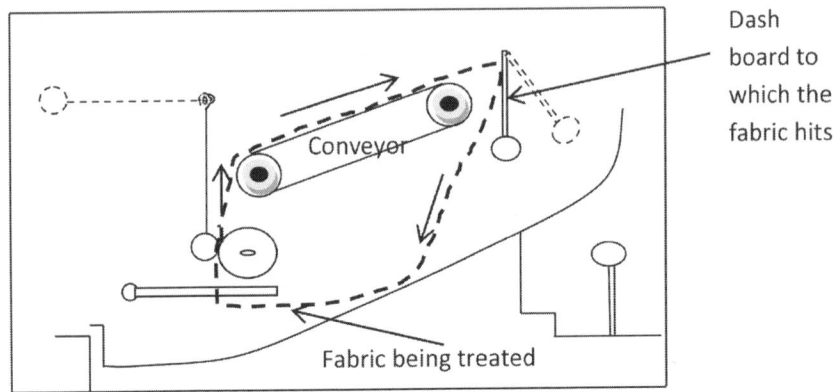

Figure 13.1   Milling operation.

Figure 13.2   Six fabrics mounted at a time.

Length of ropes may be taken from 120 to 150 m for fabric of 300 g/m² (Fig. 13.2). The cloth is fed into the machine in rope form and it is conveyed by a soft elastomer conveyor belt at a speed of over 600 m/min. An exclusive roller in compound materials pressing on the conveyor belt prevents mechanical creasing or abrasion on the cloth and assures a perfect adherence which avoids any kind of slippage. Very high processing speeds are achieved even with heavy fabrics.

The kinetic energy accumulated by the cloth propels it onto a special plate (dashboard) placed at the rear of the machine. The cloth then falls into the main tank. When it reaches the front of the machine, an air turbo system balloons and swells the fabric as well as heats it up to an adjustable temperature of about 110°C.

The mechanical action of softening and compactness is achieved by the impact of the fabric onto the special rear plate. The exclusive turbo system helps swelling the cloths (which are not bag stitched) before they reach the

Six different fabrics in rope form

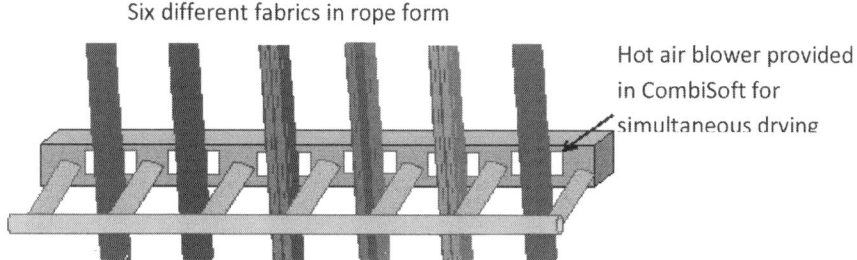

Hot air blower provided
in CombiSoft for
simultaneous drying

Figure 13.3 Hot air blower provided in CombiSoft for simultaneous drying operation.

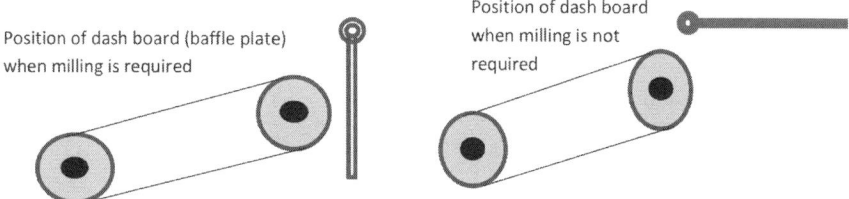

Position of dash board (baffle plate) when milling is required

Position of dash board when milling is not required

Figure 13.4 Action of dashboard.

conveyor belt and prevents any forming of creases. Furthermore, while damp treating without liquor, the action of the warm air is distributed all over a vast surface evenly and, therefore, the fabric dries faster (Fig. 13.3). A variable-speed motor allows selection of the most suitable speed for all types of fabric.

By modifying the operating parameters, such as processing time, air capacity, air temperature, humidity and cloth speed, the machine is able to produce a wide range of effects even on the same cloth type.

When the dashboard is positioned for milling, depending upon the speed of the conveyor, the fabric hits the board and milling takes place (Figs. 13.4 and 13.5). The speed of conveyor can go from 300 to 600 m/min. The operator needs to get the instructions from the supervisor, depending on the quality of the fabric.

Four types of programme are available in the CombiSoft machine; they are Heat, Modulate, Bypass and Cool. The operator can select any of the programmes, depending on the requirement by discussing with the supervisor.

a) **Heat**: In this programme, the fabric can be dried instantly and in the meantime, other programmes can also be used. Dry air is circulated over the fabric through the rear filter. The temperature of the hot air can be set at the required level.

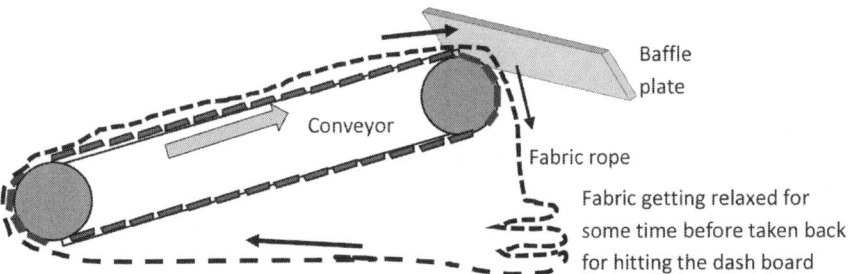

Figure 13.5   Fabric rope hitting the board (baffle plate) at high speed.

b) **Modular**: This programme removes hot air from the machine and recirculates the air. A timer is provided for this purpose which monitors the time. This programme ensures that the relative humidity inside the machine is not increased.

c) **Bypass**: According to this programme, air can be circulated without taking through the rear filter.

d) **Cool**: This programme is for keeping the cloth and the machine cool. This programme opens the machine door up by 15 cm and hot air goes out of the rear filter.

The fabric is allowed to relax without tension after getting beating as it collects in the bottom of the machine before it is lifted to pass again through the beating action.

13.5   A typical work procedure for operating CombiSoft

The operating procedure depends on the fabric being used and the degree of milling required. Following is a general guideline.

a) Take instructions regarding the fabric to be scoured.

b) Take fabrics of 100 ± 10 m and mount on the conveyor. Stitch the ends and make a continuous rope.

c) Select the angle of the dashboard. If milling is needed, keep the board near to the conveyor belt. In case of only scouring, lift the dashboard up.

d) Prepare the solution for scouring depending on the instructions given in the job card, as the blends and fibre constituents are varying from design to design.

e) To bring pressure on the fabrics, either long fibreboards or rollers can be used.

f) The fabric is allowed to relax without tension after getting beating as it collects in the bottom of the machine, before it is lifted to pass again through the beating action.

g) Depending on the activities, such as scouring and milling or only scouring, the fabric blend and construction, the time of operation changes. Get instructions for the time to be set. The duration of process may take 2–6 h.

h) After the operation is over, drain the liquor and rinse the fabric with fresh water.

i) In case of milling operation, periodically observe the fabric condition and ensure that fabrics are not overbeaten or damaged.

j) Enter the complete details of the design number, number of batches taken, total length of fabric processed, the starting time and ending time of operation, and problems observed, if any, in the production record.

k) Unload the material in the plastic cubical trolley and cover with a polythene sheet and keep in the designated place as per the instruction of supervisor.

l) Enter the details in the lot card of the design number and the process carried out and attach the same to the trolley.

13.6   Control points and checkpoints

It is essential to have clarity on the points to be controlled to achieve the targets and those to be checked to ensure the process in control. These points need to be reviewed from time to time and modified to suit the requirements of individual companies and their targets. Each mill should prepare its own 'control points and checkpoints' and display them in the work area so that the people on spot refer and follow. Following is just an illustration.

10.6.1   Control points

a) Selection of suitable scouring and milling process for the material to be processed

b) Deciding and selection of process parameters, namely:

  • Speed

- Running time for each operation
- Temperature
- Pressure
- Scoring recipe
- Liquor ratio
- Washing sequences
- Setting of dashboards

c) Deciding acceptance criteria for quality of milling

d) Deciding on chemicals and their quality to be procured

e) Employing trained and qualified employees

f) Evolving production norms

g) Evolving norms for consumption of chemicals and water

13.6.2 Checkpoints

- **Material related**

    a) The fabrics received for scouring and milling against the plan received

    b) The quantity of fabric received for scouring and milling—metres and kilograms

- **Machine related**

    a) The condition of the machine

    b) The working of various valves and controllers

    c) The condition, that is, smoothness of the dashboard surface

    d) Cleanliness of the air spray nozzles

- **Setting related**

    a) The force of fabric hitting the dashboard, that is, the speed of the conveyor

    b) The recipe prepared for scouring

    c) The distance set for the dashboard and its angle

- **Performance related**

    a) Uniformity of scouring and milling throughout the fabric

    b) The production achieved

- **Documentation related**

    a) The design number and the blend of the fabric received for milling

    b) Quantity scoured and milled

- **Work practice related**

    c) Cleaning of the machine thoroughly before loading the material

    d) Proper stitching of the ends of the fabric after feeding into the machine

    e) The speed, pressure and temperature maintained and the norms

- **Logbook related**

    a) Machines working

    b) Starting time of the running lots

    c) Activities done and to be done

- **Management information system related**

    a) Scouring and milling machine number

    b) Design number

    c) Number of metres

    d) Weight in kilograms

    e) Chemicals used

- **General**

    a) Use of safety gadgets such as gloves and gum shoes while handling chemicals

    b) Proper housekeeping around the machine so that there are no hindrances for movement of materials and operators

13.7 Dos and Don'ts

It is essential to understand clearly the activities that are necessary to get the required performance and the activities that should never be done as they can result in poor performance. Following are some examples.

13.7.1 Dos

a) Check the blend per cent of the fabric and ensure that it has significant per cent of wool.

b) Stitch the ends of the fabric securely after mounting on the conveyor belt.

c) Monitor the pH after washing and maintain at neutral.

d) Prepare the scouring liquor after referring to the instructions for the material in process, namely the percentage of wool fibres in the fabric, the count, construction and weave.

e) Discuss with the in charge and select the programme in case of CombiSoft.

13.7.2   Don'ts

a) Do not go for milling if the percentage of wool is less than 65% in the fabric.

b) Do not set the baffle board very near to conveyor.

c) Do not continue the milling operation if you notice damage to the fabric.

d) Do not increase the speed of conveyor to get faster results. It may damage the fabric.

Drum washing of synthetic fabrics

14.1 Purpose

Rotary drum washing is specially developed shrinking process for getting bulkiness and consistent grain effect on synthetics fabric. The drum washing machines are able to do processes such as sizing, scouring and creasing. The process can be performed at high temperature and high pressure. Drum washers are also used for washing garments.

14.2 Working principle

The heavy-duty horizontal drum washing machine consists of an octagonal perforated drum (Fig. 14.1) made out of stainless steel. In the inside of the drum, at the angles, a plate is provided to guide the fabrics (Fig. 14.2). The projections made at corners take the fabric along with the drum while rotating.

Fabric is loaded from front door inside a stainless steel octagonal basket (Fig. 14.3). The basket rotates in forward and reverse direction with the help of a motor. The activity is done at high temperature and high pressure.

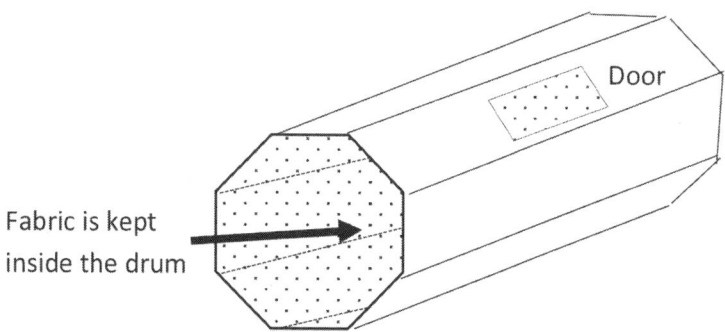

Figure 14.1 Octagonal perforated drum.

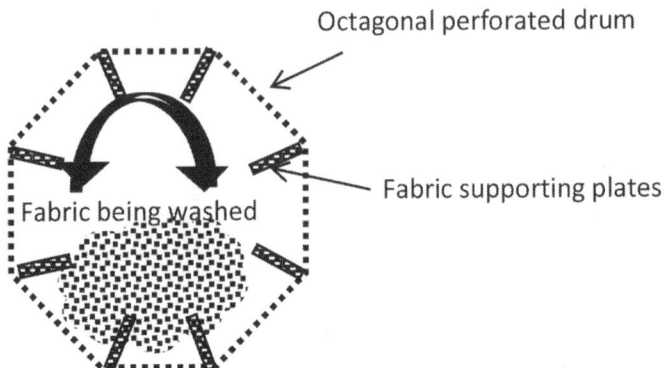

Figure 14.2   Sectional view of drum washer.

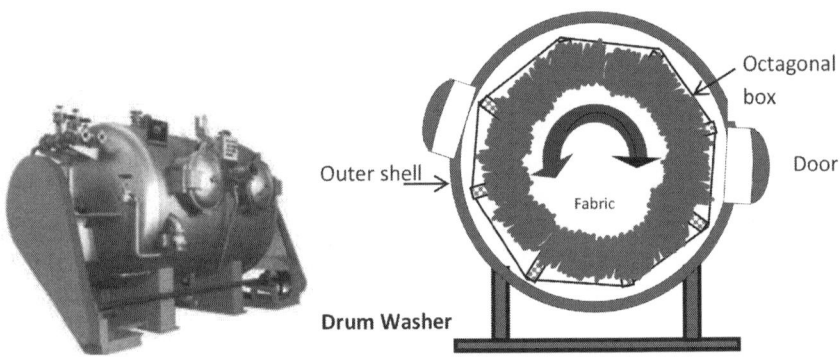

Figure 14.3   Drum washer.

In the front, two large door openings with strong door hinges are provided for loading. In the side, a stock tank is provided for adding chemicals. Normally a combination of caustic flakes and soda ash are used for scouring.

Machines are available with different capacities starting from 100up to 1000 kg. Normally, 50–65%of the capacity is loaded depending on the type of fabric.

Maximum operating pressure is up to 3.5 kg/cm^2 and maximum operating temperature is 140°C. Normally, the materials are worked at slightly lesser temperature, say 110–120°C.

Even heating throughout the process is achieved by programmable microprocessor. Automatic temperature control for steam and water is achieved by electrically operated solenoid valves. Synchronization is done for drum rotation in forward and reverse direction. Additional inspection and sight glasses are provided to view the internal process.

To ensure safety while working, valve and pressure switches are provided.

The total process takes up to 6.5 h depending on the material and the quantity put.

14.3 Typical work procedure for operating rotary drum washing machine

The operating procedure depends on the capacity of the machine, the material being washed and the processes adopted. Hence, each mill has to prepare operating procedures considering its system and requirements. Following is just an illustration.

a) Clean the machine thoroughly before starting any programme.

b) Get the details of the programme to be worked from the supervisor.

c) In case of any discrepancy either in the material or in the job card, discuss with the supervisor and get clarity.

d) Check the design number and lot number before bringing the material to the machine.

e) Drain the steam condensates from the steam line before starting the machine.

f) Load the required quantity of material.

g) Fill water up to the maximum level indicated in the side level indicator (Fig. 14.4).

h) Prepare the caustic solution in the stock tank as per the instruction of the supervisor (Fig. 14.5). Maintain concentration and add the required catalyst as directed in the job card.

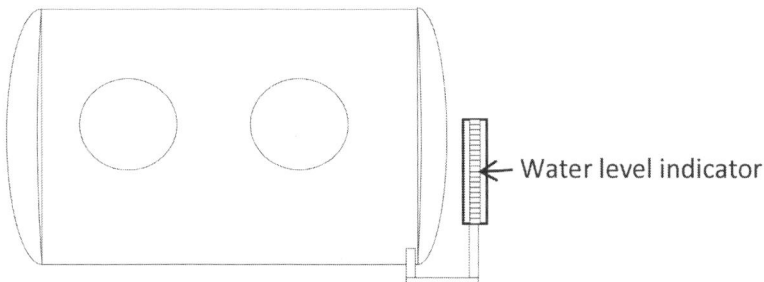

Figure 14.4   Water-level indicator.

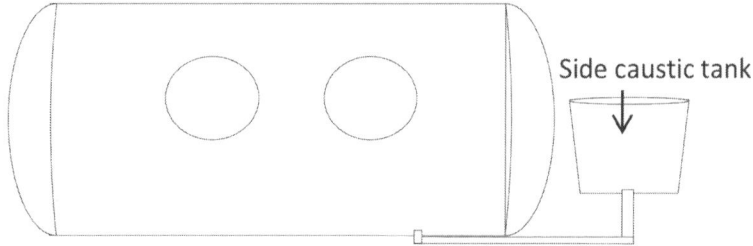

Figure 14.5 Side caustic tank.

i) Follow the job card and programme the machine for heating, the specified temperature and timings, forward running, reverse running, holding at a specified temperature, cooling and rinsing.

j) Check the pH of the fabric after rinsing before unloading the materials.

k) Enter the complete details of the design number, number of batches taken, total weight of fabric processed, the starting time and ending time of operation, and problems observed, if any, in the production record.

l) Enter the details in the lot card of the design number and the process carried out and attach the same to the trolley.

m) Unload the material, cover properly and deliver at the designated place along with all details on the process card.

14.4 Control points and checkpoints

In any process, it is essential to have clarity on the points to be controlled to achieve the targets and those to be checked to ensure the process in control. These points need to be reviewed from time to time and modified to suit the requirements of individual companies and their targets. Each mill should prepare its own 'control points and checkpoints' and display them in the work area so that the people on spot refer and follow. Following are just for an illustration.

14.4.1 Control points

a) Selection of suitable scouring and washing process for the material to be processed

b) Deciding and selection of process parameters, namely:

- Speed of rotation
- Running time for each operation

- Temperature
- Pressure
- Scouring recipe
- Liquor ratio
- Washing sequences

c) Deciding acceptance criteria for quality of drum washing

d) Deciding on chemicals to be procured and their quality requirements

e) Employing trained and qualified employees

f) Evolving production norms

g) Evolving norms for consumption of chemicals and water

14.4.2 Checkpoints

- **Material related**

  a) The fabrics received for scouring and washing against the plan received

  b) The quantity of fabric received for scouring and washing – metres and kilograms

  c) Materials loaded against the maximum capacity of the machine

- **Machine related**

  a) The condition of the machine

  b) The working of various valves and controllers

  c) Smoothness of the inner surface of the octagonal drum

  d) Cleanliness of the spray nozzles

- **Setting related**

  a) The force of fabric hitting the board, that is, the speed of the drum

  b) The recipe prepared for scouring

  c) The number of revolutions and reversals set

- **Performance related**

  a) Uniformity of scouring and washing throughout the fabric

  b) The production achieved

- **Documentation related**

  a) The design number and the blend of the fabric received for washing

  b) Quantity washed

- **Work practice related**

  a) Whether the drum washer machine was thoroughly cleaned before loading the material

  b) Whether the speed, pressure and temperature maintained are as per norms

- **Logbook related**

  a) Machines working

  b) Starting time of the running lots

  c) Activities done and to be done

- **Management information system related**

  a) Rotary drum machine number

  b) Design number

  c) Number of metres

  d) Weight in kilograms

  e) Chemicals used

- **General**

  a) Use of safety gadgets such as gloves and gum shoes while handling chemicals

14.5 Dos and Don'ts

Understand clearly what is to be done without fail and what should not be done at any cost.

Some examples are given below.

14.5.1 Dos

a) Prepare the scouring liquor after referring to the instructions for the material in the process.

b) Monitor the pH after washing and maintain at neutral.

c) Ensure that safety locks are properly secured while opening the doors.

14.5.2 Don'ts

a) Do not increase the concentration of caustic to get faster results.

b) Do not operate the machine if the pressure gauge is not calibrated or not showing the correct pressure.

c) Do not load the machines to its full capacity.

Weight reduction process of polyester fabrics

15.1 Introduction to weight reduction

Polyester is commonly used synthetic fibres for apparel purposes. However, modification of polyester fibre has been necessitated to overcome several inherent basic shortcomings in the polyester fibre properties which are not comfortable for the person wearing the polyester cloth. The drawbacks include very low moisture absorption (0.4% moisture regain) and poor dyeability due to a high degree of 'hydrophobicity' of the fibre molecular structure and compactness of polymer structure. Specifically, weight reduction is done to the heavy polyester fabric to modify polyester fibre either partially or fully by copolymerization or by treating polyester fibre with alkali/acid/organic solvents treatments.

Varieties of machines are developed for weight reduction. Softleena machine and Apollotex continuous weight reduction (CWR) machine are examples. Softleena is used for scouring and weight reduction in batches and to give chemical treatment to get special handle drape and comfort property in blended fabrics.

Weight reduction of polyester by partial alkaline hydrolysis is not only less expensive but gives better and favourable results in improved physical and comfort properties also. The weight reduction of polyester is 'saponification' of terephthalic ester (one of the major raw materials from polyester fibre) caused by sodium hydroxide under a suitable wetting agent.

Various catalysts are available such as Terrysolve, Cizapan WR and so forth. The catalysts help in partial alkaline hydrolysis causing the progressive weight reduction of polyester fibre in the surface, resulting in a small extent of loss of mass, where the other favourable properties remain almost unchanged or change for better effects. Weight reduction of polyester can be achieved at low temperature and low concentration of caustic soda by using the catalysts.

The weight reduction treatment is a major breakthrough to obtain 'skincare polyester fabric'. Various trials were conducted for controlled weight loss in polyester textiles after identifying suitable catalysts for this treatment. Following are some of the recommendations:

- To achieve 15% weight reduction, at the optimum treatment conditions at 5% NaOH, the product catalyst 'Terrysolve' when used 0.5% on weight of fabric (owf) for 60 min at 90°C obtained an optimum result as required.

- Cizopan WR gave optimum results when used in a dosage of 1–3 g/L or 2–3% owf along with 10–20 g/L of caustic soda in a jigger or jet dyeing machine at a temperature of 90–120°C for 45–60 min. This is found useful in achieving the desired level of weight reduction of polyester fabrics which could not be subjected to high temperatures and which could not be treated with a high concentration of caustic soda.

- The weight reduction of polyester fabrics while using Cizopan WR is highly accelerated. Hence, the concentration of the caustic soda, the duration of the process and the temperature should be assessed by trial before bulk production. Excessive weight reduction may result in loss of tensile strength, slippage and so forth.

15.2   Softleena machine for weight reduction of polyester fabrics

Softleena machine is specially designed for weight reduction of polyester fabrics in batch (Fig. 15.1). It has several advantages.

a) Comparing to the degradation technique of both continuous and tank type, Softleena process gives a better handle to the fabrics as it is performed with low tension because of the smooth flow of the weak alkali solution circulating repeatedly.

b) Softleena improves the handle and reduces the defect for dyeing due to perfect rinsing effect which can remove the breaking material and sizing agents left in the previous process.

c) It is suitable for small quantity of various kinds as it is easy to control the temperature and alkali concentration.

d) Softleena provides relaxing effect when treated hard twisted fabrics.

e) No variation, because of both alkali solution and the fabrics are circulated forcibly no hanger pin trace, no stretching the selvage and no overwrapping.

Figure 15.1   Softleena machine.

f)   Can save more than half of supply water, cooling water, waste water and steam due to its unique structure.

g)   The process can be reduced due to no hanger process (compared with tank type).

h)   Can decrease the time limit by the quick preparation work because of no hanger process.

i)   No accident due to breaking down of the hoist or hazardous materials with high temperature.

j)   Good working environment with closed structure due to no scattering of steam and no dropping of water to transportation.

k)   Four fabric ropes can be treated at a time in the Softleena machine.

15.3   Operating procedure for using Softleena machine

The general work procedure for weight reduction process using Softleena machine is as follows:

a)   Clean the machine thoroughly before starting any programme.

b)   Get the details of the programme to be worked from the supervisor.

c) In case of any discrepancy either in the material or in the process to be done, discuss with the supervisor and get clarity.

d) Check the design number and lot number before bringing the material to the machine.

e) Stitch the ends to make continuous rope form of the fabric.

f) Fill water in the overhead tank by operating the button 'O/H Tank Fill' (as per the requirement of process).

g) Drain the steam condensates from the steam line before starting the machine.

h) Prepare the caustic solution in the stock tank as per the instruction of the supervisor. Maintain concentration and add the required catalyst as directed in the chemical issue sheet. The effect of weight reduction is depending on the concentration of caustic solution and the ratio of catalyst added.

i) Check the pH of the fabric after rinsing before unloading the materials.

j) Enter the complete details of the design number, number of batches taken, total length of fabric processed, the starting time and ending time of operation and problems observed, if any, in the production record.

k) Enter the details in the job card of the design number and the process carried out and attach the same to the trolley.

l) Unload the material, cover properly and deliver at the designated place along with all details on the job card. Unload at room temperature to avoid crease.

Some typical process methods are explained in coming sections.

15.4 Typical process sheet for scouring on Softleena

Step no.	Process	Time (min)	Water
1	Fill water—2200 ± 100 L	8	2200
2	Loading of fabric	25	
3	Stitching	5	
4	Dosing of (a) caustic 2%		
5	While dosing starts raising temperature to 60 ± 5°C at 3°C/min	7	
6	Add (b) STNR 0.8%		

Step no.	Process	Time (min)	Water
8	Raise temperature from 60 to 95°C at 3°C/min	12	
9	Hold	60	
10	Cooling + sampling temperature up to 60°C	10	
12	Drain fully	5	
13	Refill	8	2200
14	Dosing of hydrochloric acid and oxalic acid as per slip		
15	Simultaneously raise temperature to 80±5°C	10	
16	Hold	5	
17	Fill and drain two times (2F+2D+2F) in minute	6	2000
18	Drain fully	5	
19	Refill and run 3 min	11	2200
20	Drain	5	
21	Refill (temperature should be at least 45°C)	8	2200
22	Dosing of acetic acid as per slip	3	
24	Off-load	25	
	Total time in minutes	218	
	Total water consumption in litres		10 800

15.5 Typical process sheet for weight reduction on Softleena

Step no.	Process	Time (min)	Water
1	Fill water—2200±100 L	8	2200
2	Loading of fabric	25	
3	Stitching	5	
4	Dosing of caustic 4–7%		
5	While dosing starts raising temperature to 60±5°C at 3°C/min	7	
6	Add HYPLEX 1.0–1.5%		
8	Raise temperature from 60 to 95°C at 3°C/min	12	

Step no.	Process	Time (min)	Water
9	Hold	90	
10	Cooling + sampling—temperature up to 60°C	10	
12	Drain fully	5	
13	Refill	8	2200
14	Dosing of sodium bicarbonate as per slip		
15	Simultaneously raise temperature to 80±5°C	10	
16	Hold	5	
17	Fill and drain two times (2F+2D+2F) in minute	6	2000
18	Drain fully	5	
19	Refill and run every 3 min	11	2200
20	Drain	5	
21	Refill (temperature should be at least 45°C)	8	2200
22	Dosing of acetic acid as per slip	5	
24	Off-load	25	
	Total time in minutes	250	
	Total water consumption in litres		10 800

15.6 Process recommendation for 100% polyester processing using jet dyeing machine for weight reduction (as recommended by Rossari)

15.6.1 Drumming

Objective	To open up yarn twist, thereby getting grainy effect
Machine	High temperature–high pressure (HTHP) drum
Material-to-liquor ratio (MLR)	1:4–1:6
Product	Kleenox TEP
Dosage	0.3–0.5% on weight of fabric (owf)
Pressure	2–3 kg/cm2
Hold time	45–90 min

15.6.2 Desizing

(For regular and water jet loom sizes, i.e. polyvinyl alcohol and Na salts acrylate copolymers)

Machine	HTHP jet dyeing
MLR	1:4 in U-tube and 1:8 in long tube

Process

a) Load grey and give cold wash

b) Add Kleenox BAS: 1–2 g/L

c) Adjust pH 5.5–6.0 with acetic acid

d) Run for 20 min at room temperature

e) Add soda ash: 2–4 g/L (pH 10–10.5)

f) Raise temperature to 100°C by 1°C/min gradient.

g) Hold for 45–60 min

h) Hot drain

i) Hot wash at 95°C with 0.5 g/L Kleenox BAS for 10 min

j) Cold wash and further scouring/weight reduction

15.6.3 Single-bath scouring and weight reduction

Machine	HTHP jet dyeing
MLR	1:3–1:4 in U-tube

Process

• Load material and give cold wash

• Add

   a) Geeenscour CPM: 2–3 g/L

   b) Kleenox BASM: 2–3 g/L

   c) Zylube CM (if required): 1.5 g/L

• Run for 15 min at room temperature

• Add caustic soda flakes as per weight reduction required

• Raise temperature to 120–130°C

• Hold for 30–60 min

- Hot drain
- Hot wash at 95°C for 15 min with the addition of Oligo EM 1 g/L
- Neutralization at 90°C for 20 min with
  a) Oxalic acid: 4 g/L
  b) Kleerix N: 1 g/L

15.6.4 Single-bath scouring/weight reduction and grainy effect on high twist 100% polyester qualities

Machine	HTHP jet dyeing
MLR	1:3–1:4 in U-tube

Process

- Load grey fabric and give cold wash
- Add
  a) Geeenscour CPM   2 g/L
  b) Kleenox BASM   1 g/L
  c) Kleenox TEP   0.5–1% owf
  d) Zylube CM (if required)   1.5 g/L
  e) NoFoam ND   0.25 g/L
- Run for 15 min at room temperature
- Add Caustic soda flakes as per weight reduction required
- Raise temperature to 100–130°C by 0.5°C/min and hold for 10 min.
- Maintain pressure to 2.5 kg/cm2. Hold for 60 min.
- Cool down at 90°C—drain
- Hot wash at 80°C for 10 min
- Neutralization at 85°C for 20 min with
  a) Oxalic acid—4 g/L
  b) Kleerix N—1 g/L

15.7 Continuous weight reduction process

Weight reduction can be done on continuous basis also. Apollotex CWR machine is an example (Fig. 15.2).

There are three main sections in the machine, namely caustic soda padding section, heating section and washing section. The fabric is first padded with caustic soda and then heated in the heating section. The degraded portions are washed off in the washing section.

Controlling concentration of caustic solution: This is a very important operation because if the concentration of caustic solution is higher than the specified, it can damage the fabric. The concentration of caustic solution is expressed in terms of degree Twaddle (°Tw). The concentration should be maintained between 55 and 70°Tw. The fabric density (gram per square metre; GSM) is the governing factor for deciding the concentration of caustic solution to be adopted. The supervisor checks the concentration of caustic solution before allowing it to be loaded into the machine.

Heating: The heating is done by steam at a high temperature of 60–90°C. The temperature is controlled by opening and closing the steam valves.

Mangle pressure: The padding mangle pressure is to be maintained between 3 and 4 kg/cm^2.

Follow the steps as follows while operating a CWR machine:

- Clean the machine thoroughly before starting any programme.

- Get the details of the programme to be worked from the supervisor.

- In case of any discrepancy either in the material or in the job card, discuss with the supervisor and get clarity.

Figure 15.2 Apollotex continuous weight reduction machine.

- Check the design number and lot number before bringing the material to the machine.

- Drain the steam condensates from the steam line before starting the machine.

- Prepare the caustic solution in the stock tank as per the instruction of the supervisor. Maintain concentration (°Tw) and add the required wetting agent as directed in the chemical issue slip. The effect of weight reduction is depending on the following factors:

  a) Concentration of caustic solution

  b) Concentration of wetting agent

  c) Time

  d) Temperature

- Get the fabric as per programme.

- Check the fabric GSM before putting it in the machine.

- Open the steam valve and set the temperature in the heating range.

- Set the mangle pressure.

- Start the machine after ensuring all systems are OK.

- Check the pH of the fabric after washing.

- Monitor the fabric speed which depends on weight reduction required.

- Periodically check the chamber temperature.

- Check the GSM of the fabric after the process is over and the fabric is dry.

- Enter the complete details of the design number, number of lots taken, total length of fabric processed, the starting time and ending time of the machine and problems observed, if any, in the production record.

- Enter the details in the job card of the design number and the process carried out and attach the same to the trolley and send the material to next process.

15.8   Control points and checkpoints

It is essential to have clarity on the points to be controlled to achieve the targets and those to be checked to ensure the process is in control. These points need to be reviewed from time to time and modified to suit the requirements of individual companies and their targets. Each mill should prepare its own 'control points and checkpoints' and display them in the work area so that the people on spot refer and follow. Following points are just for an illustration.

15.8.1 Control points

a) Selection of suitable scouring and weight reduction process for the material to be processed

b) Deciding and selection of process parameters, namely:

- Speed
- Running time for each operation
- Temperature
- Pressure
- Scoring and weight reduction recipe
- Liquor ratio
- Washing sequences

c) Deciding acceptance criteria for quality of weight reduced polyester fabrics

d) Deciding on chemicals and their quality to be procured

e) Employing trained and qualified employees

f) Evolving production norms

g) Evolving norms for consumption of chemicals and water

15.8.2 Checkpoints

- **Material related**

  a) The fabrics received for scouring and weight reduction against the plan received

  b) The quantity of fabric received for scouring and weight reduction – metres and kilograms

- **Machine related**

  a) The condition of the machine

  b) The working of various valves and controllers

  c) Cleaning of the spray nozzles

- **Setting related**
  a) The recipe prepared for scouring and weight reduction
  b) The number of revolutions per set
- **Performance related**
  a) Uniformity of scouring and weight reduction throughout the fabric
  b) The production achieved
- **Documentation related**
  a) The design number of the fabric received for washing
  b) Quantity washed
- **Work practice related**
  a) Cleaning of the machine thoroughly before loading the material
  b) Setting and maintaining the speed, pressure and temperature as per norms
- **Logbook related**
  a) Machines working
  b) Starting time of the running lots
  c) Activities done and to be done
- **Management information system related**
  a) Machine number
  b) Design number
  c) Number of meters
  d) Weight in kilograms
  e) Chemicals used
- **General**
  a) Use of safety gadgets such as gloves and gum shoes while handling chemicals

15.9 Dos and Don'ts

Understand clearly what is supposed to be done without fail and what should not be done at any cost. Some examples are given below.

15.9.1 Dos

a) Stitch the ends securely of the fabric after mounting on the inside reel.

b) Monitor the pH after washing and maintain at neutral.

c) Prepare the scouring liquor after referring to the instructions for the material in process.

d) Check tearing strength before and after treatment.

e) Always make a sample and decide on the parameters to be kept depending on the status of the fabric.

f) Monitor the speed of the machine, especially in continuous weight reduction process.

g) Monitor the temperature in the heating chamber in continuous weight reduction process.

h) Monitor the mangle pressure regularly in a continuous process.

15.9.2 Don'ts

a) Do not go for weight reduction process if the percentage of polyester is less than 67% in the fabric.

b) Do not go for weight reduction process for delicate fabrics.

c) Do not increase the concentration of caustic solution to get faster results.

d) Do not increase catalyst beyond the limit.

e) Do not do excessive weight reduction as it causes tendering.

f) Do not accept a job work for weight reduction unless the sample is tested and approved for weight reduction, loss of strength and change of colour.

Peaching (sueding) of fabrics

16.1 What is peaching?

Peaching or sueding is a technique that creates a soft feel while touching (Fig. 16.1). It is a mechanical finishing process in which a fabric is abraded on one or both sides to raise or create a fibrous surface. This fibrous surface improves the fabric appearance, gives the fabric a softer, fuller hand, and can mask fabric construction and subdue colouration. These improved aesthetics can increase the value of a fabric in the marketplace.

In the textile industry, the process of sueding is also known as 'sanding' or 'emerizing'. Normally, this process is done only as per buyer's requirement.

The sanding technique can be applied to any type of fibre, although it does seem to work more effectively with natural materials. It can be done on both woven as well as knitted fabrics. Even bottom weights (trouser materials) are now being sueded to get better feel.

Figure 16.1   Sueder machine.

After the fabric is woven, sections of the material are dipped in chemical compounds that permeate the fabric. The sections are then stretched taut and left to dry. Once the sections have dried, the material is brushed either by hand or with the use of automated machinery. In either case, rollers with abrasive bristles, similar in appearance to a toothbrush, are brushed across the fabric. This helps to break some of the small fibres on the exterior of the material and teases them out. Teasing the broken fibre ends creates the peached appearance and feels for the sanded fabric.

Along with the use of abrasive rollers, peached fabric can also be created by the use of chemical abrasion. With this method, the fibres are gently broken down with the use of chemical compounds rather than being sanded by bristles. Laundry abrasion is also a means of producing peached fabric. The motion within the laundering process creates a break in the outside fibres and helps to produce the soft feel that is associated with peached fabric.

Because peached fabric holds its shape very well and is soft to the touch, the material is often used in household textiles such as casual tablecloths and napkins. Kitchen curtains can also be made from peached fabric. When it comes to clothing, peached fabric is an ideal choice for casual shirts, golfing shirts and undergarments. The sanding techniques do not take away from the ability of the material to absorb and maintain colour, which means that peached fabric is available in any colour or pattern that one can imagine.

Both the finished product and the fabric treatment derive the name from a few observations about the look and feel of the material that results from the process. Looking at the material through the lens of a microscope, the material will appear to have a small layer of fuzz on the top, much like the fuzz that is found on the outside of the peach peel. Because the end result is both the look of peach fuzz and a feel that is like rubbing a hand across the fuzz on a peach, the popular name came into common use.

16.2　Sueding machines

There are two types of sueder machine, namely single-cylinder sueder machine and multicylinder sueder machine.

16.2.1　Single-cylinder sueder machine

With a single-cylinder sueder, sueding takes place when the fabric passes between the abrasive covered roll and the feed back-up roll (Fig. 16.2). The opening is adjustable within 1/10 000th of an inch on contemporary machine models. Designing a machine with such accuracy requires close tolerance

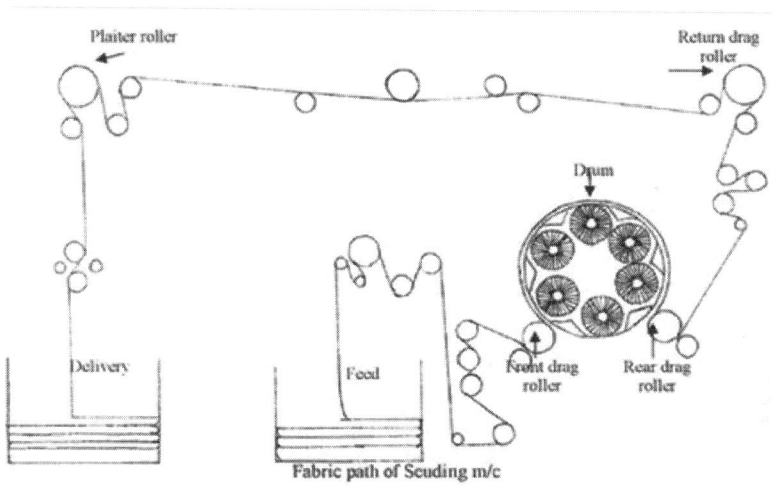

Figure 16.2 Single-cylinder sueding machine.

manufacturing and a method of maintaining this accuracy under operating conditions in a textile mill.

16.2.2 Multicylinder sueder machine

To raise a pile on fabric, the multiroll sueder relies on tension to allow the abrasive to penetrate the fabric surface. The use of several (usually five) abrasive rolls to perform this task reduces the amount of tension by spreading the work over several rolls. Each roll is independently driven and can be rotated in a clockwise or counter clockwise direction. Adjustable idler rolls between the abrasive rolls control the slack in the fabric among rolls. Entry and exit drive rolls transport the fabric and control tension. The principle of exposing the fabric to more than one abrasive action area permits the use of a finer grain abrasive.

16.2.3 Modern sueding machines

Let us have a look at some of the modern sueding machines.

16.2.3.1 Caru, Italy (The Patterson Group, Charlotte, NC)

It offers the sanding and sueding machine, model CSM/6-B. The machine introduces patented, lightweight carbon composite sueding rolls. This six-roll sueding machine features alternating current inverter drives and programmable logic controllers (PLC) control, offering numerous advantages over conventional machines, including speeding up the rolls to 2800 rpm for higher production rates. It also drives individual sueding rolls in pile and counter pile directions for special sueding effects. The PLC control assures uniform and repeatable processing. The special pneumatic paper-clamping system automatically recovers any paper loosening due to wearing and temperature and gives a constant adjustment to the paper tightness. An oscillating motion of the third and sixth rolls evens the pile and hides eventual microdefects from prior processes. All of the features of the CMS/6-B machine for dry sanding are also available on the AS/4 machine for wet sanding.

16.2.3.2 *Lafer SpA, Italy (SGA, Charlotte, NC)*

It is one of the leading suppliers of raising, shearing and sueding machines. It offers a wide range of surface finishing solutions for knitted and woven fabrics. It pioneered the double-drum technology on the raising machine and was one of the first suppliers to introduce automation on raising machines. The company based its GSMI sueding machines on the multiple roll drum technology (24 or 48 rollers), compared to conventional sueders that use only 4–8 sueding rollers. The drum technology allows greater versatility on light- to heavyweight fabrics with minimal loss of tensile strength. Streaking, which is a critical problem on conventional sueders, has been solved by the combined action of the 24 sueding rollers. The company emphasizes that the rotating drum will not cause any streaking and will actually enhance sueding uniformity.

Lafer has introduced Aqua Sand, a new type of sueding machine that widens the range of sueding effects. This can be used to process fabric in wet or dry form. Dry sueding gives a soft handle and will raise a certain pile surface (hairiness), whereas with wet sueding, the water creates a lubricating effect that gives a soft handle but without raising the pile. Subsequently, there will be a minimal change of appearance, but the fabric softness will be remarkably enhanced.

Lafer's Aqua Sand functions principally of the sueding machine (Fig. 16.3). It is supplied with an impregnation mangle at the entry and a squeezing mangle at exit so that fabrics can be processed wet or dry. It can process woven, knit and stretch fabrics made of natural, synthetic or blended fibres.

Lafer Ultrasoft and Microsand sueding machines have popular 'brush' sueding technology. They are equipped with the new sueding drum configuration allowing to use either abrasive brushes (warm effect) or Carbosint diamond rollers (cool effect) leading to no streaking and no hairiness. When using the Ultrasoft brush sueding mode, special metal plates (patented Delta system) allow to keep the fabric well spread at all times, avoiding channelling and streaking. On conventional machines without these supports,

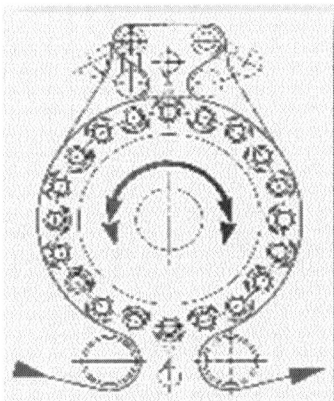

Figure 16.3   Lafer sueding machine.

the lengthwise tension causes a downward pressure that distorts the brush bristles causing channelling (lengthwise creasing), which on most fabrics will produce lengthwise streak marks and 'centre to selvedge differences'. This limits the range of fabrics that can be processed.

On the Ultrasoft, the Delta plates support the full tensioning range required for all types of fabric construction in order to maintain the ideal orientation of the brushes and avoid channelling. Higher lengthwise tension settings will produce a cleaner-cut short pile sueding effect while lower tension settings will leave a longer surface pile. Conventional systems without the Delta support plates are very much limited in the lengthwise tension settings; hence, the range of fabrics is limited.

The Lafer machine (Ultrasoft mode) is equipped with four load cells positioned in critical spots of the machine that are essential for processing even the most delicate knits or lycra-based (Ultrasoft mode available as an option on Microsand; Fig. 16.4).

A specific 'knit fabric operating software' allows treating these 'difficult fabrics' with ease.

A separate 'woven fabric operating software' (Microsand mode) leads to the best results always. Other machines cannot treat all types of fabric without accepting compromises on quality and/or production.

The patented Delta system with the Ultrasoft brush (Fig. 16.5) mode allows to cover the maximum drum angle leading to greater contact area between fabric and abrasive brush.

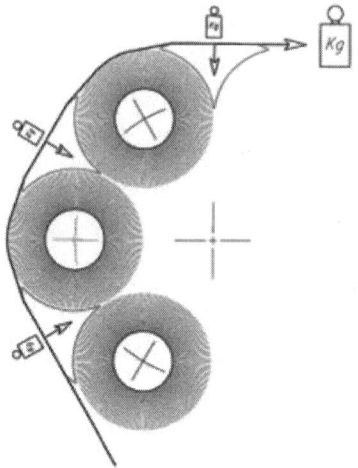

Figure 16.4　Lafer Ultrasoft.

Figure 16.5　Ultrasoft brushes.

Lafer claims that other machines without the Delta system are forced to run at slower speeds and lower tension settings in order to avoid problems (streaking, especially on knits) and in any case produce more hairiness than Lafer; one can notice a remarkably cleaner effect with Lafer.

The exclusive 'Carbosint' diamond-like emery developed by Lafer significantly reduces machine stops for emery replacements (Fig. 16.6). There are Microsand end users who have been able to exploit a full drum of Carbosint emery at 4 million metres.

On the Lafer machine, one can set various operating modes in order to create different sueding effects like rollers all clockwise, all counter-clockwise and half and half. A differential drive system with timing belts, the Sure-Grip transmission system allows to exploit the maximum energy available.

Figure 16.6   Carbosint emery rollers.

Figure 16.7   Dust collection unit.

The sueding technology using abrasive brushes or diamond emery creates very fine dust compared to conventional sandpaper which creates larger particles that are easier to eliminate. The sueding chamber is enclosed in order to create the necessary pressure to enhance the suctioning efficiency.

The sueding chamber is equipped with an exclusive vent system to help remove the sueding dust and keep the machine cleaner in order to avoid fabric contamination (Fig. 16.7). At the outlet of the sueding drum, a special finishing brush orients the pile and enhances the sueding effect on woven fabrics. On top of the machine, there is a very efficient fabric cleaning unit equipped with brush and beater connected to the suction unit.

16.2.3.3   Mario Crosta, Italy (PSP Marketing, Charlotte, NC)

It offers a wide variety of surface finishing machinery including type MC Universal and special raising and sueding machines and type SD sueding machines.

16.2.3.4   Sperotto Rimar SpA, Italy (Speizman Industries, Charlotte, NC)

It offers the Plurima wet-sueding machine. Compared to the traditional dry-sueding technique, the wet-sueding distinguishes itself in achieving new and different effects such as surface softening without pile formation perceivable to the touch and surface appearance similar to discolouration and/or aging typical of aged and/or worn fabrics. These effects are now required by the market since they are meeting the current fashion trends. The water presence in the Plurima wet-sueding machine carries out a fibre-lubricating action on the fabric. This allows more superficial abrading action, less fibrous material to be removed, a lower pile raising and the development of short and thick pile. The fabric types that proved to be particularly suitable for the wet-sueding processing are the woven fabrics from the lightest poplin to the heaviest cloth with drill-weaving construction. In the cotton sector, the application is for 100% cotton and cotton/polyester, cotton/polyamide, cotton/viscose blends and denim fabrics. In the synthetic fibre sector, typical applications are for 100% polyester, 100% polyamide and their blends with viscose/rayon and 100% viscose/rayon. In the wool sector, 100% wool, wool/viscose/rayon blends and wool/silk blends are typical applications.

16.3   Typical operating procedure for sueding machine

The operating procedure depends on the type of sueding machine, the type of sueding roller, the fabric density and weave, the peaching effect required and so on. Each mill has to prepare process sheet depending on the combinations they have. The following are general guidelines:

a) Before starting the machine, clean the inside portion of it with compressed air. It is necessary to wear a mask as there shall be a lot of microdust on the sides of the machines and also on the brushes.

b) Check the condition of the brushes and ensure proper condition and cleaning.

c) Check the guide rolls for their alignment. Clean them before starting the machine.

d) Check the air pressure and ensure it as correct.

e) Check the tension of the fabric.

f) Check the seam detector.

g) Check the exhaust fan working. Ensure that the canvas cloths are tied properly and there is no air leakage. There should not be any hole in the bag.

h) Check all dust bags. If bags are full, clean them and put new empty bags. Wear mask while cleaning the dust bags as it contains microdust, which is harmful.

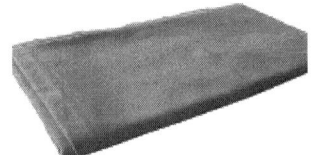

Figure 16.8 Fabric after peaching.

i) Set the fabric speed, speeds of all brushes, the setting of the guiding plates and fabric tension as per the requirement of the fabric. Refer to the process sheet for more details.

j) Inspect the fabric before peaching and in between during process. Check the peaching effect. The fabric should not become hairy and the surface should be uniform.

k) Check the fabric strength, both tear and grab test before and after peaching the fabric (Fig. 16.8).

l) Adjust the peaching parameters depending on the test result.

m) Cover the trolley with polyethylene paper at the delivery end and keep at proper place.

n) Keep proper tag on the batch.

o) Close all air pressure valve while stopping the machine.

16.4 Control points and checkpoints

It is essential to have clarity on the points to be controlled to achieve the targets and those to be checked to ensure the process in control. These points need to be reviewed from time to time and modified to suit the requirements of individual companies and their targets. Each mill should prepare its own 'control points and checkpoints' and display them in the work area so that the people on spot refer and follow.

16.4.1 Control points

a) Selection of suitable peaching process for the material to be processed

b) Deciding on the degree of peaching to be given considering the fabric construction and the raw materials used for the fabric

c) Deciding and selection of process parameters, namely:

• Speed of rollers

- Setting between fabric surface and the brushes
- Dressing of sueding brushes

d) Deciding acceptance criteria for quality of sueding of fabrics

e) Employing trained and qualified employees

f) Evolving production norms

16.4.2 Checkpoints

- **Material related**

  a) The fabrics received for sueding against the plan received

  b) The quantity of fabric received for sueding – metres and kilograms

  c) Type of fabric – woven or knitted

  d) The density (grams per square metre) and the fabric construction

  e) The type of fibres in the fabric

- **Machine related**

  a) The condition of the machine

  b) The condition of the bristles in the brush rollers – they should be sharp and the brush should have uniform diameter

  c) The condition of the emery rollers

  d) The working of various controllers

  e) The condition of the dust collection bags

- **Setting related**

  a) The setting between fabric and the brush

  b) The setting between emery and the fabric

  c) The angle at which the fabric is presented for sueding

  d) The distance of contact of fabric to the sueding media

- **Performance related**

  a) Uniformity of sueding throughout the fabric

  b) The production achieved

- **Documentation related**

    a) The design number of the fabric received for sueding

    b) Quantity sueded

- **Work practice related**

    a) Thorough cleaning of the machine before loading the material

    b) Clearing of the wastes from time to time

    c) Maintaining the speeds as per process sheet

- **Logbook related**

    a) Machines working

    b) Starting time of the running lots

    c) Activities done and to be done

- **Management information system related**

    a) Machine number

    b) Design number

    c) Number of metres

    d) Weight in kilograms

- **General**

    a) Use of safety gadgets such as gloves and masks while working

16.5 Dos and Don'ts

It is essential to understand clearly what are supposed to be done without fail and what should not be done at any cost. Some examples are given below. Display them properly in the work area so that all can follow all the time.

16.5.1 Dos

a) Wear mask while cleaning the dust bags as it contains microdust (Fig. 16.9). Check the fabric strength, both tear and grab tensile, and adjust the parameter.

b) Make a trial run before feeding the fabric in bulk for setting the peaching parameters.

c) Keep the dust collection area clean.

Figure 16.9   Wear mask.

16.5.2 Don'ts

Figure 16.10   Good bristles.   *Figure 16.11*   Fused bristle tips.

a)   Do not use the brush if the bristle tips are fused (Figs. 16.10 and 16.11).

b)   Do not allow the fabric to become weak because of peaching.

c)   Do not proceed with peaching if the fabric is found damaged or abraded.

d)   Do not decide to peach a fabric unless there is requirement from the customer.

e)   Do not use the bag if it has developed hole (Fig. 16.12).

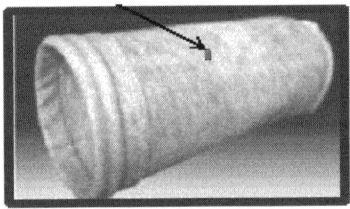

Figure 16.12   Hole in the bag.

Raising operation on fabrics

17.1 What is raising?

Raising is a process of making the embedded fibres from the fabric surface to come out. This is done to impart a soft or cushioning velvet effect on the woven fabrics. One can do only raising on a fabric or combine with different surface finishing like raising followed by shearing, sueding, emerizing and so forth.

Raising is a finishing process that raises the surface fibres of a fabric by means of passage over rapidly revolving cylinders covered with metal points or teasel burrs. The outing, flannel and wool broadcloth derive their downy appearance from this finishing process. Napping is also used for certain knit goods, blankets and other fabrics with a raised surface.

By means of this process, a hairy surface can be given to both face and back of the cloth providing several modifications of the fabric appearance, softer and fuller hand and bulk increase. This enhances the resistance of the textile material to atmospheric agents by improving thermal insulation and warmth provided by the insulating air cells in the nap. The fuzzy surface is created by pulling the fibre end out of the yarns by means of metal needles provided with hooks shelled into the rollers that scrape the fabric surface. The ends of the needles protruding from the rollers are normally hooks with 45°; their thickness and length can vary. They are fitted in a special rubber belt spiral wound on the raising rollers (Fig. 17.1). These rollers are generally alternated with a roller with hooks directed toward the fabric feed direction (pile roller) and a roller with the hooks fitted in the opposite direction (counterpile roller).

The machine also includes some rotating brushes, which suction clean the nibs in pile and counterpile directions. Normally, the trend is to have a ratio of raising rollers/pile rollers equal or one-third. The two series of rollers have independent motion and can rotate with different speed and direction; thus, carrying out different effects.

1. Roller

2. Rollers equipped with hooks

3. Fabric

4. Nib cleaning brushes

5. Fabric tension adjustment

The action of these systems is powerful and the results depend upon the effects and the type of fabric desired. The raising effect can be obtained by adjusting the fabric tension (5) or by adjusting the speed and the roller rotation direction (2). Once a certain limit is exceeded, the excessive mechanical stress could damage the fabric. It is better to pass the wet fabric through the raising machine many times (dry when processing cotton fabrics) when carrying out a powerful raising and treat the fabrics in advance with softening–lubricating agents (Fig. 17.2).

The pile extraction is easier when carried out on single fibres. It is suitable to reduce the friction between the fibres by wetting the material or, in case of cellulose fibres, by previously steaming the fabric. It is better to use slightly twisted yarns. A raising machine allows different options of independent motions:

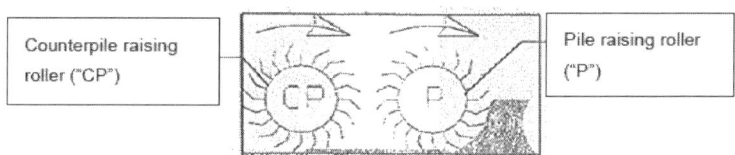

Figure 17.1   Raising rollers.

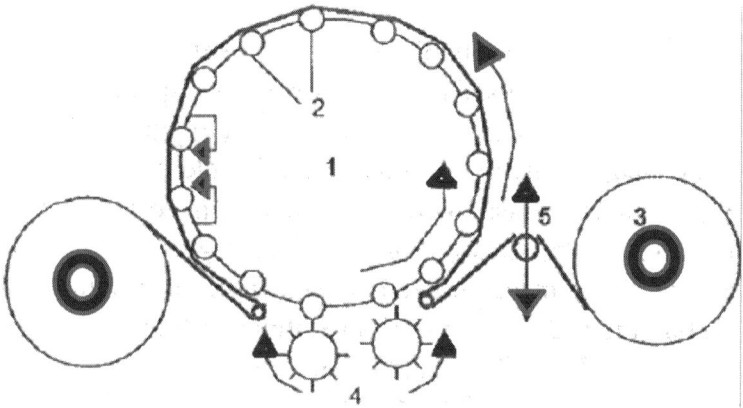

Figure 17.2   Raising (napping) machine.

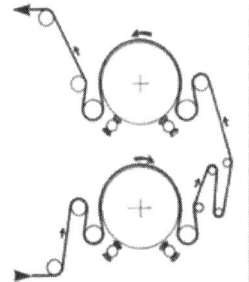

Figure 17.3   Raising the face of the fabric.

1.   Fabric moving between entry and exit

2.   Motion of large drum

3.   Motion of raising rollers

The raising intensity can be adjusted by suitably combining the above-mentioned independent motions, the tension of the textile material, the number of pile or counterpile raising rollers and their relative speeds. It is possible to obtain combed pile raising effect, 'semi-felting' effect with fibres pulled out and re-entered in the fabric, and complete felting effect. The raising machine is equipped with two overlapping drums each one featuring 24 rollers, which can process two faces or face and back of the same fabric (Fig. 17.3). The drums assembled on a standard machine can rotate separately one from the other in the fabric feeding direction or in the opposite direction by carrying out a counter rotation. In this model, all the functions are carefully monitored and controlled by a computer system; in particular, all the commands are driven by alternating power motors controlled by 'sensorless' vector inverters.

Furthermore, a series of special pressure rollers can be assembled on the feeding cylinders to prevent the fabric from sliding, thus granting an extremely smooth raising. The raising process ability lies merely in raising the desired quantity of fibre ends without excessively reducing the fabric resistance. The technique of applying the alternated use of pile and counterpile rollers is most widely used. It minimizes the loss of fibres from the fabric and the consequent resistance reduction.

Standard raising machines have been designed to work with fabrics powerfully tensioned because they are not equipped with an efficient and reliable tension control. This gives rise to the effects detailed below:

1.   The contact surface between the fabric and the raising cylinders is quite small.

2. The hook nibs work only superficially on the fabric and the raising effect is quite reduced.

3. The fabric width is drastically reduced.

The above-mentioned inconveniences have now been eliminated in the latest generation of raising machines, which reduce the number of passages and carry out the raising process by gently tensioning the fabric.

The control electric system features are normally available on all modern machines with

1. Programmable logic controller (PLC) for machine and alarms automation

2. Touchscreen to program and update all processing parameters

3. Operating conditions of each single raising process (up to one million recipes) that can be stored to facilitate the batch reproduction

17.2 Early raising machines

Raising machine was invented by Ernst Gessner in 1854, who founded Xetma Vollenweider GmbH in Aue (D) in the year 1850. The raising machine employed the pile and counterpile raising principle. This development of a universal wire raising machine with 24 rotating raising cylinders in the year 1886 was one of the most significant innovations in the field of raising (Fig. 17.4). The system, for which Ernst Gessner had obtained a patent, is the basis of all drum-type wire card raising machines existing today. Patented novelty developments from the recent past, such as the multisystem machine with its unprecedented flexibility in raising, emerizing and brush sueding as well as the twin-system double-drum machine with an entirely new drum conception are coping with these high-level requirements. Xetma Vollenweider

Figure 17.4   Universal wire raising machine.

AG, founded by Samuel Vollenweider in the year 1881 developed number of raising machines basing at Switzerland.

17.3 Modern raising machines

Let us have a look of some of the modern raising machines available in the market.

17.3.1 Xetma multisystem XR

Beside the traditional pile/counterpile procedures at the ratio 1:1 and the full felting procedure with only pile or counterpile rollers, Xetma Vollenweider developed raising technologies with different ratios of the pile and counterpile rollers (Fig. 17.5). Depending on the number of raising rollers in the drum, pile/counterpile ratios of 1:2, 1:3 or 1:4 are realized. The higher the number of counterpile rollers, the shorter and denser the raising pile turns out. This brings decisive advantages in the working of woven and knitted wears of which the base is to be intensively raised and made denser. The optimal ratio of raising rollers, specifically designed for the individual article, provides the user with higher production quantities at shorter running times.

Multisystem XR has the following salient features (Fig. 17.6):

- 24 raising rollers in working position and 12 rollers with identical or different fillets in stand-by position.

- Raising can be done on all kinds of woven, knitted and warp-knitted as well as non-woven materials.

- Attaining various finishing results by variable pile–counterpile direction ratios of raising rollers.

- Saving of raising passages by optimal raising procedure per passage.

Figure 17.5   Xetma multisystem XR.

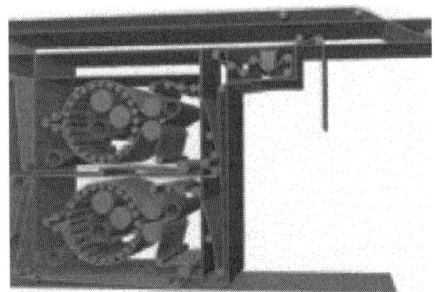

Figure 17.6   Twin system XR.

Figure 17.7   Xetma multisystem XRE.

- Automatic changing of raising procedure by push button in less than 4 min without taking material from the machine.

- Automatic adaptation of cleaning rollers to individual raising roller ratio.

- High production output for users with frequently changing articles.

- Development and testing of new effects and technologies in one machine system.

17.3.2   Xetma multisystem XRE

This is a raising cum emerizing machine with following salient features (Fig. 17.7).

a) Drum equipped with 12 raising and 12 emerizing rollers in pairs; additionally, 12 further emerizing or raising rollers in stand-by position

b) Automatic resetting between pure raising and combined raising and emerizing in less than 4 min by way of touchscreen

c) Extended flexibility by easy resetting to pure raising or emerizing system within a maximum of 1.5 h

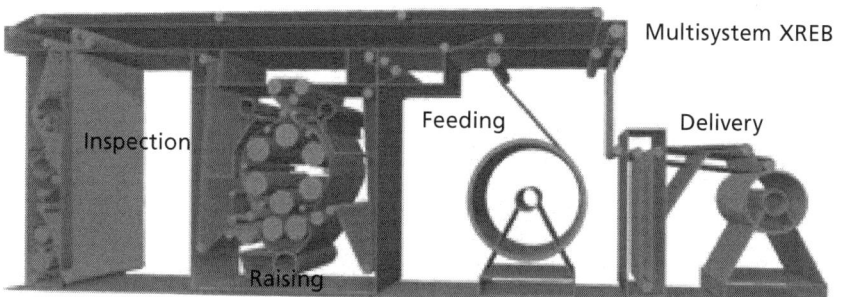

Figure 17.8   Multisystem XREB.

d)   Three technologies useable in one machine:

- XRE combined raising and emerizing
- XR raising
- XE emerizing

17.3.3   Multisystem XREB

In multisystem XERB, five technologies are combined in one machine (Fig. 17.8):

a)   XR raising

b)   XRE combined raising and emerizing

c)   XE emerizing

d)   XEB combined emerizing and brush sueding

e)   XB brush sueding

It has maximum flexibility through easy resetting between the individual technologies within 2.5 h and meets all demands for the mechanical surface finishing with only one machine system

17.3.4   Lafer raising machine

Lafer SpA is one of the largest manufacturers of raising machines in the world (Fig. 17.9). Some of their salient features are as follows:

a)   **Double or single drum configuration**: Single drum raising machines with 24, 28, 32 or 36 raising rollers and double drum machines with 24, 28 or combined 24/36 raising rollers are available in various widths in order to satisfy different production requirements. Combined raising and shearing units (GLC) are also available.

Figure 17.9   Lafer raising machine.

b) **Controls to simplify the raising process**: Unlike conventional raising machines equipped with simple control units, this machine has an exclusive system combining both PLC and industrial PC system with large touchscreen colour video with the objective of making raising a much easier process compared to other machines.

c) It has a capacity of storing 1 million raising programs.

d) On-screen troubleshooting guide with photos and instructions (over 100 alarms included) are provided.

e) The machine can be connected to a network through Ethernet card so that production data can be monitored from a remote computer.

f) An optional modem allows troubleshooting directly from Lafer headquarters.

g) Electronic fancy brush synchro control prevents possible damages to raising fillet caused by non-synchronized raising fillet cleaning brushes.

h) **Precise fabric tension control**: Fabric tension on the raising drum is one of the key parameters to achieve the desired raising effect. Fabric tension is set on the touchscreen and is controlled by the computer.

i) **Automatic minimum tension control by load cells**: Optional load cells placed at the entry and exit of each drum automatically maintain the minimum fabric tension set regardless of the raising energies. This prevents the fabric from becoming too loose at the drum entry and exit points allowing optimal raising on the entire drum and avoids bar marks, weft distortions or other inconveniences.

j) **Uniform raising**: Special pressure rollers are fitted on the entry and exit draw rollers in order to prevent fabric slippages and ensure perfect raising uniformity.

k) **Adjustable drum speed/direction**: The drum has adjustable speed and can rotate in both directions compared to the fabric. On the twin drum machine, the drum speed and rotation direction can be varied separately.

l) **Raising energies always constant**: An electronic system maintains the raising energies constant while varying the fabric or drum speed.

17.4 Typical operating procedure

The operating procedures depend on the type of machine, the processes involved, and the type of fabric and so on. The following is an example of procedures to be followed in XREB rising machine.

a) Clean the machine with compressed air and then wipe with a clean cloth before starting the work.

b) Start the air flow of the machine before switching the machine ON.

c) Start the exhaust system.

d) Check the tensions in different drives those are shown on Tension Page of the machine.

e) Check the condition of the raising fillet (Fig. 17.10) and ensure that the raising rollers are clean.

f) Check the condition of all brushes and ensure that the tips are not fused or unshaped (Fig. 17.11).

g) Check the emery required for the material to be processed (Fig. 17.12).

h) Check the condition of shearing cylinders and the ledger blades (Fig. 17.13). If they are found blunt, refer to the mechanic and get them ground.

i) Check and ensure that fabric guiding systems, seam detection systems, heating and steaming systems, brushing systems, fabric accumulation systems, batch winding systems and dust extraction systems are in order (Fig. 17.14).

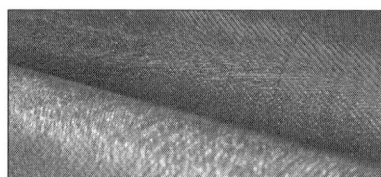

Figure 17.10 Fillet.

Figure 17.11 Brushes.

 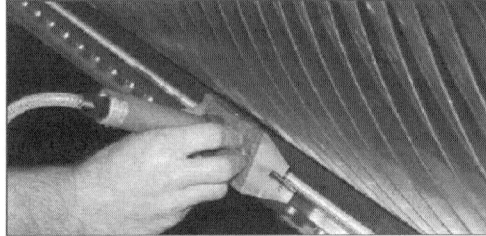

Figure 17.12   Emery.   *Figure 17.13*   Shearing cylinders and the ledger blades.

Figure 17.14   Different systems.

j) Bring the material to be raised as per the programme and keep it in place of feeding.

k) Refer to supervisor and select the menu for operating the machine depending on the degree of raising required.

l) Take the pile and counterpile configuration by discussing with the supervisor. The machine has five optional technologies namely

   i. XR raising

   ii. XRE combined raising and emerizing

   iii. XE emerizing

   iv. XEB combined emerizing and brush sueding

   v. XB brush sueding

m) Observe the fabric to ensure that rising is uniform and no damage is caused to the fabric in process.

n) Enter the details of the fabrics processed such as lot number, beam number, quality number, programme number and number of metres in the production book.

o) Cover the produced material with polythene sheet and keep at the designated place.

17.5 Control points and checkpoints

It is essential to have clarity on the points to be controlled to achieve the targets and those to be checked to ensure the process in control. These points need to be reviewed from time to time and modified to suit the requirements of individual companies and their targets considering the machines and the processes they follow. Each mill should prepare its own 'control points and checkpoints' and display them in the work area so that the people on spot refer and follow.

17.5.1 Control points

a) Selection of suitable process for the material to be processed

b) Deciding on the degree of raising for the fabric

c) Deciding and selection of process parameters, namely:

- Type of fillets
- Speed of rollers
- Setting between fabric surface and the fillets
- Dressing of brushes

d) Deciding acceptance criteria for quality of raising of fabrics

e) Employing trained and qualified employees

f) Evolving production norms

17.5.2 Checkpoints

- **Material related**

  a) The fabrics received against the plan received

  b) The quantity of fabric received—metres and kilograms

  c) Type of fabric—woven or knitted

  d) The density (GSM) and the fabric construction

  e) The type of fibres in the fabric

  f) The twist factor in the yarns used in the fabric

- **Machine related**
  a) The condition of the machine
  b) The condition of the fillets and the rollers – the fillets should be sharp and roller have uniform diameter
  c) The working of various controllers

- **Setting related**
  a) The setting between fabric and the fillets
  b) The speeds of different rollers and their direction of movement
  c) The angle at which the fabric is presented for raising
  d) The distance of contact of fabric to the raising media

- **Performance related**
  a) Uniformity of raising throughout the fabric
  b) The production achieved

- **Documentation related**
  a) The design number of the fabric received for raising
  b) Quantity of fabric raised

- **Work practice related**
  a) Thorough cleaning of the machine before loading the material
  b) Maintaining the speeds as per norms

- **Logbook related**
  a) Machines working
  b) Starting time of the running lots
  c) Activities done and to be done

- **Management information system related**
  a) Machine number
  b) Design number
  c) Number of metres
  d) Weight in kilograms

- **General**
  a) Use of safety gadgets such as gloves and masks

17.6 Dos and Don'ts

Understand clearly what is supposed to be done without fail and what should not be done at any cost. Some examples are given below.

Dos

a) Check the condition of the raising fillet and the tip of the brushes.

b) Clean the machine thoroughly with compressed air before starting.

c) Check the condition of emery cloth, brushing and beating devices, cleaning system and dedusting devices before starting the operation.

d) Check the seam quality before feeding material to raising machine.

Don'ts

a) Do not use the fillet or brush if tips are damaged (Fig. 17.15).

b) Do not run the machine if holes are found in the fabric after raising.

c) Do not run the machine if the raising is not uniform throughout the width of the fabric.

d) Do not take the material for rising if the seams have loose threads (Fig. 17.16).

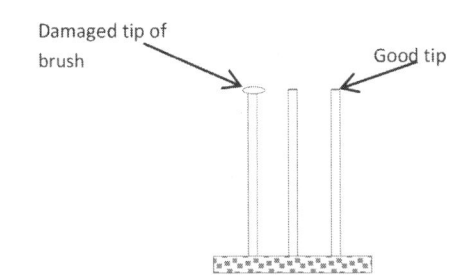

Figure 17.15 Condition of bristles.

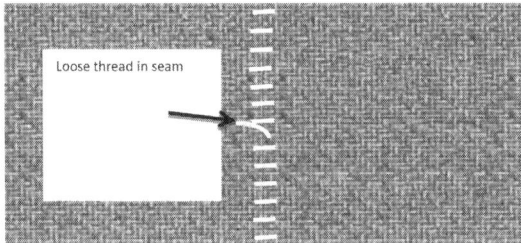

Figure 17.16 Loose thread in seam.

18.1 What is shearing?

In textile manufacturing, the cutting of the raised nap of a pile fabric to a uniform height to enhance appearance is referred as shearing. It is an operation of mowing the raised fibres and making a lawn-type surface with uniform height. The activity of the removal of protruding ends of threads, knots and hairs from the surface of a fabric and making even the length of pile to improve fabric's appearance is also termed as shearing. In animal husbandry, shearing is the cutting off of the fleece of sheep and other wool-bearing animals, using special shears. Here, we discuss on cutting of the raised nap to give a uniform height to the protruding naps.

Shearing machines operate much like rotary lawn mowers, and the amount of shearing depends on the desired height of the nap or pile. Shearing may also be applied to create stripes and other patterns by varying surface height.

Both woven and knitted fabrics undergo shearing. The process is carried out on two types of machines: lengthwise shearing machines (the major type) and transverse shearing machines, which are used to cut the fabric ends. The working element is a shearing mechanism consisting of a cylinder to which spiral blades are attached, a flat steel blade, a table and guiding rollers. The rapidly rotating cylinder and the stationary flat blade form a scissors mechanism that cuts the fabric as it passes through the shearing mechanism.

Machines with outputs from 27 to 81 m/min are used to shear coarse cotton, linen, light woollens and artificial and natural silk fabrics. Machines with outputs of 8–24 m/min are used for heavy woven and knitted fabrics.

18.2 Different applications of shearing

The function of a shearing machine is to cut the pile (or the yarn), present on a fabric surface, at a constant level of height. This operation can be applied to a wide range of fabrics and for a wide range of applications. The shearing machine can process a wide range of woven as well as knitted fabrics. However, knitted fabrics cannot be processed in tubular form, on the shearing machine. The effects of the shearing machine in fabrics can be varied:

a) **Shearing the pile of a raised fabric**: This function is used to shorten and equalize the height of a raised fabric, to create more space for a subsequent raising or to have a final effect.

b) **Shearing the pile of an acrylic fur**: Acrylic furs need a complete finishing line that is composed of triggering machines, polishing machines and finally of shearing machines. The shearing machine is placed in the middle and at the end of the finishing process to cut the pile during its finishing and to give the final effect.

c) **Pattern shearing**: If we replace the cutting stand with a roller that bears a pattern design, only those parts of the fabric that correspond to the pattern will be sheared. A very nice effect with a design 'sculpted' on the pile will be the result.

d) **Cutting the yarn of a terry towel or of knitted velour**: A knitted fabric made with a terry machine will have loops on its surface; if we cut the loops of this fabric, a very nice effect of velvet will be the outcome. In addition, other effects like acid burning are giving extra value to the finished fabrics.

e) Other applications cover the fields of cleaning a surface, for example, removal of any hairiness in grey fabrics before printing or in worsted fabrics.

18.3　How does a shearing machine work?

This operation is performed by a device called shearing unit (Fig. 18.1). The shearing unit is composed of two sharpened elements that work in a similar way to scissors.

The first element is a roller, with a certain number of blades (usually from 10 to 24 but can be less or more) fixed on it with a helical displacement. This roller is rolling on a flat blade; the contact between the helical blades and ledger blade is giving the cutting function. The rotation of shearing roller on the flat blade creates a continuous cutting effect. A point where the flat blade and helical blades meet is called the cutting point.

The fibres cut by the cutting operation are removed by a vacuum system which is composed of a suction box, placed in the back of the shearing roller.

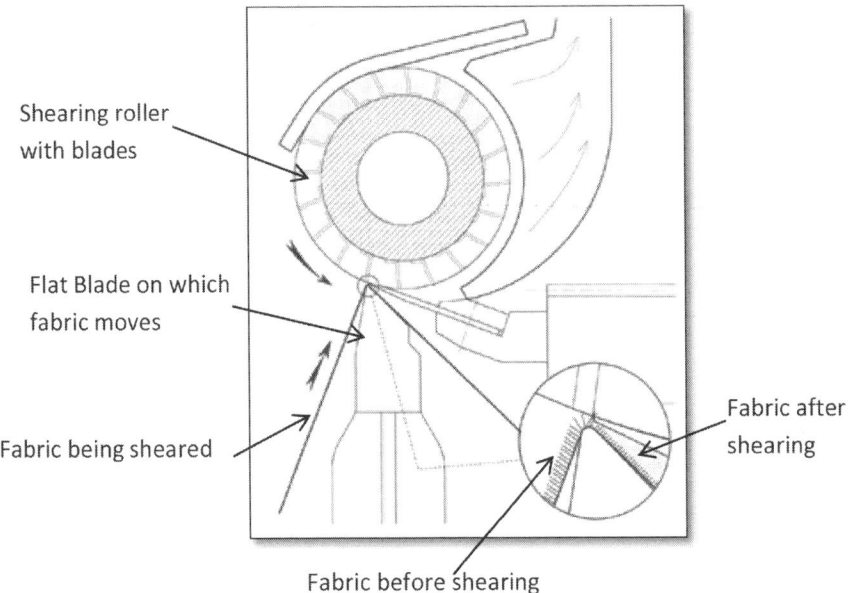

Shearing roller
with blades

Flat Blade on which
fabric moves

Fabric being sheared

Fabric after
shearing

Fabric before shearing

Figure 18.1   Shearing operation.

The fabric is brought to the cutting point by the cutting stand. The cutting stand may have different shapes according to the kind of fabric used for the process.

The distance between the cutting point and the cutting stand is called shearing height. After shearing, the ensemble of fabric and pile will have the same height as shearing height. One needs to consider the fabric thickness while working out the desired height of the pile.

A key element for the good functioning and the optimal result of the shearing operation is given by the position of each of these components with each other:

•   The shearing roller and the ledger blade

•   The shearing unit (shearing roller and ledger blade) with the cutting stand

To do this, the shearing unit is equipped with a series of hand wheels and adjusting screws that are setting all the reference positions of the various components.

As said previously, the shearing unit is like scissors; in order to have a good cut, for scissors, it is enough that the two blades are well sharpened and pressed in contact one with each other. For the shearing unit, this is a bit more complicated and to get better cut efficiency, one has to make sure that this contact between the two blades is constant on all width. The key point for a good shearing efficiency is the perfect contact between the flat blade and the shearing roller.

The shearing process tends to wear the contact point between these two elements so the shearing quality will be worse after some time; therefore, periodic setting of the shearing unit and sharpening the blades are important.

18.4   Modern shearing machines

Lafer SpA is one of the largest manufacturers of raising and shearing machines in the world. Let us have a look at some salient features as claimed by the manufacturer.

The 'Diablo' version allows shearing polar fleece and terry velour to double the speed of conventional machines with completely automated adjustments (Fig. 18.2).

a)   **Exclusive anti-vibration shearing cylinder**: The shearing cylinder is of a unique design, with a special anti-vibration technology, that ensures a perfect rotation in order to achieve an optimal cut. The cylinder revolves at a speed adjustable from 500 to 1500 rpm.

b)   **Maximum shearing precision**: The shearing precision is 0.03 mm, compared to most other machines with a precision of around 0.1 mm. This gives great advantages of better control, shearing effect and quality.

c)   **Automatic shearing and velveting height control based on the fabric thickness**: This device optimizes shearing by ensuring the same pile height regardless of the actual thickness of the fabric. Sensors

Figure 18.2   Lafer Diablo shearing machine.

continuously monitor the fabric thickness (sensitivity 0.03 mm) and the shearing and velveting parameters are automatically adjusted when the seam passes by adding the degree of shearing and velveting to the fabric thickness.

d) **Shearing rest and ledger blade alignment**: The alignment of the shearing rest and ledger blade is maintained constant when the shearing height is varied by the 'compass' movement.

e) **Automatic fabric tension control**: The fabric tension set among the draw rollers is automatically monitored by highly sensitive load cells, making it possible to shear in continuous mode and with minimum tension for any kind of woven, knit and stretch fabrics. In order to prevent the fabric from slipping or dragging on the velveting brush when a strong action is set, the draw rollers have a wide contact angle of the fabric to ensure the correct grip.

f) **Automatic lubrication**: It is critical to maintain the shearing cylinder evenly lubricated in order to ensure uniform shearing. The Lafer system can be programmed for frequency and quantity of oil to the felt and oil to be distributed. When the oil finishes, the computer will give an alarm.

g) **PC-based machine management system**: Lafer automated the functions of the shearing machine and has continuously introduced innovative solutions to make operation easier and more efficient.

h) **Controls of functions**: The Lafer shearing machines are driven by alternating current motors controlled by sensorless vectorial inverters and the electronic control system is equipped with:

- Programmable logic controller for the machine automation and alarms;

- Operator interface by industrial PC using colour touchscreen technology for setting and storing all operating parameters. Shearing parameters of each fabric can be memorized (over 1 million recipes can be stored); hence, the desired effects can be easily and quickly reproduced.

18.5   Typical operating procedure

The operating procedure depends on the type of machine, type of fabric, pile height, shearing height required and so forth (Fig. 18.3). The process should be designed and come in the standard operating procedures written by the concerned head of processing, considering their working conditions and variables. The following is just an illustration.

Lafer Shearing Machine   Lafer CMS, 220 cm, shearing machine

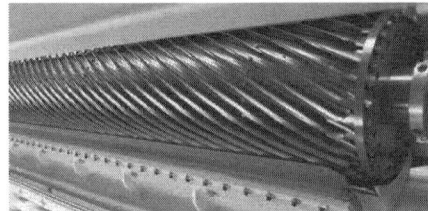

Exhaust system   Shearing roller

Figure 18.3   Lafer shearing machine

1. Clean the machine with compressed air and wipe with a clean cloth before starting the work.

2. Start the air flow of the machine before switching the machine ON.

3. Start the exhaust system.

4. Check the tensions on different drives of the machine.

5. Check the condition of the shearing blades.

6. Check the condition of the brushes.

7. Check and ensure that fabric guide plates, fabric guides, fabric thickness sensor, seam detection systems, fabric accumulation systems, fabric delivery systems and dust extraction systems are in order.

8. Bring the material to be sheared as per the programme.

9. Understand the sequence, that is, one-side shearing or both sides shearing.

10. Refer to the supervisor and select the menu for operating the machine considering the degree of shearing to be done.

11. Decide the direction of rotation of the shearing blade, that is, in line with the movement of the fabric or opposite to the movement of the fabric.

12. Keep the machine speed as needed; normally, the speed is 10 m/min.

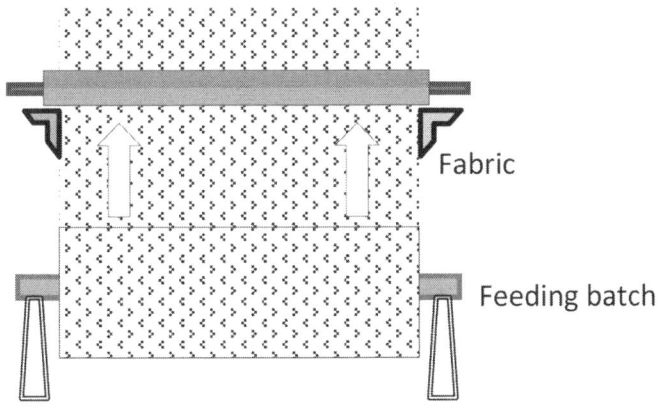

Figure 18.4 Aligning the batch.

13. Load the fabric into the machine and run.

14. If the feeding is in batch, align the batch so that the fabric fed is in line with the shearing system (Fig. 18.4).

15. If the feeding is in loose fabric form, then keep a man for controlling the folds in the fabrics.

16. If shearing is needed on both the sides, reverse the fabric and feed again.

17. Check the quality of seams before feeding the materials into the machine for shearing. If loose threads are there, then get them attended before taking them for shearing.

18. Observe the fabric and ensure that shearing is uniform and no damage is caused to the fabric in the process.

19. Enter the details of the fabric sheared like lot number, batch number, quality number, programme number and number of metres produced.

20. Cover the produced material with a clean polythene sheet and keep at the designated place.

21. Clean the dust bags and put fresh bag by checking the dust collected in the bag.

18.6   Control points and checkpoints

It is essential to have clarity on the points to be controlled to achieve the targets and those to be checked to ensure the process in control. These points need to be reviewed from time to time and modified to suit the requirements of individual companies and their targets. Each mill should prepare its own 'control points and checkpoints' and display them in the work area so that the people on spot refer and follow. Following are just for illustration.

18.6.1 Control points

a) Selection of suitable process for the material to be processed

b) Deciding on the level of shearing for the fabric considering the density

c) Deciding and selection of process parameters, namely:

- No of blades and their sharpness

- Speed of shearing rollers

- Setting between fabric surface and the shearing blade

- Dressing of brushes

d) Deciding acceptance criteria for quality of shearing of fabrics

e) Employing trained and qualified employees

f) Evolving production norms

18.6.2 Checkpoints

- **Material related**

    a) The fabrics received against the plan received

    b) The quantity of fabric receive—metres and kilograms

    c) Type of fabric—woven or knitted

    d) The density (grams per square metre) and the fabric construction

    e) The type of fibres in the fabric

    f) The twist factor in the yarns used in the fabric

- **Machine related**

    a) The condition of the machine

    b) The condition of the blades and the rollers—the blades should be sharp and rollers should have a uniform diameter

    c) The working of various controllers

- **Setting related**

    a) The setting between the fabric and blades

b) The speeds of different rollers and their direction of movement

c) The angle at which the fabric is presented for shearing

d) The distance of contact of fabric to the shearing media

- **Performance related**

  a) Uniformity of shearing throughout the fabric as per the pattern, if specified

  b) The production achieved

- **Documentation related**

  a) The design number of the fabric received for shearing

  b) The quantity of fabric sheared

- **Work practice related**

  a) Cleaning of the machine before loading the material

  b) Maintaining the speeds as per norms

  c) Disposing of the wastes from time to time

- **Logbook related**

  a) Machines' working

  b) Starting time of the running lots

  c) Activities done and to be done

  d) Problems faced while working

- **Management information system related**

  a) Machine number

  b) Design number

  c) Number of metres

  d) Weight in kilograms

- **General**

  a) Use of safety gadgets such as gloves and masks.

18.7 Dos and Don'ts

It is essential to understand clearly what are supposed to be done without fail and what should not be done at any cost. Some examples are given below.

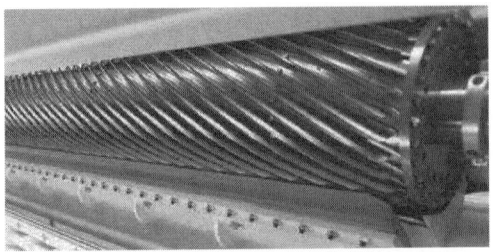

Figure 18.5   Shearing roller.

18.7.1   Dos

a)   Check the condition of the blades of the shearing roller (Fig. 18.5).

b)   Clean the shearing blades with compressed air.

c)   Check the conditions of brushes and dust extraction unit before starting the operation.

d)   Clean the machine thoroughly with compressed air before starting.

e)   Engage an extra person for controlling the folds while feeding fabric in loose form.

18.7.2   Don'ts

a)   Do not take the material for shearing if the seams are improper and have loose threads (Fig. 18.6).

b)   Do not operate if the shearing blades are loaded with fluff (Fig. 18.7).

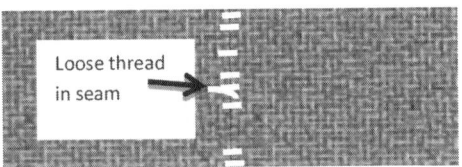

Figure 18.6   Loose thread in seam.

Figure 18.7   Shearing blades.

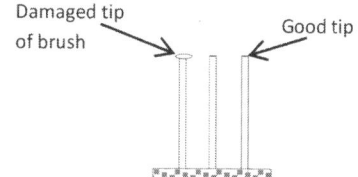

Figure 18.8 Tip of brush.

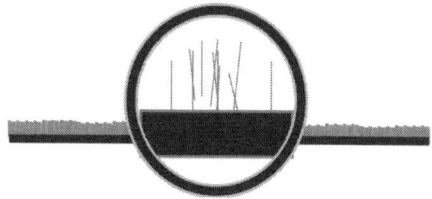

Figure 18.9 Uneven shearing seen by magnifying glass.

c) Do not use the brush if the tips are damaged (Fig. 18.8).

d) Do not run the machine if the shearing is not uniform (Fig. 18.9).

e) Do not engage an extra person in the back while feeding fabrics in a batch form.

19.1 What is decatizing?

Decatizing, also known as crabbing, blowing and decating, is the process of making permanent a textile finish on a cloth so that it does not shrink during garment making. Though used mainly for wool, the term is also applied to processes performed on fabrics of other fibers, such as cotton, linen or polyester. Crabbing and blowing are minor variations on the general process for wool, which is to roll the cloth onto a roller and blow steam through it. By this process, high-quality wool and wool-blend fabrics receive a permanent fixing. The wool fabrics acquire important basic qualities such as a flowing drape, crease resistance and a discreet glaze finish. Decatizing takes place through the application of steam under pressure.

The decatized wool fabric is interleaved with a cotton, polyester/cotton or polyester fabric and rolled up onto a perforated decatizing drum under controlled tension. The fabric is steamed for up to 10 min and then cooled down by drawing ambient air through the fabric roll. The piece is then reversed and steamed again in order to ensure that an even treatment is achieved.

Commonwealth Scientific and Industrial Research Organisation Division of Wool Technology, Sydney Laboratory investigated the effects of decatizing temperature, prior rotary pressing and fabric initial regain on the mechanical properties of decatized fabrics of various constructions and dyeing stages after pressure decatizing. Undyed and plain weave fabrics were found to be the most sensitive to decatizing treatment. Higher decatizing temperature consistently increased extensibility and bias extension and decreased bending rigidity. Rotary pressing had a marked influence on the mechanical properties of fabrics subsequently set by pressure decatizing; the resulting decrease in tensile properties and increase in bending rigidity are explained by simple mechanisms. They also studied the magnitude of changes in the warp and weft extensions, bias extension and bending rigidity with setting temperature and rotary pressing in relation to various fabric types.

19.2 Purpose of decatizing

The main purposes of decatizing are:

- Improvement of the texture through added humidity
- Moderation or setting of the glaze or sheen
- Dye-fast finish
- Wet fixing in length and width

19.3 Decatizing wrappers

The fabric is covered with a wrapper while decatizing. The functions of wrappers are as follows:

- Separation of the fabric layers to avoid pressure marking
- Transfer of surface pressure to reduce the thickness of the fabric
- Conducting the steam (heat and humidity in and through the fabric)
- Gentle, crease-free, low-tension transport of the fabric when unrolling or rolling up the reel
- The type and the surface of the wrapper affect the feel and sheen effect

The wrappers can be used several hundred times. Just a knot in the surface of the wrapper would be impressed into the woollen material and result in B-grade production quality. Since the wrapper is used several hundred times, many pieces of material would be affected by this fault. The same applies to yarn irregularities, stains and holes. Therefore, it is clear that the manufacture of decatizing textiles demands a high degree of sensibility with regard to production quality. The decatizing wrapper can be used on kier decatizing machine and finish decatizing machine.

There are different types of wrappers available:

a) **Satin decatizing wrappers**: The satin decatizing wrappers are made out of ring-spun 100% cotton or special cotton/synthetic blends. In the development of the wrappers, importance is given to a long lifetime which will result in more economic decatizing.

b) **Napped satin decatizing wrappers**: These wrappers are made from special cotton/synthetic blends and are used to produce low-lustre woollens.

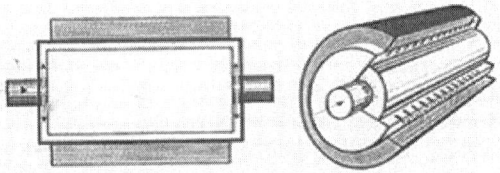

Figure 19.1 Decatizing drum.

c) **Molton decatizing wrappers**: The molton wrappers are manufactured from pure cotton (specially cultivated) or special cotton/synthetic blends and are used in boiler and finish decatizing.

d) **Underwindings:** These are used to protect the wrappers and also to prolong their lifetime. Through their use, a better and more uniform distribution of steam in the decatizing boiler is ensured. The maintenance requirement is reduced since the perforated roller, being protected by the underwinding, needs no additional cleaning. Polyester as well as aramid underwindings with hand-combed ends are used in continuous decatizing. Underwinding made of acryl or aramid fibres is ideal. These are wound directly on the perforated roller with three to five wrappings. The advantages are:

- The decatizing wrapper is protected and its life is prolonged

- A better and more uniform distribution of steam and

- The maintenance requirement is reduced because the holes in the perforated roller remain free and therefore do not have to be cleaned.

In continuous decatizing, one can have a choice between underwindings in 100% polyester or in 100% aramid fibre fleece. To avoid pressure marks and to increase strength, the underwindings can be alternatively combed out at the ends and/or reinforced on one side.

To protect the Machons in use on the decatizing machines in the continuous decatizing process from premature colouring through dye residue coming from the material undergoing finishing, an extremely thin fleece with very good strength is used.

e) **Spreading strips**: These help avoid end pressure marking.

f) **Satin decatizing wrappers**: The satin decatizing wrappers are mainly made of special cotton/synthetic blends whereby in the development of the product great importance needs to be given for a long lifetime performance.

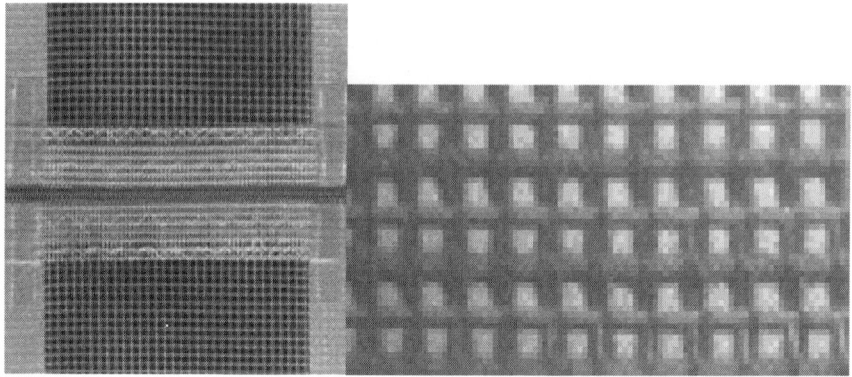

Figure 19.2   Spreading strips.

Calculation of the wrapper length in boiler decatizing depends on the amount of wool cloth that needs to be treated in one go and the parameters of the decatizing boiler.

Felts are used in continuous decatizing machines. The felts are offered with the varied technical parameters. The interior perimeter and the width of the endless needle felts should fit the dimensions of the machine. The surface weight of the needle felts which is put to use varies according to the type of machine.

Differing materials, material thickness and surface weight result in different effects on the surface of the fabric or knitted material which is receiving treatment. Needle felt belts are generally made from wool, special polyester fibres, aramid as well as polyester–aramid blends. These are suitably hydrolysis-resistant and avoid the destruction of the blanket at extreme temperatures and conditions of humidity. At the same time, these feltings also constitute the base material for silicone blankets. According to customer's requirement, needle felting belts can be made with different technical parameters from several components.

19.4   Decatizing machines

There are different types of wool decatizing machines such as batch decatizing machines, continuous decatizing machines, wet decatizing machines and dry decatizing machines. There are three main types of decatizing, namely kier decatizing, continuous decatizing and finish decatizing.

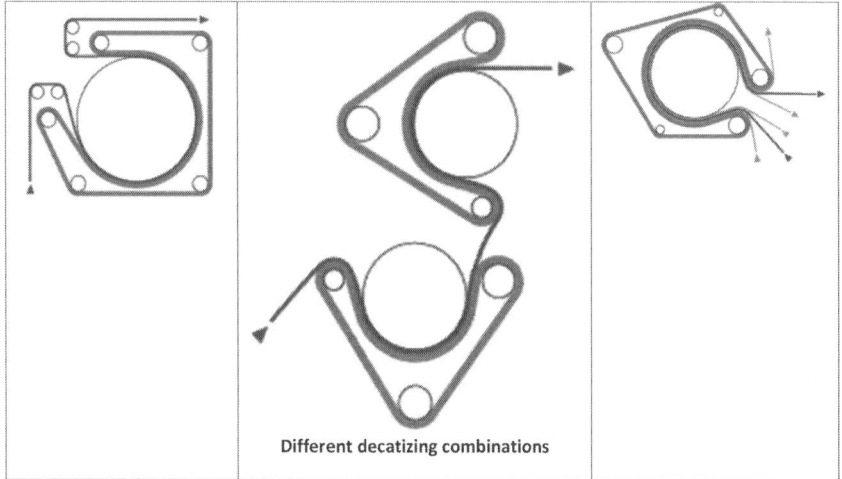

Figure 19.3 Different decatizing combinations.

19.4.1 Kier decatizing

The kier decatizing is a discontinuous process with a purpose of achieving maximum dimensional stability, volume, texture as well as the appearance of the wool and wool-mix fabrics.

A discontinuous decatizing machine is used for the process of woollen fabrics in order to give stabilization as well as handle (or surface smoothness) to the worsted and wool-blended fabrics. The process itself consists of a steaming cylinder (also called decatizing cylinder) onto which pieces of woollen fabrics are rolled up inside a wrapper (either satin or molton) and to put them together in a kier decatizing machine. Choosing a molton wrapper or a satin wrapper will depend on the requested effect for the woollen fabrics.

These machines are generally equipped with two steaming cylinders; it enables the processed fabrics to be unrolled from one cylinder while the other

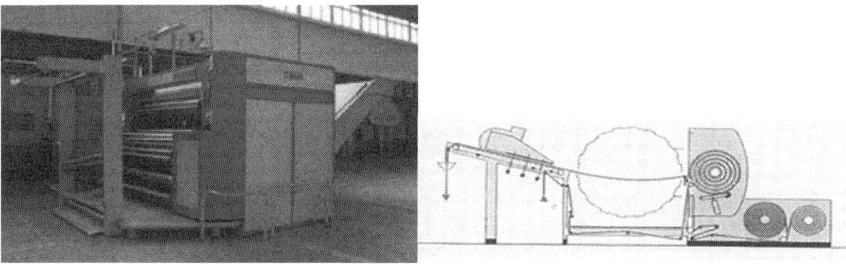

Figure 19.4 Kier decatizing.

wrapper is inside the kier decatizing machine for the finishing process. Two satin wrappers per machine are requested.

19.4.2 Continuous decatizing

A continuous decatizing machine comprises a perforated decatizing cylinder supplied with steam. A fabric web to be decatized is passed around the cylinder and is pressed against the cylinder by means of backing cloths. Pressure is applied to the backing cloths by means of strips which are impermeable to steam.

This process improves wool and wool-mix materials, especially with regard to texture structuring and the appearance of the product. During this process, either steam not under pressure and in a cooled-down state is brought into contact with the material or controlled humidity is introduced into the treating area and over the material and is then converted into steam by pressure and the effect of heat.

In the continuous decatizing process, needle felts or silicone blankets are used. To protect the needle felts underwindings, fleeces are employed.

The calendering machines for continuous decatizing of fabrics and knitted materials are usually equipped with endless needle felting. The surfaces of

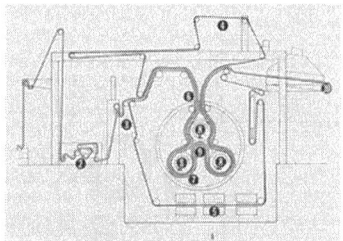

Figure 19.5   Continuous decatizing.

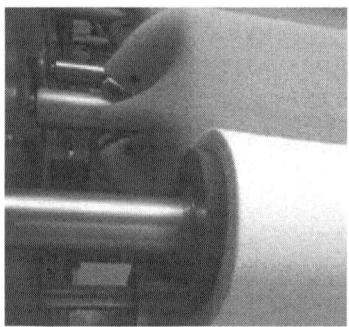

Figure 19.6   Needle felts.

Figure 19.7   Silicone blankets.

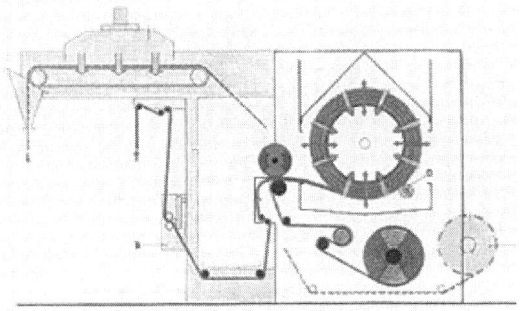

Figure 19.8   Finish decatizing.

felts need to be very smooth in order to rule out marking on the fabrics or knitted material. Only special polyester fibres, which are hydrolysis-resistant, are used to avoid destruction of the blanket under extreme temperature and moisture conditions.

The endless decatizing felts and the Machon include non-rusting eyes and cord.

The endless belt and silicone blankets used have a significantly longer life. Owing to the absence of overlapping connecting pieces, the endless composition makes a finish without the possibility of any marking. The hand-ground, rub-resistant surface of the belts helps for a long-lasting, optimal decatizing quality. Belt with high grip surface is made of 100% polyester to work up to 150°C and with air permeability of 1400 ft³/min.

As a protection and to extend the life of the Machons on the decatizing machinery, a very uniform light fleece weighing approximately 30 g/m² is used. This is wound around the Machon in several layers and should be continually changed.

19.4.3   Finish decatizing

The setting of the finished material in the boiler decatizing process is followed either by shrinking plus finish decatizing or only the latter. The purpose of this treatment is to remove the glaze (which is not set) on the surface of the material and then by this process to create a more attractive, dye-fast fibre sheen. Additionally, the texture structure and the shrink fastness of the finished fabric are favourably influenced. For finish decatizing wrappers, molton or napped satin as well as underwindings and spreading strips are recommended.

19.5   Some commercially available decatizing machines

19.5.1   KD Minimat HQ model manufactured by Biella Shrunk Process (Italy)

KD Minimat HQ is a discontinuous decatizing machine, used to process woollen fabrics in order to give stabilization as well as handle (or surface smoothness) to the worsted and wool-blended fabrics. The process consists of a steaming cylinder (also called decatizing cylinder) onto which pieces of woollen fabrics are rolled up inside a wrapper (either satin or molton) and to put them together in a kier decatizing machine. Choosing a molton wrapper or a satin wrapper will depend on the requested effect for the woollen fabrics.

These machines are generally equipped with two steaming cylinders; it enables the processed fabrics to be unrolled from one cylinder while the other wrapper is inside the kier decatizing machine for the finishing process. This is the reason for two satin wrappers per machine.

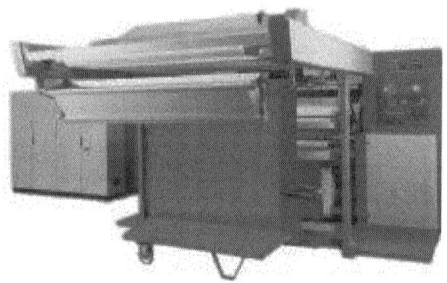

Figure 19.9   KD Minimat HQ.

Satin wrappers, also called decatizing satin, are with a standard width of 1.77 m at the lengths vary from about 180-m long for old and small machines to 1000 m; this length is relevant to the diameter of the kier decatizing machine.

19.5.2 KD Gigante—machine for 'KD' permanent finishing and for atmospheric decatizing

KD Gigante has a single unit for permanent KD treatment and an atmospheric decatizing with the deep well pump. The feature of KD Gigante constitutes a crucial turning point in the field since it concerns the size of the machine and also the basics of treatment and technical solutions. The size of the autoclave and decatizing rollers with internal tank for reduction of oxygen are the main feature in addition to the exclusive 'roller magazine' which contains three technical cloth rolls which can be easily and rapidly replaced or alternated on the machine allowing an extreme versatility on different treatments.

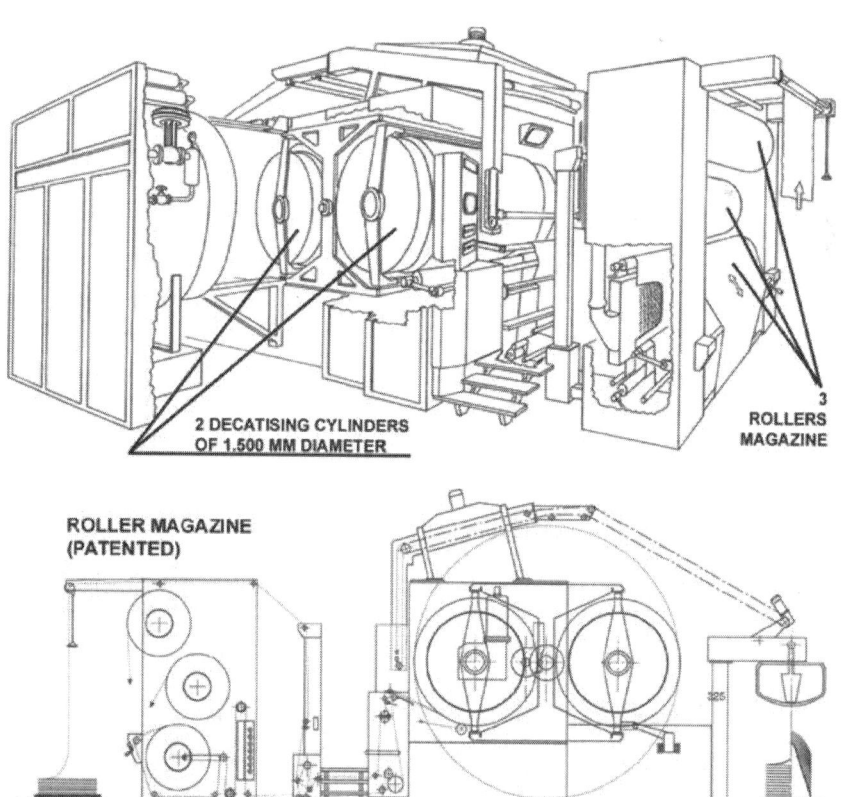

2 DECATISING CYLINDERS
OF 1.500 MM DIAMETER

3
ROLLERS
MAGAZINE

ROLLER MAGAZINE
(PATENTED)

Figure 19.10 KD Gigante.

Main technical characteristics of KD Gigante are as follows:

- Decatizing cylinder diameter—1500 or 900 mm
- Autoclave diameter—1600 or 2100 mm
- Number of wrappers—four (two satins and two molletons or two different kinds of satin)
- Rolling/unrolling speed—0–100 m/min
- Wrapper length—750 m
- Fabric length—600 m
- Production—1800 m/h
- Steam consumption—150 kg/h
- Installed power—105 kW
- Steam feeding—6 bar minimum

KD Gigante has some parts which are their exclusive patents. They are as follows:

a) **High-quality system**: extra-large decatizing cylinders containing a sealed tank to reduce the internal air volume and reduce the steam treatment time.

b) **Roll magazine**: with up to three technical clothes for fast replacement.

c) **PC-controlled modulating steam reduction valve**: 'run-up' cycle granting a progressive and delicate steam inlet phase which reduces the thermal and physic shocks to the fabric, particularly suitable for the voluminous finish and soft handle.

d) **Cold cycle**: special automatic system to detect eventual steam leakages during the first phase of the autoclave cycle.

e) **Steam quality control device**: fully automatic electric steam overeaters to maintain constant steam supply quality irrespective of environmental conditions; it reduces the steam crossing time phase through the roll.

f) **Combined pressure—tension system**: PC-controlled winding wrapper tension and compacting cylinder action. Variables-free programming in order to compensate the fabric thickness loss during the treatment; it grants a perfect finishing uniformity.

g) **Fabric outlet transport device**: new solution consisting of two suspended arms providing an accurate 'fabric transfer' from the cylinder to the discharge belt.

h) **Wrapper tensioning device**: this solution consists in an intermediate 'braking cylinder' placed in between the roller magazine cylinder and the decatizing cylinder; it permits to minimize the mechanical stress and increase the life of the technical cloth.

19.5.3 Thermo Duplex 90—continuous decatizing machine

Thermo Duplex 90 is a continuous decatizing machine; it provides a finishing treatment with advantages over the traditional batch decatizing machine, that is, greater output and treatment uniformity.

Thermo Duplex 90 provides better setting performances due to the patented steam quality control device which generates overheated steam fed through the additional decatizing field. An electronic thermometer detects the real steam temperature and compares it with the preset required value; this system allows keeping the steam temperature constant and, if required, allows working with higher temperature (overheated steam) granting higher setting effect, particularly suitable for wool and synthetic fibres. The special decatizing field construction also grants a permeable belt limiting wear and tear by which it has working life two to three times longer compared to standard solutions.

Main technical characteristics of Thermo Duplex 90 are as follows:

*   Steam consumption—350/400 kg/h

*   Working speed—0–35 m/min

Figure 19.11   Thermo Duplex 90.

- Working width—1.700 mm
- Installed power—44 kW

19.5.4 Continuous pressing and setting machines Contipress GPP 400

Kettling and Braun Contipress GPP 400 system does continuous pressing and setting at high surface pressure without lengthening.

Wovens, felts and knits made of different kinds of material, namely wool blends and wool as well as synthetic fibres blended with cotton, viscose and so forth, can be processed for pressing, compacting and smoothening of the fabric. The machine gives different effects such as pressing, pretreatment for kier decatizing, press board effect, decatizing effect and bright lustre finish. The Contipress is replacing the disadvantageous rotary press while processing wool. The Contipress process ensures zero stretch during pressing, as there are no relative movements. This can be used beyond the field of outerwear, for example, for automobile upholstery. The process fully meets the requirements made upon the improvement of appearance, touch and tailorability as well as to wear and pilling resistance. A special pressure belt presses the fabric with adjustable surface pressure against an effect cylinder. Both elements are heated independently from each other. On an incomparably longer way of pressure influence, pressing is carried out quicker and more intensively than with linear pressure.

The pressure belt, developed over many years, does not produce any marking on the fabric. It has a long lifetime and is resistant against chemicals and temperatures of up to 200°C. There is the possibility of moistening the fabric, especially in combination with Hygrocor, and thus an intensive decatizing as per the superfinish principle under extreme surface pressure

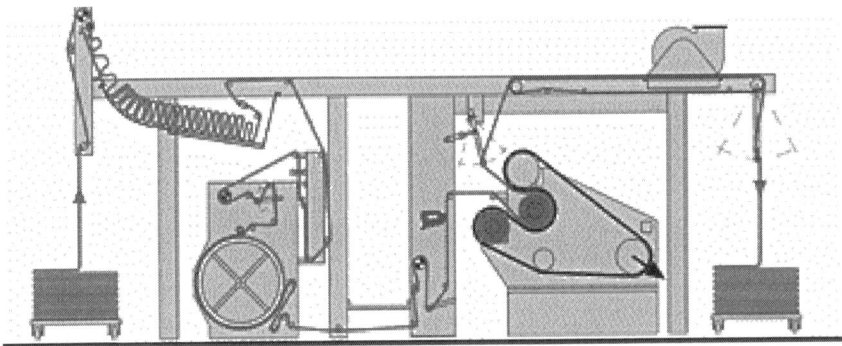

Figure 19.12   Contipress GPP 400.

and high fabric temperatures are obtained. The decisive factors for the setting process, namely humidity, temperature and surface pressure are optimally available at the same time.

19.5.5 Continuous-decatizing-machines—Superfinish GFP 800, system Kettling and Braun

This is a continuous pressing, setting and decatizing machine with the GFP effect: softer, more flexible and better shaping. This can be used for wool and wool blends as well as synthetic fibres, for example, in the automotive field, especially for blends with viscose and cotton.

Superfinish GFP 800 is a continuous decatizer working with the evaporation system. Important feature of this machine is the defined and adjustable moisture application at machine entry. By the action of temperature and pressure, the moisture applied is converted into steam. An impermeable pressure belt prevents escaping of the steam. Fabric runs throughout the machine at absolutely minimum tension. By heating pressure belt and effect cylinder and through the surface pressure produced by the tensioning roller, high temperature conditions are achieved. The fabric temperature, which can go up to 140°C, can be achieved by increasing the pressure. Conventional continuous decatizers work with the suction principle by which temperatures of only up to 105°C can be achieved. Touch, lustre, volume, degree of fixation and elasticity can be influenced within wide bounds through humidity, temperature and pressure whereby the height of temperature is a decisive factor.

Superfinish GFP 800/900 with final decatizing unit gives additional possibilities by influencing fabric effect. An intensive final steaming changes the fabric condition to extremely soft, mat and voluminous.

We will discuss some specialized shrink processes such as Formula 1 and Suprema developed by Biella in forthcoming chapters.

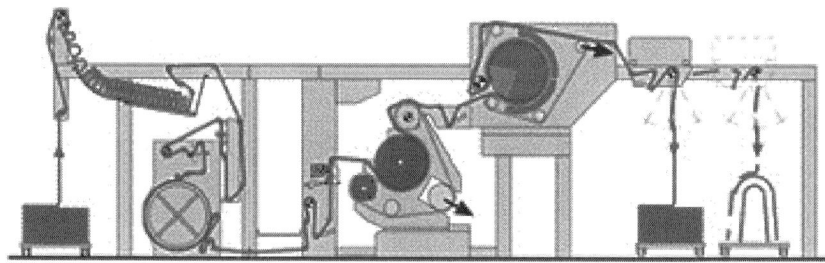

Figure 19.13 Superfinish GFP 800—continuous decatizing machine.

19.6   Typical operating procedure of a continuous decatizing machine

The operating procedures depend on the type of machine used, the fabric being processed and the level of decatizing needed. Hence, each mill has to prepare operating procedures suiting to its environment and condition. Following is an example for operating Sperotto Rimar Multidecat continuous decatizing machine.

The machine should be cleaned with compressed air and then wiped with a clean cloth before starting the work. All sensors and photocell units are to be cleaned with dry cotton cloth.

a)   Discuss with the supervisor and bring the materials to be decatized. Ensure that the width of the fabric is at least 10 cm less than the width of the blanket.

b)   Switch On the power.

c)   Drain the water from the compressed air line and then open the valve.

d)   Open the steam condensate trap and remove the condensates fully, and then open the steam valve slowly.

e)   Open the heat exchanger delivery manifold valve.

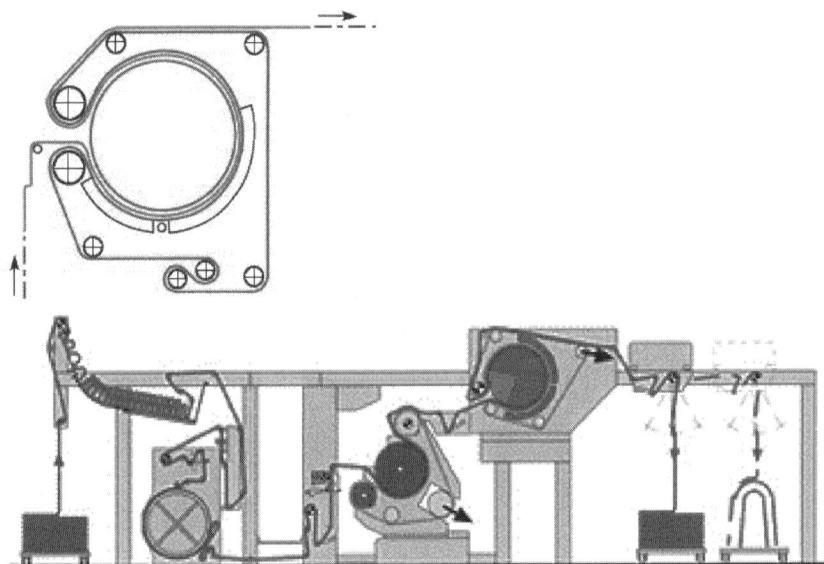

Figure 19.14   Sperotto Rimar Multidecat continuous decatizing machine.

f) Keep open the steam trap attached to manifold for 1–2 min and ensure that all the condensate is drained out. Then close the valve.

g) Check the quality of decatizing blanket and ensure it to be good.

h) Check and ensure that the quality of the conveyor is in good condition.

i) Start the finishing blanket of the machine and then start the fans of the machine.

j) Start the steam feeding canvas selector.

k) Check and ensure that all the safety systems in the machine are working properly.

l) Start the machine.

- Operation of the machine is in two modes, namely manual and automatic.

- By pressing button F1, the machine will be started in manual mode. You can change the machine parameters while the machine is working, and the parameters will get automatically updated.

- By pressing button F2, you can start the machine in automatic mode. You can feed up to 30 programmes in the machine. These entire programmes can be designed by the technician depending on the type of fabric being processed. The operator needs to select the programme depending on the quality being processed.

m) Check the finish of the fabric and decide on the programme to be run.

n) Cover the produced material and keep in the designated place as suggested by the supervisor.

o) Write the details of production such as batch number, lot number, quality or design number, quantity produced along with the date and time in the production record.

p) In case any abnormalities are found in the fabric, inform the supervisor immediately, and process the material after getting clear instructions.

q) In case of any problem in the machine, either electrical or mechanical, call the concerned person and get it attended.

19.7 Dos and Don'ts

It is essential to understand clearly what are supposed to be done without fail and what should not be done at any cost. Some examples are given below.

19.7.1   Dos

- Check the condition of the blanket before starting the machine.
- Clean the machine thoroughly with compressed air before starting.
- Use reliable wrapper.
- Ensure that the blanket width is less by at least 10 cm compared to cylinder width.

19.7.2   Don'ts

- Do not run the machine if the blanket has become thin or irregular.
- Do not allow the temperature to go above 140°C although the machine is capable of running up to 150°C.
- Do not stretch or overfeed the material.
- Do not accept the fabrics having higher width. The width of the fabric should be at least 10 cm lesser than the width of the wrapper

Formula 1—KD Biella Shrunk Process

20.1 What is Formula 1?

Formula 1 shrink process was designed by Miss K D Biella to carry out several finishing processes on fabrics like pressing/calendering before decatizing, paper press and superfinish. The machine has two cylinders wound by a single belt. These machines represent a new landmark in the finishing field by their extraordinary power and versatility. They have rapidly obtained a worldwide approbation both in the wool and synthetic fabric sectors. Through customized finishing lines suitable to meet the different customers' needs, several finishing processes can be carried out. The best definition for these machines is 'all in one' (Figs. 20.1 and 20.2). It is in fact possible to realize the following treatments:

- Pressing/calendering before KD
- Paper press imitation

Figure 20.1   Multipla Formula 1.

- Last 'finish' decatizing to improve the SiroFast formability parameters

- Single-passage permanent chintz

- Intense wet setting (crabbing)

As the machine does multiple operations, it is also popular as Multipla Formula 1.

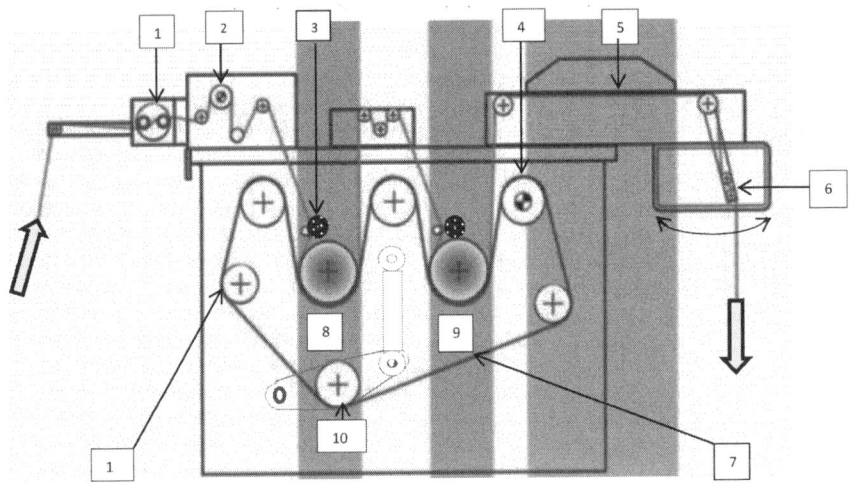

Figure 20.2   Line diagram of Multipla Formula 1

Legend:
1. Double stretcher cylinder with springs
2. Feeder cylinder
3. Cylinder cleaning device with brush and high prevalence suction blade
4. Motorized cylinder
5. Fabric cooling group with a conveyor belt
6. Folder at variable and increasing length
7. High resistance silicone belt (KD Biella specific)
8. Heated effect cylinder no. 1
9. Heated effect cylinder no. 2.
10. Lever arm for the belt-tensioning 'power multiplier system'
11. Belt-tensioning cylinder

20.2   Working principle

a)   The pre-humidified fabric passes around a steam heated cylinder transported by an impermeable belt. Under this condition, partial water evaporation takes place, the self-generated steam increases its pressure and is kept in contact with the fabric so as to create a microenvironment saturated with slightly overheated steam.

b) A belt-tensioning system, named 'power multiplier', allows the fabric to reach a uniform pressure of up to 11.5 kg/cm².

c) The machine is equipped with two effect cylinders wound by a single impermeable belt enhancing its performance in terms of versatility and productivity. The double effect cylinders are technologically and economically convenient. In fact, the process performed by this machine cannot be matched even by combining two distinct traditional machines; with Multipla, the fabric is subjected to five rapid sequence variations of temperature, pressure and humidity, which grant incomparable results.

d) The chromed cylinders' surface is different from the silicone belt surface, creating a different finishing effect between the face and the back of the cloth (shine/dull).

e) With the twist option, the fabric is overturned between the first and the second effect cylinder in order to obtain the same handle and the same finish effect on both the faces.

f) The temperature can be set independent for each cylinder that allows a careful control on the progressive and uniform humidity evaporation of the fabric; the maximum temperature is 200°C.

g) Material is damped before entering the basic Multipla machine by using a specially made Igrofast humidifier (Fig. 20.3).

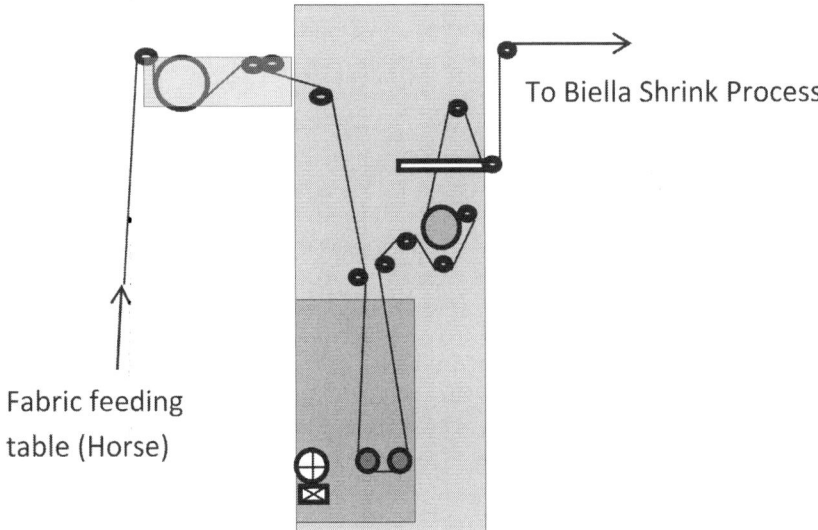

To Biella Shrink Process

Fabric feeding
table (Horse)

Figure 20.3 Igrofast humidifier unit.

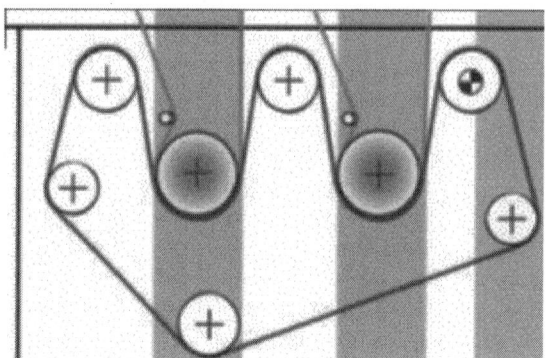

Figure 20.4 Decatizing part.

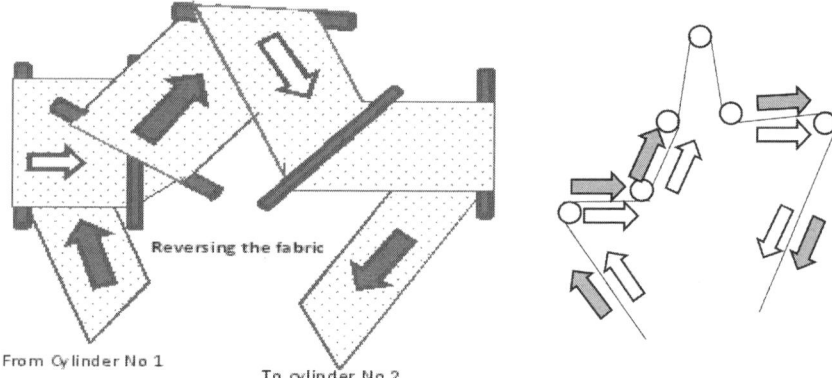

Figure 20.5 Twist and full touch arrangement.

h) Material goes between a high resistance silicone belt and the cylinder two times in the Multipla basic unit (Fig. 20.4). The cylinders are maintained with different temperatures, with the maximum limit of 200°C.

i) While material is taken from first cylinder to the second cylinder, it is reversed so that the fabric surface which was in contact with the cylinder will now be in contact with the silicone belt (Figs. 20.5 and 20.6).

j) Different effects in finishing can be obtained by altering the belt tensions. At very low pressures, the self-generated steam of the fabric swells and blows up the fabrics (Fig. 20.7).

Fabric reversing – Twist and Full touch

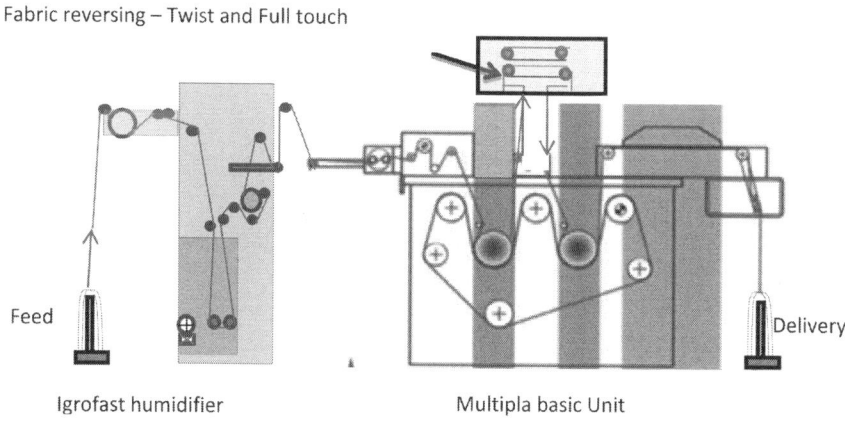

Feed

Igrofast humidifier

Multipla basic Unit

Delivery

Figure 20.6 Movement of fabric.

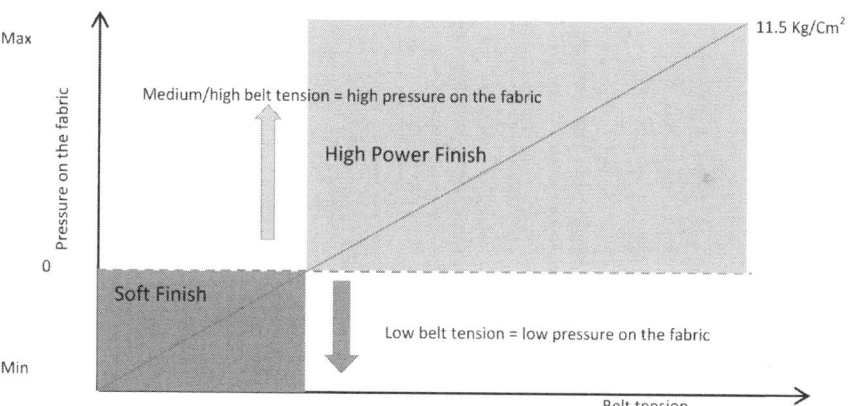

Figure 20.7 Effect of belt tension and pressure on the fabric.

20.3 Typical working procedure

Following guidelines are to be followed while operating a Multipla machine. The exact process sheet shall be prepared by the concerned head of the process by considering the fabric being treated.

a) Clean the machine with compressed air and then wipe with a clean cloth before taking any material for working in the machine.

b) Open all the three taps and slowly open the steam.

c) Drain out the water from the air compressor line and then open the air valve.

d)   Open the valve near the Igrofast humidifier unit.

e)   Switch on the machine and heating programme and allow it to heat for 30 min.

f)   Close all the three steam traps.

g)   Check the silicone belt and the conveyor belts before starting the machine.

h)   Before feeding the material, enter the required controls in the machine programme. Take the programme in writing from the supervisor before feeding it into the machine.

i)   If you find any discrepancy in the programme, machine or in the material, refer to the supervisor and get the things clarified and corrected. Write the complete details of the discrepancy, decisions taken and the actions in the production report book.

j)   If you have any problems in the chemicals or in the electrical items, inform the supervisor and get them attended. Write the details in the production report book.

k)   Discuss with the supervisor regarding the samples to be sent for testing to the testing laboratory and arrange to send the sample accordingly.

l)   Cover the produced material with a polythene sheet and keep at a designated place.

m)   After completing the programme or while stopping the machine, activate the 'cooling' programme.

n)   Open the machine tap and close the steam.

o)   Stop the cooling programme when the temperatures in both the cylinders reach between 35 and 40°C.

p)   To cool the belt, spray water in small quantities, but do not stop the belt when it is wet. Run the machine till the belt becomes completely dry.

q)   Enter the complete details of the materials worked with like the beam number, quality number, lot number, programme number and the quality problems in the production report.

r)   Clean the belt at least once in a week with water and a cotton cloth to remove the chemical depositions.

s)   Check the belt tension and the performance of 'power multiplier' periodically.

20.4 Control points and checkpoints

It is essential to have clarity on the points to be controlled to achieve the targets and those to be checked to ensure the process in control. These points need to be reviewed from time to time and modified to suit the requirements of individual companies and their targets. Each mill should prepare its own 'control points and checkpoints' and display them in the work area so that the people on the spot refer and follow. Following are some examples.

20.4.1 Control points

a) Selection of process parameters for the material to be processed

b) Deciding on the level of pressing/calendering for the fabric, considering the material

c) Deciding and selecting process parameters, namely,

- Length of the fabric to be wrapped at a time
- Steaming time and temperature
- Distance between the two fabric ends being fed
- Cutting at the point of stitches

d) Deciding acceptance criteria for the quality of decatizing of fabrics

e) Employing trained and qualified employees

f) Evolving production norms

20.4.2 Checkpoints

- **Material related**

a) The fabrics received against the plan received

b) The quantity of fabric received—metres and kilograms

c) The density (grams per square metre) and the fabric construction

d) The type of fibres in the fabric

e) The twist factor in the yarns used in the fabric

- **Machine related**

a) The condition of the machine

b) The working of various controllers, such as sensors and photocell units

    c)   Quality and condition of the wrapper being used

    d)   Quality and condition of the conveyor

- **Setting related**

    a)   The pressure between the fabric and the flannel

    b)   The speeds of different rollers and their direction of movement

    c)   The steam pressure and timings of the autoclave

- **Performance related**

    d)   Uniformity of decatizing throughout the fabric as specified

    e)   The production achieved

- **Documentation related**

    a)   The design number of the fabric received for shrink processing

    b)   Quantity of the fabric shrink processed

- **Work practice related**

    a)   Thorough cleaning of the machine with compressed air and then wiping with a clean cloth before starting

    b)   The speeds maintained and the norms

    c)   Cutting of the stitched portions without wasting good fabric

    d)   Maintaining a gap of 15 m while feeding different fabrics

    e)   Keeping 30–40 m of blank space at the start and end of each fabric roll

- **Logbook related**

    a)   Machines working

    b)   Starting time of the running lots

    c)   Activities done and to be done

- **Management information system related**

    a)   Machine number

    b)   Design number

    c)   Number of meters

    d)   Weight in kilograms

- **General**
  - a) Use of safety gadgets
  - b) Working of safety systems

20.5 Dos and Don'ts

It is essential to understand clearly what are supposed to be done without fail and what should not be done at any cost. Some examples are given below.

20.5.1 Dos

a) Clean the machine thoroughly with compressed air and then wipe with a clean cloth before taking any material for working in the machine.

b) Before opening the air valve, ensure that water is drained out from the pipes.

c) Feed the cloth only after completing minimum 30 min of heating.

d) Stop the cooling programme after both the cylinders have come down at the temperature between 35 and 40°C.

e) Before opening the air valve, drain out water from the pipe.

f) Clean the belt at least once in a week with water and a clean cotton cloth.

g) Check the quality of the silicone belt.

h) While stopping the machine, activate the 'cooling' programme.

i) Cool the belt by spraying water in small quantities when the belt is running.

20.5.2 Don'ts

a) Do not keep any material in the front pocket of your shirt while inspecting the belt as there are chances of it falling down and damaging the machine parts.

b) Do not stop the silicone belt if it is wet; keep it running till it becomes completely dry.

c) Do not select the programme without referring to the design card and the process in charge.

Suprema—KD Biella Shrunk Process

21.1 Purpose and concepts

The shrinkage of textile goods, such as wool and cotton, on laundering has long been one of the major problems requiring the attention of the textile technologists. Various methods of both mechanical and chemical treatments are developed for making fabrics shrink-free. Suprema—KD Biella shrunk process is one such process that is used for making wool fabric and blends shrink-free giving smooth decatizing finish with dimensional stability (Figs. 21.1–21.5).

The autoclave decatizing treatment carried out on this machine imparts excellent dimensional stability and permanent finish (Fig. 21.6). Worsted or woollen cloths of pure wool, mixed wool/polyester fibres or other synthetic fibres including raised cloths can be treated.

21.2 Principle of operation

Shrink finish is provided by wrapping the fabric in a wrapper and treating with steam under pressure. Fabrics along with a wrapper are rolled in open width onto a perforated covered cylinder and subjected to steam in an autoclave.

Figure 21.1 Feeding zone in Suprema Biella shrunk process.

Three operations are conducted simultaneously by rolling the fabric along with a wrapper, treating the roll with steam and unrolling the treated cloth and delivering using a plaiting device.

Two versions, that is, KD1300 and KD1600 are available in Suprema 95 KD shrunk process.

21.2.1   Main characteristics of 'KD Suprema 95'

a)   Stainless steel AISI 316 large diameter decatizing cylinders—'high quality' system. The cylinders have a diameter of over 35% of the autoclave diameter.

b)   The carousel with three cylinders and three transport trolleys. Each cylinder has a trolley with consequent dead times reduced to minimum.

c)   The cylinders are tubes and contain a tank which reduces the air volume, guaranteeing quicker treatment and the best treatment conditions.

d)   Maintenance free inverter-controlled AC motors.

e)   System of continuous alignment wrapper centreing with no mechanical stress during rolling operation. The centreing is obtained by checking the decatizing cylinder position during the cylinder-locking phase.

f)   Wrapper tension and compacting cylinder pressure are variable according to the curves of increasing values to obtain the desired effects: combined pressure–tension system managed by the means of a computer. This provides a constant tension to the cloth after the treatment and not during the rolling.

g)   Computer control system with touchscreen ergonomic panel and Windows-based software, more than 1000 different cycles can be programmed.

h)   'Run-up' cycle with motorized reduction valve, drastically reducing the cycle initial thermal shock, improve the treatment uniformity. This is suitable for very soft and mat-finishing effects.

i)   Treatment uniformity: after 400 m of charge the values of Δe read on the spectrophotometer shall range from 0.2 to 0.3.

j)   'Cold' control cycle finds potential steam leaks during the first phase of cycle. If irregularities are found that would compromise treatment quality, an alarm would go on.

k) Steam quality control device assures steam consistency all year irrespective of external conditions of input steam quality. This is provided as an optional system.

l) New cycles for 'whites' and 'in temperature' especially for ecru clothes and delicate colour fabrics.

m) Cuttle can be 90 or 180 cm.

n) Automatic search that stops machine where cloth is to be inserted.

o) Satin wrapper metering device, to check wrappers life.

p) Metering device to record daily production for each shift.

q) Rolling speed of up to 100 m/min, continuously variable.

r) Step by step display of the program cycle with an automatic diagnostics system, remote control possibility (optional).

s) Versatility: varying the steam pressure, the dwell time in the autoclave, the wrapper tension and the treatment cycle, it is possible to get infinite types of cloth finishing.

t) Double cooling system—water ring pump for suction from inside cylinder and cooling of the cloth at discharge with both suction and ventilation.

21.2.2 Technical characteristics

Main technical characteristics	KD 1300 model	KD 1600 model
Decatizing cylinder diameter	670 mm	900 mm
Autoclave diameter	1.310 mm	1.610 mm
Number of wrappers	2	2
Rolling/unrolling speed	0–100 m/min	0–100 m/min
Wrapper length	Up to 550 m	Up to 900 m
Fabric length	Up to 450 m	Up to 750 m
Production	1.800 m/h	Over 2.000 m/h
Steam consumption	100 kg/h	100 kg/h
Installed power	35 KW	35 KW
Steam feeding	6 bar minimum	6 bar minimum

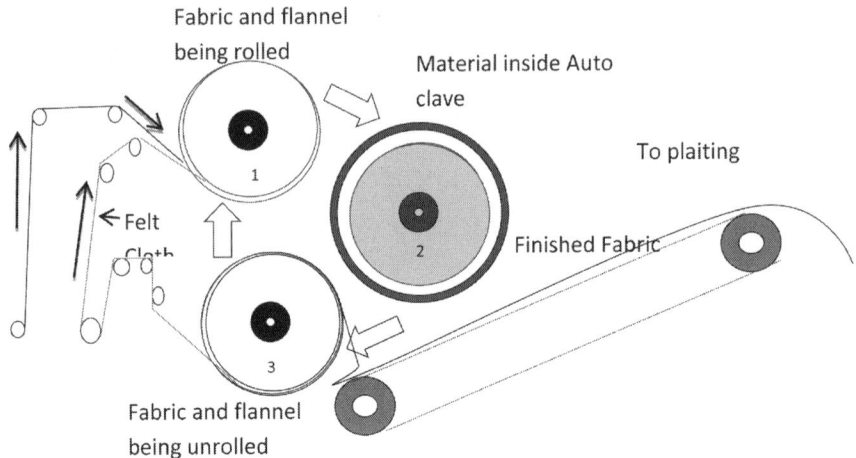

Figure 21.2   Side view.

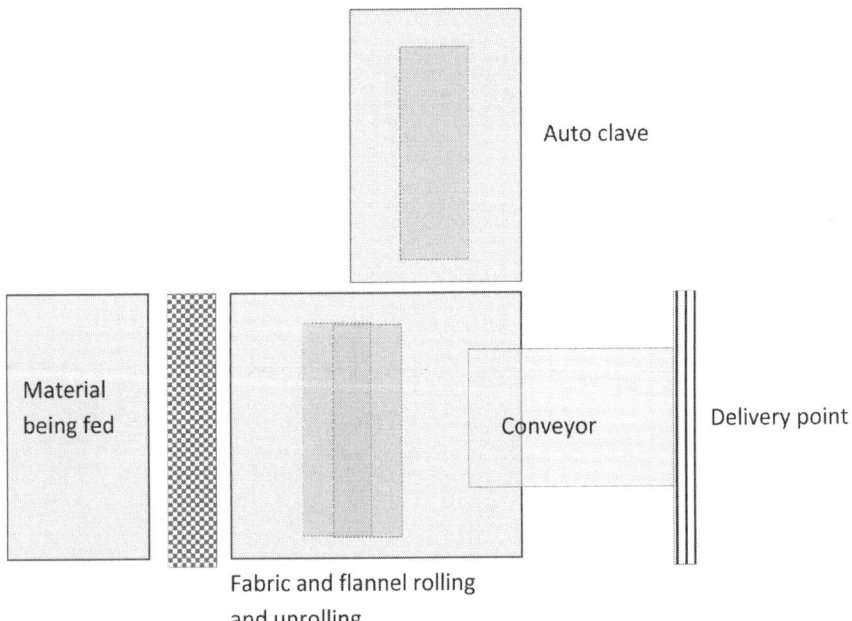

Figure 21.3   Plan view of Suprema—Biella shrunk processing machine.

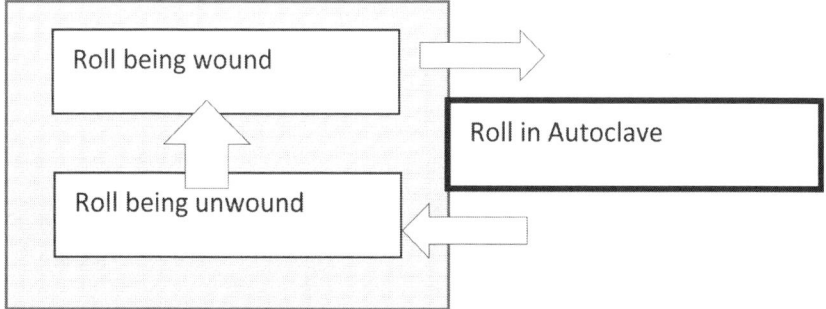

Figure 21.4 Front view.

Figure 21.5 Delivering portion.

Figure 21.6 Autoclave.

21.3 Typical operating procedure

Although the process sheet is prepared separately for each batch depending on the fabric being processed, the following steps need to be followed in a Suprema—K. D. Biella shrunk process Machine.

- Clean the machine with compressed air and then wipe with a clean cloth before starting the work. Clean all sensors and photocell units with clean, dry cotton cloth.

- Discuss with the supervisor and bring the materials to be shrunk processed/decatized. Ensure that the width of the fabric is at least 10 cm or less than the width of the blanket.

- Switch on the power, drain the water from the compressed airline and then open the valve. Open the steam condensate trap and remove the condensates fully, and then open the steam valve slowly.

- Get the programme for running the cloth by the supervisor along with the details of lot number, beam number and the total length of material.

Stitch

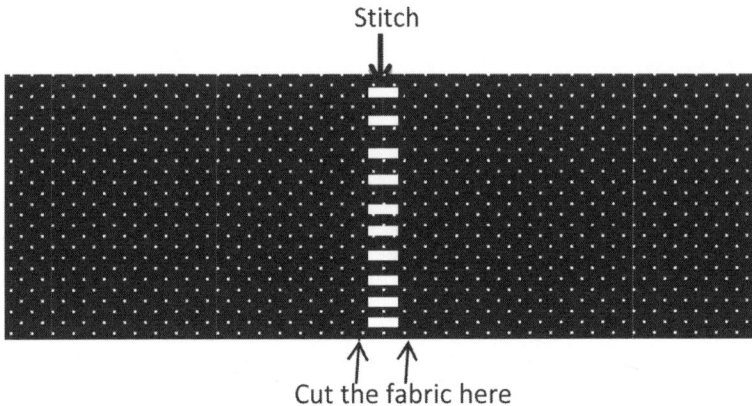

Cut the fabric here

Figure 21.7   Point to cut the fabric.

- Keep the steam trap open for 1–2 min and ensure that all the condensate is drained out and close the valve.

- Bring the material and keep it aligned with the feeding arrangement. Ensure that the cloth does not move to any corner or folds are not formed while running.

- Run the programme suggested by the supervisor which controls interconnected properties as per fabrics.

- While feeding the material, cut the fabric with scissors at the place of stitching and feed inside the roll (Fig. 21.7).

- Ensure that while feeding the cloth in between the wrapper there are no crease or fold of any type.

- Feed the fabric between two layers of the flannel (Fig. 21.8).

- While rolling the cloth with wrapper, ensure that from 30 to 40 m are kept blank without fabric both in the start as well as at the end of the roll. Keep 15–30 m gap between each piece of fabric after cutting the stitched portion (Fig. 21.9).

- While feeding the material ensure that the material is fed from a single point.

- Ensure that the treated material is collected without any crease (Fig. 21.10).

- Check the water level at suction pump.

- Check the quality of wrapper and ensure it is good (Fig. 21.11).

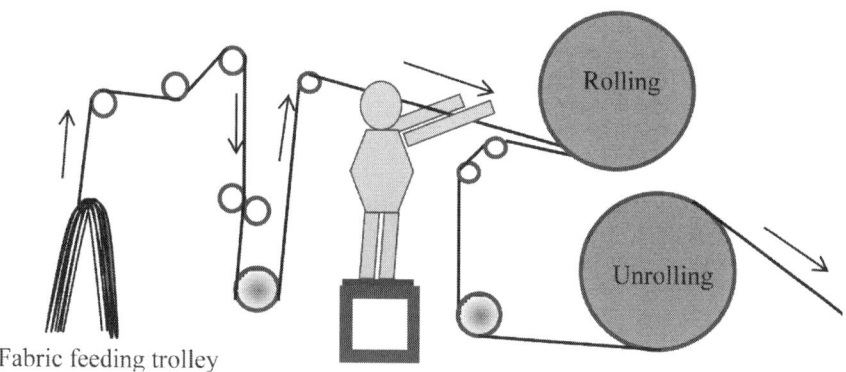

Fabric feeding trolley

Figure 21.8   Flannel.

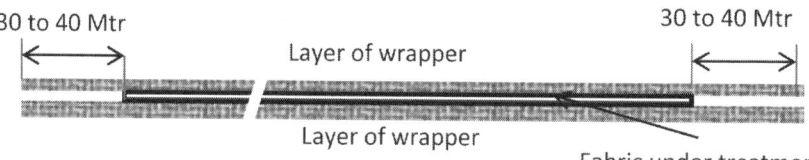

30 to 40 Mtr   30 to 40 Mtr

Layer of wrapper

Layer of wrapper

Fabric under treatment

Figure 21.9   Rolling with wrapper.

Figure 21.10   Treated fabric collected on conveyor with synthetic mesh and conveyed forward.

- Check and ensure the quality of the conveyor is in good condition.
- Check and ensure that all the safety systems in the machine are working properly
- Start the machine.

Figure 21.11   Wrapper.

- Operation of the machine has three stages. The fabric is first wound along with a wrapper cloth on perforated drum in no 1 position. Then, it moves to no 2 position for steaming treatment, and after steaming, it comes to no 3 position for unrolling and delivering the fabric. When the wrapper and fabrics are being wound on perforated drum in position no 1, the perforated drum in position no 2 wrapped with wrapper and fabric shall be in the autoclave undergoing the steaming operation, and perforated drum in position no 3 shall be unwinding the wrapper which is taken up by the drum no 1 and at the same time delivering the fabric out on to the conveyor of the delivery zone. The drums interchange the positions in a cycle.

- Check the finish of the fabric and decide on the changes in programme, if needed.

- Cover the produced material and keep in the designated place as suggested by the supervisor.

- Write the details of production such as batch number, lot number, quality or design number, quantity produced along with the date and time in the production record.

- In case any abnormalities are found in the fabric, inform the supervisor immediately, and process the material after getting clear instructions.

- In case of any problem in the machine, either electrical or mechanical, call the concerned person and get it attended.

21.4   Control points and checkpoints

It is essential to have clarity on the points to be controlled to achieve the targets and those to be checked to ensure the process in control. These points need to be reviewed from time to time and modified to suit the requirements of individual companies and their targets. Each mill should prepare its own 'control points and checkpoints' and display them in the work area so that the people on spot refer and follow. Some examples are given below.

21.4.1   Control points

a)   Selection of suitable process parameters for the material to be processed

b)   Deciding on the level of decatizing and shrinking for the fabric considering the material

c)   Deciding and selection of process parameters, namely

  •   Length of fabric to be wrapped at a time

  •   Steaming time and temperature

  •   Distance between two fabric ends being fed

  •   Cutting at the point of stitches

d)   Deciding acceptance criteria for the quality of decatizing of fabrics

e)   Employing trained and qualified employees

f)   Evolving production norms

21.4.2   Checkpoints

•   **Material related**

  a)   The fabrics received against the plan

  b)   The quantity of fabric received—metres and kilograms

  c)   The density (GSM) and fabric construction

  d)   The type of fibres in the fabric

  e)   The twist factor in the yarns used in the fabric

•   **Machine related**

  a)   The condition of the machine

  b)   The working of various controllers such as sensors and photocell units

c) Quality of wrapper being used

d) Quality of conveyor

- **Setting related**

  a) The pressure between fabric and the flannel

  b) The speed of different rollers and their direction of movement

  c) The steam pressure and timings at autoclave

- **Performance related**

  a) Uniformity of decatizing throughout the fabric as specified

  b) The production achieved

- **Documentation related**

  a) The design number of the fabric received for shrunk processing

  b) Quantity of fabric shrunk processed

- **Work practice related**

  a) Thorough cleaning with compressed air and then wiping with a clean cloth before starting the machine

  b) Maintaining the speed as per norms

  c) Cutting the stitched portions properly without wasting good fabric

  d) Maintaining the gap of 15 m while feeding different fabrics

  e) Keeping 30–40 m blank space at the start and end of each fabric roll

- **Logbook related**

  a) Machines working

  b) Starting time of the running lots

  c) Activities done and to be done

- **Management information system related**

  a) Machine number

  b) Design number

  c) Number of metres

  d) Weight in kilograms

- **General**
  a)   Use of safety gadgets such as gloves and masks

21.5   Dos and Don'ts

Understand clearly what is supposed to be done without fail and what should not be done at any cost. Some examples are given below.

21.5.1   Dos

a)   Cut the cloth at stitching and feed inside the roll. If the fabrics are not cut at stitching, we get stitch marks on the fabric and the fabric gets rejected.

b)   Check the condition of the wrapper and conveyors before starting the machine.

c)   Clean the machine thoroughly with compressed air before starting.

d)   Periodically check and ensure that the cloth does not move to any corner or folds are not formed while running.

e)   Ensure that the wrapper width is less by at least 10 cm compared to cylinder width.

f)   Leave a gap of 30 m in the wrapper while starting a new design.

g)   Leave at least 15 m between two cloth pieces fed to the machine.

h)   Switch on the power, drain the water from the compressed airline and then open the valve.

21.5.2   Don'ts

a)   Do not allow stitches and folds in the fabric being fed. They remain as permanent mark on the fabric.

b)   Do not run the machine if the wrapper has become thin or irregular.

c)   Do not accept the fabrics having higher width. The width of the fabric should be at least 10 cm or less than the width of the wrapper cloth.

d)   Do not overlap two fabrics while being fed.

Mechanical shrink-proofing for fabrics

22.1   What is shrink-proofing?

Shrinkage is the contraction in the dimension of the fabric due to any reason. Normally, cotton and woollen fabrics shrink after washing. Cotton fabrics suffer from two main disadvantages of shrinking and creasing during subsequent washing as it swells in water affecting shrinking. When the cotton fabric is put in water, swelling and rearrangement of internal forces take place. The fibres will become free from tension and come to the original tensionless state. Further, the mechanical stress, strain and tension are released during spinning, weaving and so forth and it causes the fabric to shrink.

Creasing is overcome by the resin finishing, whereas the shrinking is prevented by special mechanical finishing, namely sanforizing or zero–zero preshrinking, comfit or steam relax shrinking.

Sanforizing and zero–zero finishing is used mainly for cotton and cotton-rich blends. Comfit is a process of treatment used for polyester and its blend fabrics. It is a method of stretching, shrinking and fixing the woven cloth in both length and width. It gives soft and smooth feel in fabrics. Steam relaxing is a process of shrink-proofing delicate fabrics containing lycra and wool.

The purpose of sanforizing or zero–zero finishing is to give shrink-proof finish to cotton fabrics. Sanforized fabrics should not shrink more than 1%. There are a number of mechanical finishing treatments to make fabrics shrink-proof.

22.2   Sanforization

Sanforization is a process of treatment used for cotton fabrics mainly and most textiles made from natural or chemical fibres, patented by Sanford Lockwood Cluett (1874–1968) in 1930. It is a method of stretching, shrinking and fixing

Figure 22.1   Sanforizing machine.

the woven cloth in both length and width, before cutting and producing to reduce the shrinkage which would otherwise occur after washing. Fabrics bearing this trademark will not shrink more than 1% because they have been subjected to a method of compressive shrinkage involving feeding the fabric between a stretched blanket and a heated shoe.

The cloth is continually fed into the sanforizing machine (Fig. 22.1). It is moistened with either water or steam. A rotating cylinder presses a rubber band against another heated rotating cylinder. The rubber band briefly gets compressed and afterwards shrinks to its final size. The cloth to be treated is transported between rubber band and heated cylinder and is forced to follow this brief expansion and recontraction and thus gets shrunk. The bigger the pressure applied to the rubber band, the bigger the shrinking afterwards.

The process can be described by the following schematic diagram (Fig. 22.2).

Fabric (F) passes through the skyer (S) or another moistening device and is moistened with water and/or steam. This will lubricate the fibres and promote shrinkability within the fabric. Normally, a fabric must be moistened in such a way that every single thread achieves a moisture content of approximately 15%. This allows compression of the fabric with very little resistance.

When the fabric passes through the clip expander (C), fabric gets the required width. The clip expander also transports the fabric to the most important part of the machine, the rubber belt unit (indicated by curved arrow in above figure). In the close-up of figure, the endless rubber belt (R) can be seen. By squeezing rubber belt (R) between pressure roll (P) and rubber belt cylinder (RB), an elastically stretching of the rubber belt surface is obtained. The more the rubber belt is squeezed, the more the surface is stretched. This point of squeezing is known as the pressure zone or the nip point.

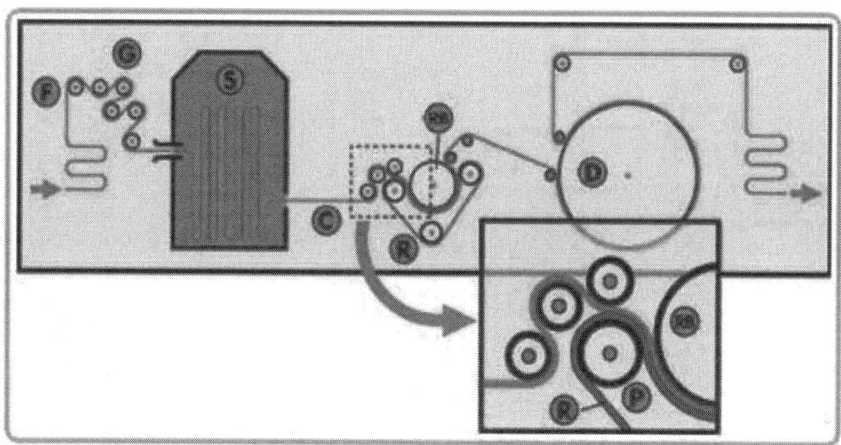

Figure 22.2 Schematic representation of sanforizing machine.

Fabric (F) is next fed into the pressure zone. When leaving the pressure zone, the rubber belt recovers itself and the surface returns to its original length, carrying the fabric with it. The effect of this action is a shorting of the warp yarn which packs the filling yarns closer together. At this moment, shrinkage occurs. After compaction within the rubber belt unit, the fabric enters the dryer (D). Here, the fibres are locked in their shrunken state by removing the moisture from the fabric.

After the compressive shrinkage process is completed, another sample of the fabric is taken. This sample is also wash-tested. The final result of this test must meet the sanforized standard, in length and width, before it may carry the sanforized label.

22.2.1 Procedure adopted for sanforizing

The operating procedure is decided by the type of fabric being sanforized. The following are general guidelines:

1. Clean the machine and its surroundings before starting any work on the machines.

2. Check the rubber blanket (Fig. 22.3) for crack, holes and so forth.

3. Check the rubber blanket squeezer properly. Verify whether it is properly working or not. Clean it before starting the machine.

4. Check the Palmer felt properly.

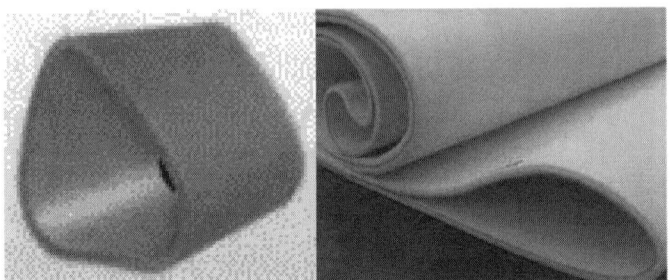

Figure 22.3   Rubber blanket.

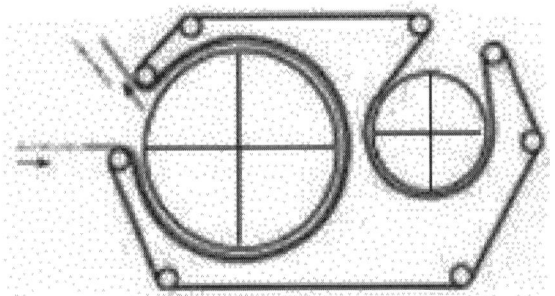

Figure 22.4   Drum and the Palmer felt.

5. Check the steam pressure of rubber blanket cylinder and Palmer felt cylinder (Fig. 22.4).

6. Check the rubber blanket water spray. Check water spray nozzle; if not found clean, clean immediately.

7. Check cooling drum working.

8. Check guiders.

9. Check the cleaning of stitching machine. It should be cleaned daily.

10. Take programme from shift supervisor.

11. Process control parameters should be strictly followed.

12. Change batch in machine-running position with minimum time.

13. Check the guider entry and exit side.

14. Set shrinkage as per programme.

15. Check the online residual shrinkage.

16. Check width of fabric before and after sanforizing (Fig. 22.5).

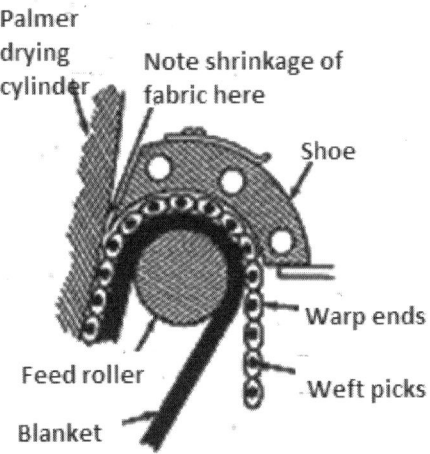

Figure 22.5 Sanforizing unit.

17. Inspect the fabric at the delivery horseback trolley.

18. Check the fabric tension at the delivery end (fabric should run without tension).

19. Cover the horseback trolley with polyethylene paper at the delivery end and keep at proper place.

20. Cut sample in exact metres to lab as specified (not more or less) as the tests to be performed (skew, shrinkage, width, tear and tensile strength, shade, CS, pilling, fastness, etc.).

21. Cover the horseback trolley with polyethylene paper after sanforizing the fabric.

22. Keep proper tag on the batch.

23. In case of any breakdown, inform engineering department through the shift in charge or supervisor to attend the same.

24. If any noise is coming from any part of the machine or leakage is found at any part, the operator informs to shift in charge/supervisor.

25. Check damages such as crease, chapti crease, rubber impression, dagi (stain), water dagi, skew and bowing.

Procedure for terminating the production of sanforizing machine

1. Water spray should work continuously till it gets cooled.

2. Close steam valve and air pressure valve.

3. Rubber belt cylinder and Palmer felt cylinder should be moved slowly until it comes down at room temperature.

4. Release the air pressure to keep pneumatically loaded squeezing roller and expander pressure roller in relaxed mode.

5. After the rubber blanket cylinder, temperature comes down at room temperature, switch off and close the water line valve.

6. Clean the rubber blanket and cylinder during cooling.

7. Cover the rubber belt and Palmer felt after it gets cooled properly.

8. Drain the water from all cylinders (Palmer felt/rubber blanket steam cylinder).

9. Cover all batches with a polyethylene sheet.

Maintenance schedule of sanforizingmachine

1. Clean the machine on weekly basis.

2. Do the oiling and greasing as per the instruction of engineer.

3. Clean rubber blanket cylinder, rubber cooling spray pipe, cooling drum and so forth during weekly cleaning.

4. Check the rubber hardness and fabric shrinkage before deciding for grinding.

22.2.2 Dos and Don'ts for sanforizing

It is essential to understand the essential activities which are supposed to be done and the activities that should never be done while operating a sanforizing machine. Following are some examples.

Dos

a) Change the batches in running machine.

b) Switch off and close the water valve only after the drum comes to room temperature.

c) Clean the rubber blanket and cylinder during cleaning.

d) Cover the fabric on horseback trolley with a polythene sheet.

Don'ts

a) Do not heat the drum and blanket when it is not rotating

b) Do not use the rubber blanket if cracks have developed

c) Do not overshrink the fabric

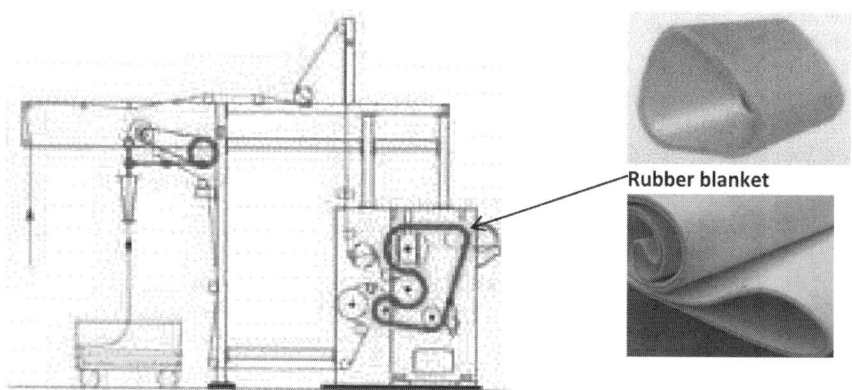

Figure 22.6 Comfit machine.

22.3 Comfit

Comfit is a process of treatment used for polyester and its blend fabrics. It is a method of stretching, shrinking and fixing the woven cloth in both length and width. It gives soft and smooth feel in fabrics.

The cloth is continually fed into the comfit (Fig. 22.6). A rotating cylinder presses a rubber band against another heated rotating cylinder; therefore, the rubber band briefly gets compressed. The cloth to be treated is transported between rubber band and heated cylinder and is forced to follow this brief expansion and recontraction and thus gets soft and smooth finish.

When the fabric passes through the endless rubber belt by squeezing rubber belt between pressure roll and rubber belt cylinder, we obtain an elastically stretched surface of the rubber belt. The more the rubber belt is squeezed, the more the surface is stretched. This point of squeezing is known as the pressure zone or the nip point.

Fabric is then fed into the pressure zone. When leaving the pressure zone, the rubber belt recovers itself and the surface returns to its original length carrying the fabric with it. The effect of this action is a shorting of the warp yarn which packs the filling yarns closer together. At this moment, shrinkage occurs. After compaction within the rubber belt unit, the fabric enters the cooling cylinder.

22.3.1 Procedure for operating comfit machine

The standard operating procedure depends on the type and construction of fabric being processed. Following are general guidelines:

1. Clean the machine and its surroundings before starting any work on the machines.

2. Check the rubber blanket for crack, holes and so forth.

3. Check the rubber blanket squeezer properly. Verify whether it is properly working or not. Clean it before starting the machine.

4. Check the steam pressure of rubber blanket cylinder.

5. Check the rubber blanket water spray. Check water spray nozzle; if not found clean, clean immediately.

6. Check cooling drum working.

7. Check guiders.

8. Check the cleaning of stitching machine. It should be cleaned daily.

9. Take programme from shift supervisor.

10. Process control parameters should be strictly followed.

11. Change batch in machine-running position with minimum time.

12. Check the guider entry and exit side.

13. Check width of fabric before and after comfit.

14. Inspect the fabric at the delivery horseback trolley.

15. Check the fabric tension at the delivery end (fabric should run without tension).

16. Cover the horseback trolley with polyethylene paper at the delivery end and keep at proper place.

17. Cut sample in exact metres to lab as specified (not more or less) as the tests to be performed (skew, shrinkage, width, tear and tensile strength, shade, CS, pilling, fastness, etc.).

18. Cover the horseback trolley with polyethylene paper after comfit the fabric.

19. Keep proper tag on the batch.

20. In case of any breakdown, the shift in charge or supervisor sends a breakdown memo to the engineering department to attend the same.

21. If any noise is coming from any part of the machine or leakage is found at any part, the operator informs to shift in charge/supervisor.

22. Check damages such as crease, chapti crease, rubber impression, dagi (stain), water dagi, skew and bowing.

Terminating the production of comfit

a) Water spray should work continuously till it gets cooled.

b) Close steam valve and air pressure valve.

c) Rubber belt cylinder should be moved slowly until it comes down at room temperature.

d) Release the air pressure to keep pneumatically loaded squeezing roller and expander pressure roller in relaxed mode.

e) After the rubber blanket cylinder, temperature comes down at room temperature, switch off and close the water line valve.

f) Clean the rubber blanket and cylinder during cooling.

g) Cover the rubber belt after it gets cooled properly.

h) Drain the water from all cylinders.

i) Cover all batches with a polyethylene sheet.

Maintenance schedule of comfitmachine

a) Clean the machine on weekly basis.

b) Do the oiling and greasing as per the instruction of engineer.

c) Clean rubber blanket cylinder, rubber cooling spray pipe, cooling drum and so forth during weekly cleaning.

d) Check the rubber hardness before deciding for grinding.

22.4 Steam relaxing

Steam relaxing is the process of stabilizing the fabric and removing its swaying and shrinking properties by making use of steam and an overfeeding device. Fabric is driven by means of one conveyor belt inside the steaming chamber. The fabric circulates free of tensions and with its natural and best shrinking conditions. Synchronization between the conveyor belt and the feeding cylinder contributes to achieve the maximum fabric capacity inside the chamber as well as optimal shrinking conditions.

Steam relaxing is used specifically for very delicate fabrics containing lycra and wool.

22.4.1   What is relax shrinking?

Relax shrinkage is a shrinkage induced by the relaxation of strains present in a textile. The strains are of a temporary nature that can be relaxed to a varied degree, for example, by steam pressing or by immersion in water.

Relaxation shrinkage in wool fabrics is normally measured after relaxation of the fabric in water or steam. Values obtained for relaxation shrinkage after wet relaxation depend on the manner in which the fabric is dried. If the wet fabric is dried directly in an oven, the values for relaxation shrinkage are low, but if the fabric is allowed to dry under ambient conditions before oven-drying, values are high. Steaming procedures produce relaxation shrinkage values that are considerably lower than those obtained with wet relaxation.

The purpose of this finishing process is to give relaxed shrinkage to delicate woven fabrics, especially those containing all types of lycra and wool while retaining the fabric properties.

22.4.2   Process of relax shrinking

The process of relax shrinking incorporates the stabilization of the fabric and removal of its swaying and shrinking properties by using steam and an overfeeding device. Fabric is driven by means of one conveyor belt inside the steaming chamber. The fabric circulates free of tensions and with its natural and best shrinking conditions. Synchronization between the conveyor belt and the feeding cylinder contributes to achieve the maximum fabric capacity inside the chamber as well as optimal shrinking conditions.

Commercially available extractors and dryers are equipped with spreading and overfeeding devices to regain width and return the stretched loops to a more normal state. Spreading must be accompanied with overfeeding in order to be effective. Relaxation drying allows those excessive tensions to be released since the fabric is dried under little or no restraints.

Stabila Relaxing machines from Sperotto Rimer, Italy, is used for completely relaxing and shrinking the fabrics (Fig. 22.7). The processing of woollen blends is done on steam relaxation principle. This is used to give relaxed shrinkage to delicate woven fabrics, especially those containing all types of lycra and wool while retaining the fabric properties.

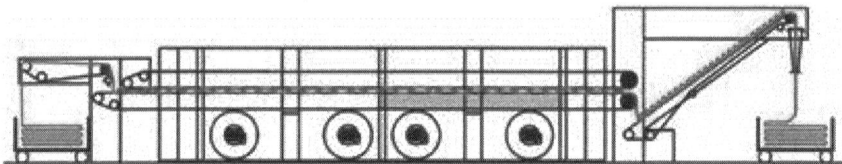

Figure 22.7   Schematic diagram of Stabila Relax shrinking machine.

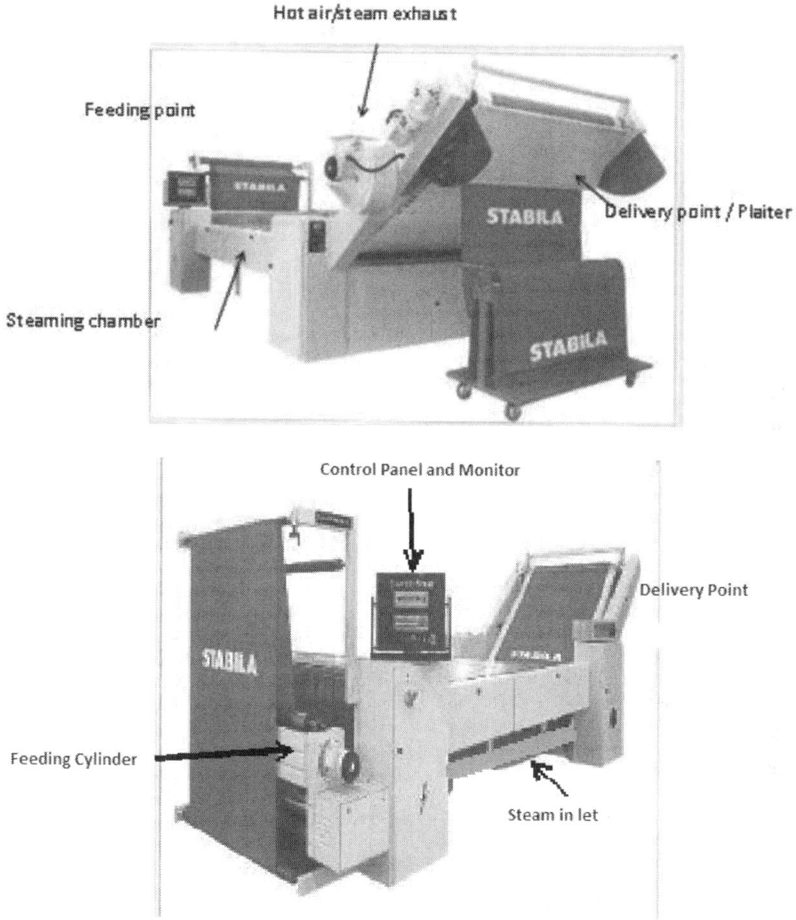

Figure 22.8   Stabila Relax shrinking machine.

Normal features of a Stabila machine consists of 2000-mm roller width for steaming, shrinking and stabilization of fabric, automatic microprocessor, high-entry, brushing units, tensionless conveyor belt, two steaming zones, exit with conveyor belt and folder (Fig. 22.8).

Most relaxation dryers are based on the belt principle, where the fabric is placed between two belts and then passed through the drying zone. The bottom belt supports the fabric but allows for shrinkage, while the top belt prevents any stretching. In some cases, the bottom belt can be vibrated for additional mechanical action. Airflow is normally directed down and up through the fabric to give a ripple/wave effect. Once the drying is completed under relaxed conditions, those excessive tensions which have occurred during prior processing have been released. If the fabric has been spread with overfeeding prior to relax drying, width shrinkage occurs first. This may prevent the length from shrinking initially, but as the drying progresses, both width and length shrinkages occur. At the exit of the relaxation dryer, the fabric width will be inconsistent and may not be completely wrinkle-free. Thus, calendering or compacting is necessary to provide a uniform, finished roll for cutting and sewing.

22.4.3 Operating procedures for steam relaxing machine

The following steps need to be followed in a steam relaxing machine similar to Stabila:

a) Clean the machine with compressed air and then wipe with a clean cloth before starting the work. Clean all sensors and photocell units with a clean dry cotton cloth.

b) Drain water from the compressor pipelines and then open the air valve and then switch on the machine.

c) Drain condensates from steam pipes and then open the steam valve slowly. Ensure that steam pressure is at 6 bar.

d) Press 'Heatingin Auto F11' to heat the steaming chamber. After the machine reaches full temperature, open the F6 valve.

e) When the heating process is on, you will get a message 'Heating in Progress' on the screen.

f) After the heating process is complete, you will get a message on the screen as 'Machine Heated, Start Steam'. Press F6 and start steam.

g) While heating process is on, run the feeder clamps. If heating process is done by keeping the feeder stationary, the conveyor felt is likely to lose its life.

h) Set the machine parameters like speed and overfeed for the material to be processed by discussing with a supervisor.

i) Keep tunnel temperature at $105 \pm 5°C$.

j) Write the complete details of the material being processed in the production book.

k) Check the fabric panel from time to time and write the findings in the production book.

l) For checking overfeed, make marking on 1-m fabric in the back and check the length after drying on the delivery side. Enter the details in the production book/job card.

m) In case of any abnormality, inform the concerned supervisor and get it rectified.

n) In case of any electrical or mechanical problem in the machine, get it rectified by the concerned engineering person. Write the details in the production book.

o) Cover the processed material with a polythene sheet and keep at the designated place.

22.4.4 Control points and checkpoints

It is essential to have clarity on the points to be controlled to achieve the targets and those to be checked to ensure the process in control. These points need to be reviewed from time to time and modified to suit the requirements of individual companies and their targets. Each mill should prepare its own 'control points and checkpoints' and display them in the work area so that the people on spot refer and follow.

Control points

a) Selection of suitable process parameters for the material to be processed

b) Deciding on the level of shrinking for the fabric considering the material

c) Deciding and selection of process parameters, namely

- Steaming time and temperature
- Overfeeding in case of Stabila
- Moisture content before shrinking

d) Deciding acceptance criteria for quality of shrinking of fabrics

e) Employing trained and qualified employees

f) Evolving production norms

Checkpoints

- **Material related**

  a) The fabrics received against the plan received

  b) The quantity of fabric received—metres and kilograms

  c) The density (grams per square metre) and the fabric construction

  d) The type of fibres in the fabric

- **Machine related**

    a) The condition of the machine

    b) The working of various controllers such as sensors and photocell units

    c) Quality of rubber blanket being used

    d) Quality of conveyor

- **Setting related**

    a) The pressure between fabric and the rubber blanket

    b) The speeds of different rollers and their direction of movement

- **Performance related**

    a) Uniformity of shrinking throughout the fabric as specified

    b) The production achieved

- **Documentation related**

    a) The design number of the fabric received for shrink processing

    b) Quantity of fabric shrink processed

- **Work practice related**

    a) Whether the machine was thoroughly cleaned with compressed air and then wiped with a clean cloth before starting

    b) Whether the speeds were maintained as per norms

- **Logbook related**

    a) Machines working

    b) Starting time of the running lots

    c) Activities done and to be done

- **Management information systems related**

    a) Machine number

    b) Design number

    c) Number of metres

    d) Weight in kilograms

- **General**

    a) Use of safety gadgets

22.4.5 Dos and Don'ts for relax shrinking

It is essential to understand clearly what is supposed to be done without fail and what should not be done at any cost. Some examples are given below.

Dos

a) Change the batches in running machine.

b) Switch off and close the water valve only after the drum comes to room temperature.

c) Clean the rubber blanket and cylinder during cleaning.

d) Cover the fabric on horseback trolley with a polythene sheet.

e) Drain the condensed water from time to time, especially in case of Stabila.

f) Synchronize the speed between conveyor belt and feeding cylinder to get required overfeed in Stabila.

g) Check the overfeed by marking 1 m in the feeding in Stabila.

Don'ts

a) Do not heat the drum and blanket when it is not rotating.

b) Do not use the rubber blanket if cracks have developed.

c) Do not allow condensation inside the chamber in Stabila.

d) Do not increase the speed of the machine to get higher production.

e) Do not stretch the material, especially in relax shrinking.

22.5 Jet air relaxation drying

The jet air relaxation dryer is designed for both woven and knitted fabrics of either open-width or tubular form. The models, available in the common nominal widths, are of modular design.

JetAir 5000 has single fabric passage, fabric inlet from the front, fabric delivery to the rear and fabric feed in either open-width or tubular form with one or more webs.

Fabric types that can be processed are knitted fabrics (in either open-width or tubular form), woven fabrics, microfibres and carpets (coated and non-coated); thermo-bonding is also possible. Bonding of non-woven fabrics using the thermo-fusion process is also possible in this machine.

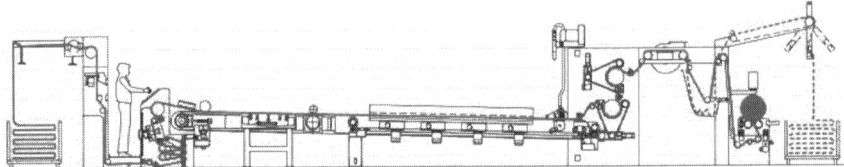

Figure 22.9   Schematic diagram of Santacompact CK.

The Santacompact CK provides the mechanical basis for requirements on open-width knitted fabrics, by securing the results of preceding shrinking and relaxing drying and also adjusting the fabric to the minimum possible residual shrinkage values through sensitive additional compressive shrinking in the compacting calender, while treating it extremely gently and preserving maximum volume (Fig. 22.9).

He important features are as follows:

a)   Width control and equalization of open-width knitted fabric through distortion-free, controlled pinning and width management in an equalizing frame, optionally equipped with individual chain adjustment system

b)   Condensate-free steam box with compensation for different widths, completely corrosion-resistant version

c)   Tensionless and distortion-free pinning with the short, direct and crease-free feed into the compacting unit

d)   Specially designed elastic felt belt in both compacting units

e)   Heatable smoothing and shrinking rollers for the fabric smoothing

Optionally:

a)   Fabric winding on cardboard cores with manual or automatic changing facility with inspection table

22.6   Relax shrinking of knit fabrics containing Lycra

The addition of 3–10% elastane fibres to cotton knitwear increases stretchability and elastic properties. As a result, the shape of the garment is retained despite using and washing numerous times. The finishing of such articles requires complex processes.

Unfixed, grey knitwear containing elastane is very crease sensitive. For this reason, the grey fabric is often prefixed on the stenter frame. During fixing, silicon- and mineral-based knitting oils evaporate and enter the waste air from the stenter frame. This is environmentally undesirable emission.

The fixing of the grey fabric also often causes problems for setting the technical parameters. The compaction necessary to achieve the weight per square metre cannot be generated using stenter frame shrinkage. If the knitwear is further treated after fixing using the extension process, it is then often sewn to form a tube. Although tubular heat setting systems are available, they have failed to become established due to their lack of flexibility. Additionally, the problem of emissions and smoke still occurs.

Benninger has developed a process, which permits controlled relaxation and simultaneous removal of silicon and knitting oils from grey fabric. Further processing is possible without heat setting. The process is suitable for finishers processing knitwear both continuously open-width and discontinuously in rope form.

A combined dwelling and drum washing unit is developed by Benninger for knitted fabrics containing lycra. The grey fabric is saturated in an impregnating unit with soda, suitable wetting agents and emulsifiers. It is then placed on a roller bed. Conventional relaxation units (conveyer belts or under liquor compartments) bear an acute risk of crease marking of unfixed knitwear containing elastane.

The key differences of the Benninger relaxation unit are (Fig. 22.10):

a) The dwell time can be adjusted from 0.5 to 2 min to suit a specific fabric. On delicate articles, the packing density is so low that there are only a few loops in the accumulator.

Figure 22.10   Benninger roller bed section for free relaxation.

b) The fabric is continuously sprayed with washing or chemical liquor. This forms a protective film of water to prevent lay marks.

c) The roller bed continuously moves the loops of fabric. In this way, the fabric is turned and lay marks cannot even form.

d) It is possible to process particularly delicate fabrics by tight strand with low tension. The silicon and knitting oils are still washed out adequately.

The washing of the relaxed knitwear takes place on an open-width drum washing machine. This phase is particularly important as the fabric shrinks during relaxation and unfixed fabric curls particularly strongly. To wash out the applied chemicals at the selvedges too, the fabric is spread out using expanding rollers and transferred over a short distance to the Trikoflex washing drum. Although the inlet side of the impregnating unit and the roller bed plaiter are designed for a large speed range, the total number of washing drums depends on the production speed. The smallest possible production unit, so-called all-in-one concept, includes the processes in a very small space such as impregnation, dwelling and washing out. This plant concept is suitable for product quantities of 2–4 metric t/day and represents a particularly economical machine variant.

The knitwear is then further processed either open-width or in rope form; here, the first variant is to be preferred to meet today's quality and cost demands. Further processing in open width guarantees a very smooth fabric appearance with less hairiness at 15–20% lower production costs. The timing of the heat setting process is dependent on the article. In the majority of cases, it is recommendable to perform heat setting directly after the relaxation process; however, at the latest before the cold pad-batch dyeing.

If further processing is performed discontinuously, it is possible either to heat set before or after dyeing. Practical experience has shown that a significant portion of fabric can be fed directly to the jet without intermediate drying. This variant is possible because a certain amount of intermediate stabilization takes place during the hot water treatment. The heat setting process is then performed economically after dyeing. In these cases, the relaxation of rope fabric is particularly interesting because it is not necessary to slit and subsequently sew the tube.

A low-tension prewash is needed for the removal of spinning oil. The continuous pretreatment concept used by Benninger involves an impregnation, emulsification and wash process on Trikoflex drum wash compartments (Fig. 22.11). During the emulsifying phase, the chemical-saturated knitwear is placed in loops on to a dwelling system. The micro-movement of the

Figure 22.11   Benninger Trikoflex prewash and relaxing.

individual loops ensures that the knits are continually loosened and promotes free relaxation, and at the same time, prevents the formation of creases. In this way, the bidirectional shrinkage of the material is excellent. The hydro-shrinkage is significantly stronger and longer lasting than thermal shrinkage, for example, using hot air. Whether with or without subsequent thermal fixing process, the stretchability and elastic recovery of the material are retained, even following several household wash cycles and wear. The fit of the garments is retained. In the prewash process, the emission of silicone and mineral knitting oils during the fixing process of unwashed knitwear is avoided. The black smoke in the exhaust air from the stenter frame has become a thing of the past.

The high washing efficiency in Benninger washing machine is due to the unique washing principle of the Trikoflex drum. By its multilayer construction, it is possible to wash the fabric web on both sides with good mechanical washing effect. This results in a reduction in water and energy consumption of up to 50%.

22.7   Compacting tubular knitted fabrics

Compacting of knit fabrics is done to get required dimensional stability with lowest residual shrinkage values. Let us have a look at Swastik Tubular Compactor TC-400 (Fig. 22.12).

The following treatment steps are carried out on the Swastik Tubular Compactor (Figs. 22.13 and 22.14):

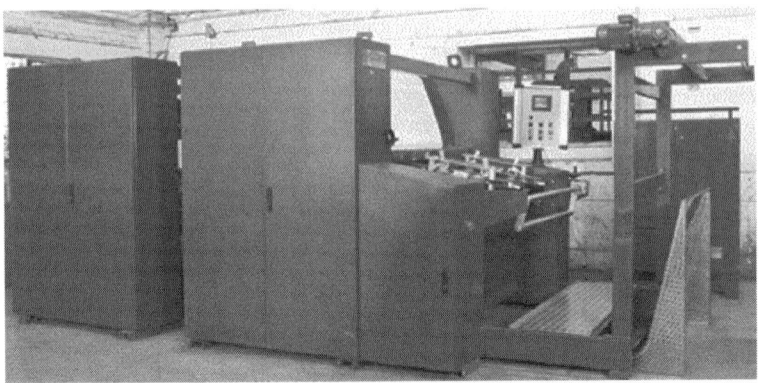

Figure 22.12   Swastik Tubular Compactor.

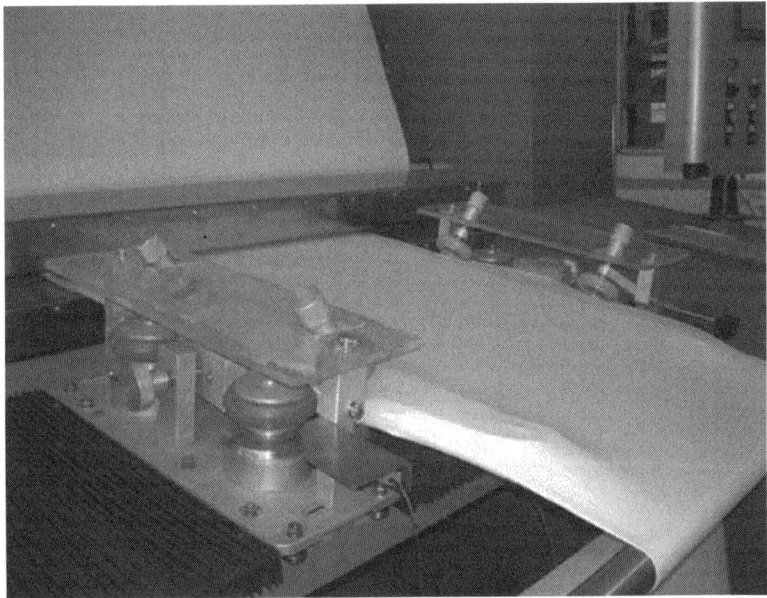

Figure 22.13   Width control.

1. Width control through a stepless adjustable special tubular fabric spreader driven by variable speed motor for distortion-free fabric guidance.

2. Steaming with a condensate-free steam box which is easily operated and completely made from stainless steel (SS).

3. Compacting through two Nomex felt belts.

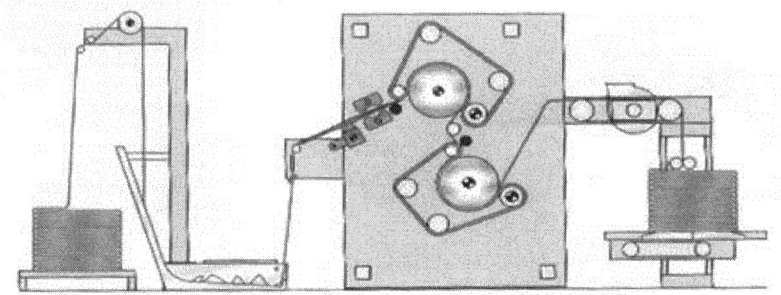

Figure 22.14   Schematic diagram of Swastik Tubular Compactor.

4. Calendering while passing between the felt belt and the heated shrinking rollers.

5. Precision plaiting with automatic platform level adjustment controlled by folded fabric height. Alternatively, a fabric rolling system can be provided.

The fabric is fed through the guiding system and stretcher then takes the fabric through the steam box onto the felt of the twin compacting units.

At the fabric delivery, the machine is equipped with a precision plaiting device with its platform. The height of the platform is controlled automatically and is adjustable according to the plaited fabric height.

22.8   Open-width compactor

Open-width compactor, for example, OC 400 is suitable for open-width knit fabrics to achieve exact dimensional stability and a soft feel (Fig. 22.15). The machine generally consists of a feeding frame with the centring device and driven scroll rollers, an equalizing stenter frame with overfeed roller and brush pinning arrangement.

The entry section of pin frame is provided with edge spreaders infra-red (IR) in-feed device, an SS-fabricated steaming unit for uniform moistening of the fabric. The steaming device has SS sliding shutters that allow steam to flow only as per the width of the fabric.

A low-contact gluing and drying unit is provided with an SS trough. Four selvedge drying units with IR emitters are placed on either side of the machine. The delivery side section consists of edge dryer, selvedge trimmer, a suction device, exit roller and width adjustment device, and the drive to the chain are housed in an exit box.

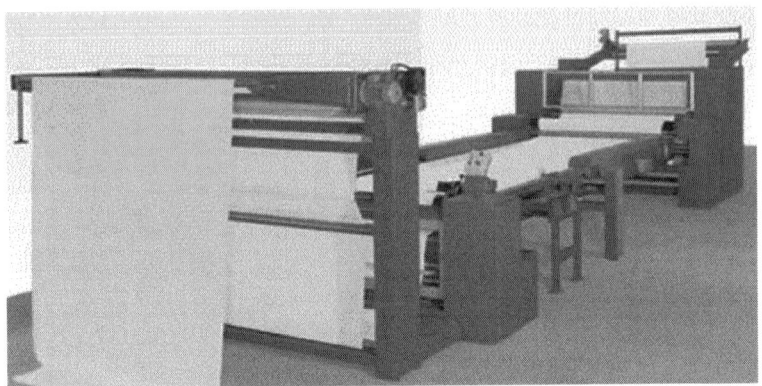

Figure 22.15   Open-width compactor.

The compacting unit consists of two felt compacting units, each of them consisting of a Nomex felt approximately 20-mm thick, a steam-heated chrome-plated centre roller of diameter 400 mm, a rubber-covered roller driven by variable-frequency drive, a compacting pressure roller, a felt-tensioning roller and a felt-centring roller. Each unit is provided with a special antifiction sheet-type shoe controlled by an electrical actuator to control the compressive shrinkage. A fabric-cooling roller is provided after second felt to cool the fabric by means of chilled water circulation. Fabric tension through the machine is controlled with the help of sensitive load cells and variable frequency drive with programmable logic controller and touchscreen.

Calendering of fabrics

23.1   What is calendering?

Calendering is pressing the moist fabric at the nip of high pressure mangles at an increased temperature to add lustre to fabrics. It improves aesthetics of a fabric by giving lustre and smooth feel. It is a finishing process where fabric is folded in half and passed under rollers at high temperatures and pressures. A series of hard pressure rollers are used that form or smooth a sheet of material.

Calendering is a mechanical finish. It is done for imparting softness and shining of the fabric or stiffness according to the nature of the chemical. This process is done after the chemical finishing as per buyer's requirement. These are heavy machines and the speed of the machine is 10–100 m/min (Fig. 23.1).

At least two rollers are required for calendering and the numbers can go up to 10. Alternately, one roller is made of steel and the other is made of softer material such as wool paper, cotton fibre and corn husks. The steel rolls may be equipped to be heated by gas or steam. The fabric passes rapidly between the rolls and then wind up on the back of the machine.

Figure 23.1   Fabric calendering machine.

The calendering finish is easily destroyed and does not normally last well. Washing in water destroys it and also does wear with time.

23.2 Objectives of calendering

The objectives of calendering can be listed as follows:

a) To upgrade the fabric hand and to impart a smooth silky touch to the fabric

b) To give different visual and feel effects on the fabric

c) To improve the opacity of the fabric

d) To compress the fabric and reduce its thickness

e) To impart different degree of lustre to the fabric

f) To reduce yarn slippage

23.3 Components of calendering machine

Normal parts of a calendering machine are as follows (Fig. 23.2):

a) **Heavy steel roll:** It is used to give smoothness and lustre. The temperature provided to steel roller is about 32–200°C with the help of electric heater. As we increase the temperature shining will increase. Steel rolls are used only for cotton, cotton viscose blends and polyester cotton.

b) **Cotton roll:** For soft finish the fabric is passed through cotton bowl. This roller is made up of cotton.

c) **Reclon roll:** When fabric passes through Reclon and steel rolls, smooth and lustre effects are generated. When it runs on steel, Reclon and cotton rolls, dull effect is generated. When it runs between steel and Reclon rolls, gloss effect is generated.

d) **Cooling drum:** To cool down the fabric, water circulates inside the cooling drum.

e) **Seam detector:** Function of seam detector is to bypass the seam.

f) **Metal detector:** Metal detector is installed to detect metal particles. If any metal particle is there, it might create holes in the fabric.

g) **Antistatic rod:** Antistatic rod is used to remove static charges.

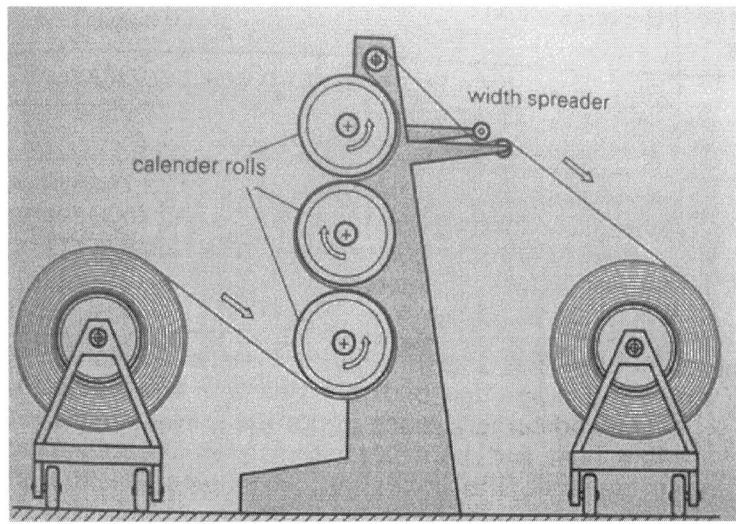

Figure 23.2 Principle of calendering.

h) **Small Winding:** It is used when short width fabrics are to be calendered. We can also run short width fabric with long width fabric.

i) **Oscillating roll**: This is used to avoid selvedge overlapping on batcher.

23.4 Different types of calendering

Several different finishes can be achieved through the calendering process by varying different parts. The main types of finishes are beetling, watered, embossing, Schreiner, swizzing and so forth.

23.4.1 Beetling

Beetling is a finish given to cotton and linen cloth which makes it look like satin (Fig. 23.3). Pounding of linen or cotton fabric is done which gives a flat, lustrous effect. This process produces a hard, flat surface with high lustre and also makes the texture less porous. In this process, the fabric, dampened and wound around an iron cylinder, is passed through a machine in which it is pounded with heavy wooden mallets. The fabric goes over wooden rollers and is beaten with wooden hammers.

Only table linen is put through beetling and not the dress linen. When applied to cotton fabrics, beetling gives it the feel and appearance of linen. This process permanently flattens the yarns of the fabric on which it is applied.

Figure 23.3   Beetling.

23.4.2 Watered

The watered finish, also known as moiré, is produced by using ribbed rollers. These rollers compress the cloth and the ribs produce the characteristic watermark effect by moving aside threads as well as compressing them. This leaves some of the threads round while others get compressed and become flat.

Watered finish calendering produces the two moiré effects known as 'moiré antique' and 'moiré anglaise'. This is a purely a physical phenomenon. Two ribbed rollers polish the surface and make the fabric smoother and more lustrous. High temperatures and pressure are used as well, and the fabric is often damped before being run through the rollers. The end result is a peculiar lustre resulting from the divergent reflection of the light rays on the material, a divergence brought about by compressing and flattening the warp and filling threads in places, forming a surface which reflects light differently. The weft threads also are moved slightly.

23.4.3 Embossed

In the embossing process, the rollers have engraved patterns on them (Fig. 23.4), and the patterns become stamped onto the fabric. The end result is a raised or sunken pattern, depending on the roller. This works best with soft fabrics and on nonwoven. Textile embossing is generally done on smooth surfaces. This is done by pressing the fabric under high pressure and at high temperature in a special backing creating a three-dimensional effect giving a unique look to the garment.

Embossing calender is similar to the Schreiner calender, but the bowls of this calender are much bigger (38–45 cm in diameter).

The top chilled iron roll has a design engraved on it and is heated, by which, damask effect can be produced on cotton fabric (Fig. 23.5). It lasts longer on

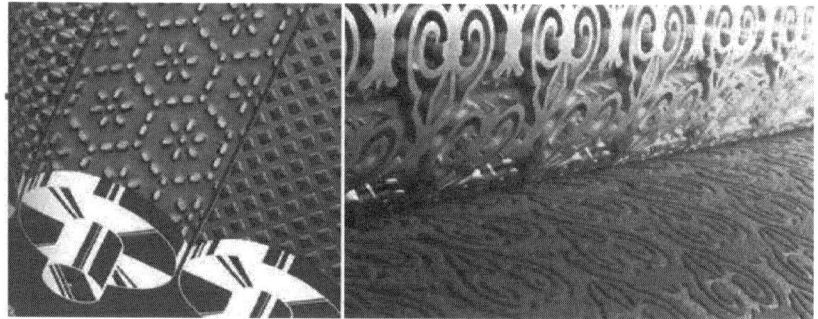

Figure 23.4 Embossing rollers.

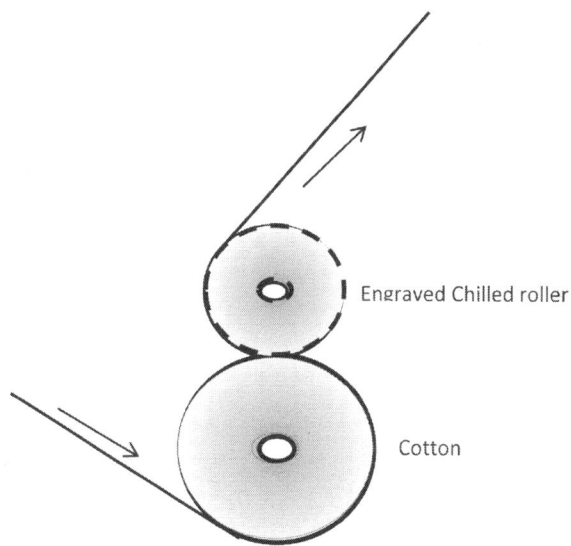

Engraved Chilled roller

Cotton

Figure 23.5 Calendering rollers for embossing effect.

mercerized fabric than unmercerized fabric. In addition, a high degree of lustre is produced on the surface of ordinary swiz-calendered fabric, flattened and smoothened for the processing. The lustre is due to the polishing effect of the roll engravings under the influence of heat and pressure. The degree of lustre can be increased by passing the fabric once or twice through a friction calender prior to embossing. The degree of lustre may also be modified as in friction calender finishes by heating the chilled iron roll, pressure at the nip, speed of the machine, per cent moisture present in the fabric and filler content of the fabric.

It is essential to condition and regulate the moisture content of the fabrics prior to embossing. If excess damping is done, a hard papery hand and

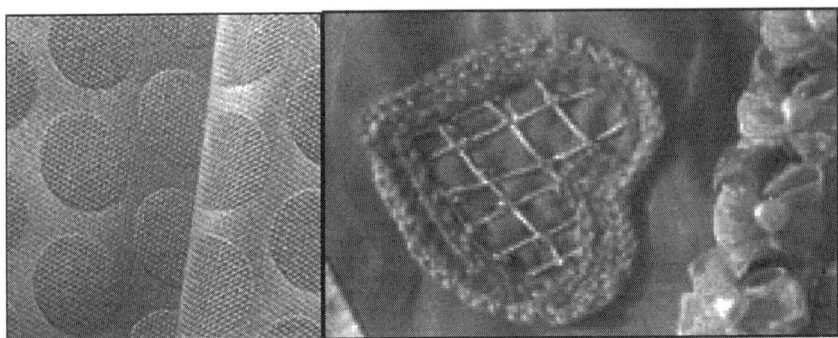

Figure 23.6 Embossed fabrics.

appearance is produced unlike the silky soft hand of the properly embossed fabric (Fig. 23.6). In actual practice, the goods are damped to give slightly more than the normal moisture content and finished on the embossing calender, bringing the moisture content below the normal and afterwards conditioning them to the normal moisture content.

In general, embossing the design engraved on the surface of a hard steel bowl is transferred onto the fabric as a temporary finish. The engraving (as in Schreiner) may be in line cut at an angle with the weft.

Embossing is different from the regular engraving or printing, in which plates are pressed against the surface to create an imprint. In embossing, the pressing raises the surfaces adding a unique and appealing effect to the fabric. Bleached, dyed or printed fabrics singed with a preliminary finish and chemically unmodified fabric are used for embossing. Working pressure depends on the quality of the cloth, the embossing design and different pressing areas of the design. This is durable, with an average life of 50–60 washings.

Embossing in textiles is mainly used in nonwovens such as napkins, diapers, tissue papers and so forth. Apart from this, it is also used in apparels such as t-shirts which gives a very trendy look to the outfit. Additional effects on embossing can be done by decorating the embossed surface with embroidery or screen printing.

There are different types of embossing such as:

a) **Blind emboss:** When the embossed image and the fabric surface are the same, it is termed as blind emboss.

b) **Tint emboss:** Here, a pastel foil or pearl is used.

c) **Single level emboss:** Where the image area is raised to one flat level, it is termed as single level emboss.

d) **Multilevel emboss:** In multilevel emboss, the embossed image is raised to different levels to give a depth to the embossing.

e) **Printed emboss:** In printed emboss, the embossed part registers with a printed image.

f) **Registered emboss:** In registered emboss, the printed image is embossed to give a raised look.

g) **Glazing:** This is a polished emboss used on dark coloured surface. More heat and pressure are applied during embossing which give a shine to the fabric surface. This method is most commonly used for contrasting designs.

Embossing has applications in advertising and marketing industry, where it is widely used for making promotional materials. In home textiles, it is used in decorating curtains, drapes, bed spreads, cushion covers, table mats and many more. In fashion segment, embossing finds way in embellishing t-shirts, caps and casual wears. Logos are pressed in the apparel with embossing techniques. It can be used on fleece, leather, denim, bags and appliqués.

Good quality embossing changes the look of a fabric and gives a breathtaking appearance. It provides a fabulous depth and texture to the outfit. With a subtle and sophisticated appearance, embossing is an innovative technique for creating fashionable garments.

23.4.4 Schreiner

Similar to the watered process, in the Schreiner process the rollers are ribbed and the ribs are very fine with as many as 600 ribs per in under extremely high pressure. Under a pressure of 4500 lb (2045 kg), these lines are pressed into the fabric. The result is that the round threads are pressed flat, but the lines break up the flat surfaces into little planes that reflect the light much better than an ordinary flat surface would. This peculiar light reflection gives the cloth the quality of a very high lustre. Heating the rolls makes this lustre more lasting. The effect is very beautiful. Mercerized cotton finished in the Schreiner finish rivals silk in appearance. Cloth finished with the Schreiner method has a very high lustre, which is made long-lasting by heating the rollers.

A typical Schreiner calender has strong frames carrying two bowls (7.5 cm in diameter) the top one being of special, fine-grained steel, which is engraved with the required number of lines and is heated by gas (Fig. 23.7). The bearings of the top bowl are cooled by water.

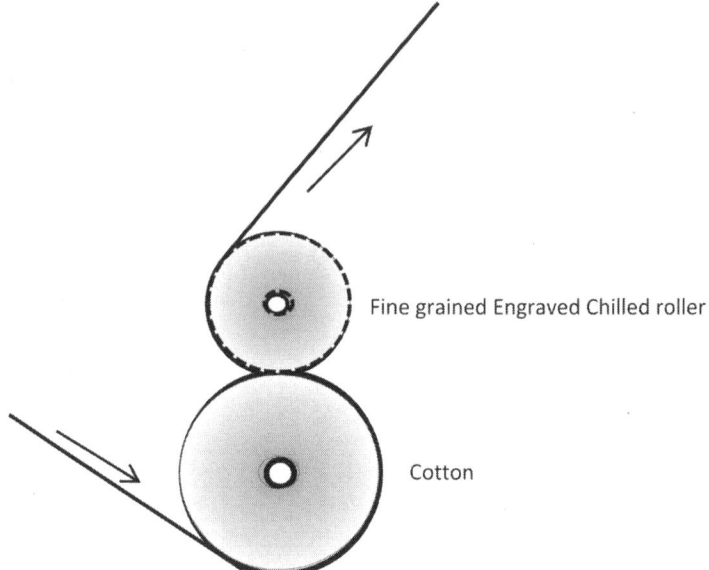

Figure 23.7   Schreiner calender.

The upper roller is in contact with the lower cotton bowl, when the machine is running and is separated while not being used in order to prevent damaging the cotton bowl. The top bowl is fixed in position and the lower one is moved up or down. The required pressure is exerted by hydraulic rams positioned underneath the bearing of the bottom bowl. Very high pressure is exerted on the fabric while it passes through the nip. The calender is provided with a relief valve allowing the bowls to be separated instantly. Skewing arrangement is provided for the bearings of the lower bowl by which the bowls can be set at an angle to the top roll. An enhanced brilliance can be obtained by using the skewing arrangement.

An accumulator with adjustable weight regulates the hydraulic pressure supplied by means of a pump, driven from the calender, which works automatically with the accumulator. The relief valve gives direct communication between the hydraulic cylinders and the accumulator closes the passage and discharges the pressure by returning the water to the pump tank. This valve is operated by a hand lever placed in the convenient reach of the calender operator. The calender is fitted with let-off apparatus, swivel stave rails and batching up or plaiter at the exit end.

A typical two-roll Schreiner calender of John Verdulin Machine Corporation (USA) has 325 cm (130 in) wide rolls, which can exert a pressure of 130 t. It is a modified pressure machine, incorporating a direct current drive system for both the main and winder drives, the latter being operated separately to enable the operation to get any degree of tension desired. The winder drive prevents stretching of the material where it is undesirable. When stretch is not wanted, the winder drive can be adjusted to produce a slight overfeed, keeping the material in a relaxed state. If starch is needed, the speed of the winder drive may be trimmed to give the desired degree of tension. To obtain zero tension, the speeds of the let-off and take-ups are equalized. Scorching and scuffing of the middle cotton roll are prevented by the heated steel rolls. In the conventional three-roll Schreiner calender operation through a slip clutch, the steel rolls are heated to 450°C in contact with the cotton bowl while it is motionless.

Butterworth Universal Calender is designed for rolling, frictioning and Schreinering. A special clutch permits it to be adjusted from one to the other. An off-nip drive automatically functions whenever the nip is open so that the speed of the cotton bowl is maintained. Jack screws are provided in the hydraulic rams to adjust roll parallelism. This is particularly valuable in the case of cotton bowl, which can wear out unevenly. Roller bearing is cooled and lubricated with circulating oil to reduce the danger of bearing damage and to substantially reduce the friction. The end thrust from the rolls is taken by a specially designed roller bearing. Roll pressure may be independently adjusted on either side of the calender. The rolls are designed in comparatively large diameter to reduce deflection under pressure. The large diameter roll provides not only a better quality of finish but also reduces the tendency to displace the material of the cotton bowl towards the end, resulting in longer life of the cotton bowl. The rolls are available in the range 135–225 cm with loads of up to 60 t.

23.4.5 Swizzing effect

Swizzing means simply running the cloth through all the nips and then either plaiting or batching. This operation closes interstices of the cloth and gives it a smooth appearance. Normally, 7-bowl calender is used (Fig. 23.8); however, a 10-bowl calender is used when the production is very large. Prior to calendering, softening and filling agent should be applied (as the finishing mix) to produce lustre. In this machine, all the bowls rotate at the same speed.

In this operation, if a cold calender is used, smooth and some surface lustrous are produced. A hot calender produces smooth and lustrous surface. When all the bowls of a universal calender are used, the fabric acquires a smooth appearance and gloss without the high-glaze characteristic of a friction calendered fabric (Fig. 23.9).

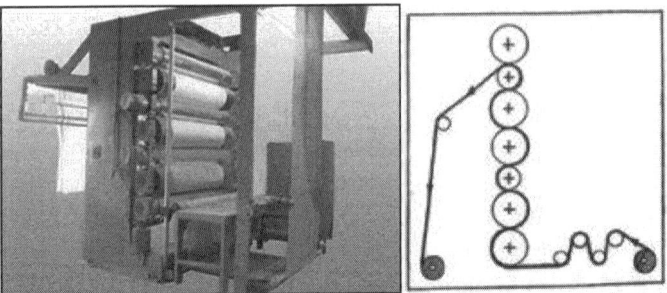

Figure 23.8   Seven-roll calendering machine.

Figure 23.9   Multiple calendering.

23.4.6   Multifunctional up and down double-sided press calendering machine

In Cumins (Hong Kong) Multifunctional up and down double-sided press calendering machine, there are multiple choices for upper and bottom soft roll. The nylon roll is normally preferred by customers. The fabric can be simultaneously pressed by the upper and bottom soft roll making a wrap angle when passing the middle heating roll, resulting in extended heating time. By passing the cooling roll on its way out, along with proper shrinkage techniques, the fabric that goes through double-sided calendering feel thin and looks glossy.

23.4.7   Friction calender

In friction calender, plain calender rollers are used with hot metallic roller in the entry, synthetic roller in the middle and cooling roller at the delivery (Fig. 23.10). The machine is composed of a soft roll in the middle and two

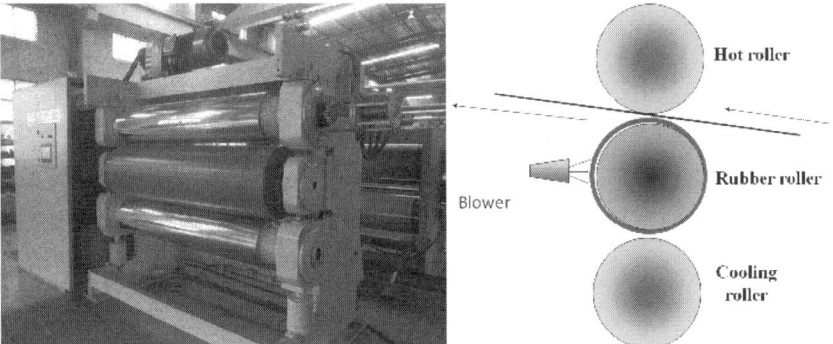

Figure 23.10 Friction calender.

hard rollers on both the sides. The upper hard roll makes polished friction rubbing the fabric along with the middle soft roll, while the bottom hard roll forms hard rolling point with the middle soft roll. The fabric first passes the hard rolling point and then the friction point. The surface speed of the friction roll is faster than the soft roll, ensuring gloss appearance of fabric due to the friction between them. This also remarkably reduces the gap between threads or fibres. The speed ratio between friction roll and fabric can reach 1.2:1. For the friction roll, the friction range can be changed and the temperature usually is from 100 to 160°C.

23.5 Typical working procedure for a friction calender

The standard operating procedure depends on the type of fabric, the finish required and the chemical finishes used earlier. The concerned technical person should prepare the process sheet. Following are general guidelines.

1. Understand the process to be done and the type of fabric

2. Check all roll (cotton, nylon, steel) of calender for crack, holes and so forth

3. Clean the rolls with dry cotton cloth before starting the machine

4. Check the working and positioning of metal detector

5. Check the heating system for getting correct temperature

6. Check the blower. Ensure working of blower on continuous basis

7. Check air pressure and ensure it as per requirement

8. Check the guider entry and exit side

9. Take the parameters by discussing with the supervisor as the speed, pressure and temperatures vary depending on the type of fabric being processed

10. Example of process parameter

	Cotton	Polyester viscose
Calender pressure	45 kg/cm^2	110 kg/cm^2
Temperature	100°C	125–130°C
Speed	45–50 m/min	12–22m/min

11. Set the speed depending on the type of material being produced, viz., cotton, linen, polyester, polyester-viscose, ploy-wool and so forth.

12. Feed the fabric as per plan

13. Change the new batch on the running machine with minimum time. Do not stop the machine

14. Inspect the fabric at the delivery end

15. Check the fabric tension at the delivery end.(fabric should run without tension)

16. Cover the batch with polyethylene paper at the delivery end and keep at proper place

17. Keep proper tag or batch card on the batch

18. Check for damages such as crease, chapti (flat) crease and stains.

23.5.1 Terminating the production of calender machine

1. Blower fan should work continuously till it gets cooled

2. Close heating system and air pressure valve

3. Steel roll, nylon roll, cotton roll should be moving slowly until it comes down to room temperature

4. Once temperatures of the steel, nylon, cotton rolls come down to room temperature, switch off the electrical and close all lines

5. Clean the steel, nylon, cotton rolls during cooling

6. Cover the rolls after they get cooled down properly

7. All batches are to be covered with polyethylene paper

23.6 Control points and checkpoints

It is essential to have clarity on the points to be controlled to achieve the targets and those to be checked to ensure the process in control. These points need to be reviewed from time to time and modified to suit the requirements of individual companies and their targets. Each mill should prepare its own 'control points and checkpoints' and display them in the work area so that the people on spot refer and follow. Following are just illustrations.

23.6.1 Control points

a) Selection of suitable calendering process parameter for the material to be processed considering the customer requirements.

b) Deciding and selection of process parameters, namely:

- Speed of top and bottom rollers and their ratio
- Temperature
- Pressure
- Pretreatment before calendering

c) Deciding acceptance criteria for quality of calendering

d) Employing trained and qualified employees

e) Evolving production norms

23.6.2 Checkpoints

- **Material related**

  a) The fabrics received for calendering against the plan received

  b) The quantity of fabric received for calendering—metres and kilograms

- **Machine related**

  a) The condition of the machine

  b) The surface of steel rollers and synthetic/cotton/rubber rollers

  c) The working of various valves and controllers

  d) Working of metal detector

  e) Proper working of heating system

  f) Check air pressure and ensure it as per requirement

  g) Check the guider entry and exit side

- **Setting related**

    a) The recipe prepared for pretreating before calendering

    b) The temperature set

    c) The speed set for each roller

    d) The pressure set

- **Performance related**

    a) Uniformity of calendering throughout the fabric

    b) The production achieved

- **Documentation related**

    a) The design number of the fabric

    b) Type of calendering done

    c) Quantity calendered

- **Work practice related**

    a) Thorough cleaning of the machine before starting

    b) Maintaining the speed, pressure and temperature as per norms

- **Logbook related**

    a) Machines working

    b) Starting time of the running lots

    c) Activities done and to be done

- **Management information system related**

    a) Machine number

    b) Design number

    c) Number of metres

    d) Weight in kilograms

    e) Chemicals used

- **General**

    a) Use of safety guards such as nip guards, safety valves.

23.7 Dos and Don'ts

It is necessary to understand clearly what is supposed to be done without fail and what should not be done at any cost. Some examples are given below.

23.7.1 Dos

a) Decide the type of calendering depending on the fabric and the requirement of the customer.

b) Check the condition of the synthetic rollers and steel rollers and decide on buffing (Fig. 23.11).

c) Check the guider entry and exit side and see that fabric does not come to the sides.

d) Check the working of various valves and controllers and ensure they are perfect before starting the machine.

e) Ensure that blower fan is working continuously.

f) Check the performance of calender by visual observation.

g) Check the working of the metal detector and ensure it as perfect.

h) Ensure that all rolls such as steel, roll, cotton rolls should be moving slowly until it comes down to room temperature.

i) Clean all rolls during cooling.

j) Cover the rolls after they are fully clean.

k) Cover the calendered batch with polythene sheet.

23.7.2 Don'ts

a) Do not use a roll which has impression on it (Fig. 23.12).

Figure 23.11 Synthetic roller.

Damaged synthetic roller

Figure 23.12   Damaged synthetic roller.

b)   Do not switch off the blower fan or else the synthetic roller gets de-
shaped due to heat. Switch off the blower fan only after the rollers
are fully cool.

c)   Do not stop the machine with heating 'on' and pressure 'on'.

d)   Do not expose the calendered fabrics to humid weather.

Printing of textile materials

24.1 Introduction

Textile printing is the process of applying colour to the fabric in definite patterns or designs. It is a part of wet processing, which is carried out after pretreatment of fabric or after dyeing. It is done for producing attractive designs on the fabric.

In properly printed fabrics, the colour is bonded with the fibre so as to resist washing and friction. It happens when dyes and pigments are applied properly on fibres. A strong bonding is produced between dye and fibre.

Textile printing is related to dyeing, whereas in dyeing, the whole fabric is uniformly covered with one colour. In printing, one or more colours are applied to the fabric in certain parts only, and in sharply defined patterns. Printing is therefore called as localized dyeing. The dyes and pigments are applied locally or discontinuously. Colourants used in printing contain dyes thickened to prevent the colour from spreading by capillary attraction beyond the limits of the pattern or design.

24.2 Steps in printing

The normal steps in fabric printing may be given as follows:

a) Fabrics are pretreated before printing

b) Printing paste is prepared by using printing ingredients. The performance of printing depends on a well-prepared paste

c) Making an impression of the print paste on the fabric by any method of printing

d) Drying the printed fabric

e) Steaming the fabric for fixing the printing paste

f) Neutralizing the fabric after the treatment process

Some examples of flowcharts for printing different fabrics and styles are given below:

24.2.1 Printing flowchart of 100% synthetic fabrics

Synthetic fabric printing is easier than natural fabric printing. Pretreatment is not required for synthetic fabric as it does not contain impurities. The printing flowchart depends on the fibre. Following is the basic chart (Fig. 24.1).

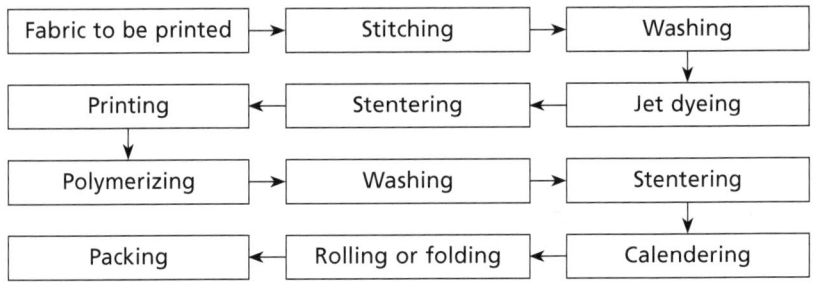

Figure 24.1   Flowchart for printing synthetic fabrics.

Synthetic fibres are produced by polymerizing different monomers. After printing, finishing is done for adding some properties. Therefore, it requires understanding of the complete sequence and chemicals.

24.2.2 Printing flowchart of 100% cotton fabrics

Cotton printing is the commonly used process. Cotton being a natural fibre contains a number of impurities. Hence, cotton fabrics need to be pretreated like scouring before printing. After printing, finishing is done to improve the outlook of the printed fabrics. The general flowchart is as follows (Fig. 24.2):

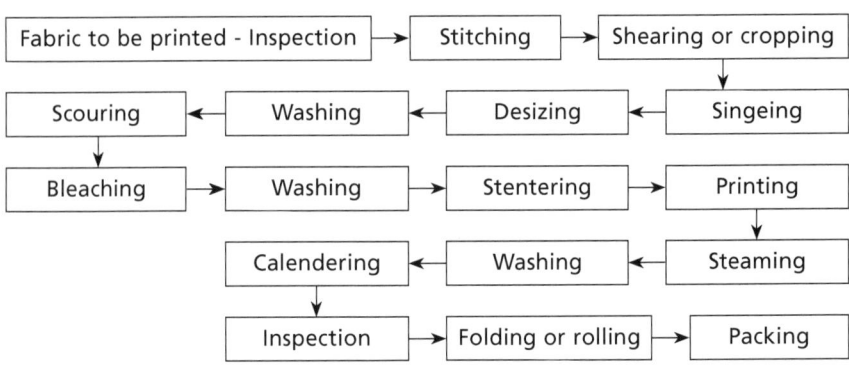

Figure 24.2   Flowchart for printing cotton fabrics.

The above flowchart is for non-mercerized fabrics. In case of mercerized fabrics, mercerizing, neutralizing and washing are done between washing and stentering process, that is, after scouring and bleaching.

24.2.3 Printing process flow of plastic solution (high-density printing process)

Following is the sequence for printing with high-density printing paste (Fig. 24.3).

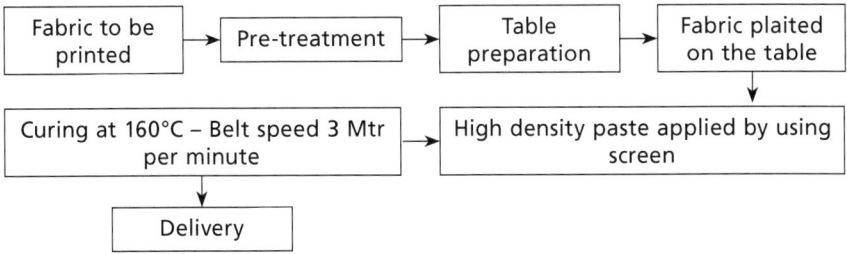

Figure 24.3 Flowchart for high-density printing paste.

Printing with the plastic solution is used for flock printing, foil printing, pigment printing and discharge printing.

24.2.4 Sequence of discharge printing process on cotton: discharge printing style

Printing is done by direct discharge style. Among all the styles, white discharge and colour discharge are common. Discharge means removing specific coloured area by another colour or reducing by a bleaching agent.

Discharge printing is also called as extract printing. By this printing process, colour is destroyed by one or multiple colours. Discharge printing can be done on cotton. Synthetics and blends are not suitable for this style of printing. The dyes used for printing must be suitable for discharging. The suitable recipe needs to be prepared. Following is a general flowchart (Fig. 24.4):

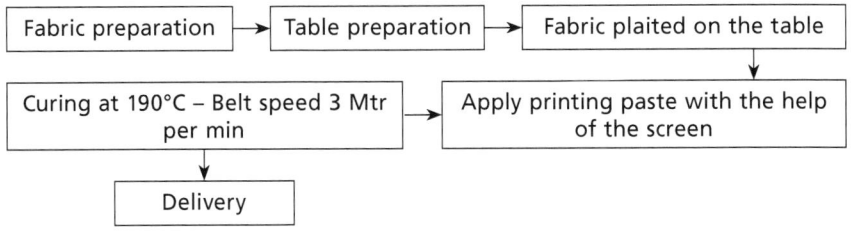

Figure 24.4 Process flow of discharge printing.

24.2.5   Process flow of emboss printing: pub printing process

Emboss printing is done for special logo printing or decorative purpose. It is done by embossing the printing paste on the textile material.

Sample recipe → rubber paste—49%, pub/emboss—49% and fixer—2%

Following is the general process (Fig. 24.5):

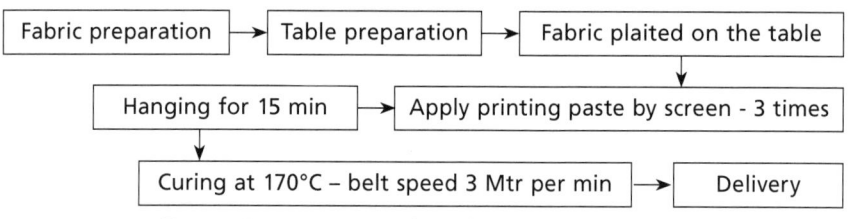

Figure 24.5   Process flow for emboss printing.

24.2.6   Sequence of glitter printing process on textile materials

Glitter is a transfer printing process. By this, different types of decorative designs are produced on T-shirt, polo shirt and other garments. After pretreatment of the fabric printing is done, printing glitter paste is applied to the fabric by screen printing process. After printing, curing is done at a high temperature (Fig. 24.6). Curing should be done slowly otherwise it may affect the printing performance.

Recipe used: rubber paste 70%, glitter 28% and fixer 2%

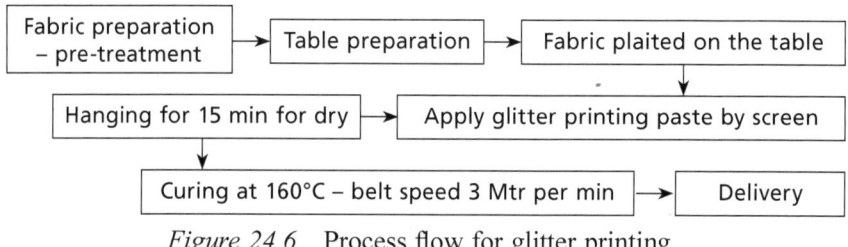

Figure 24.6   Process flow for glitter printing.

24.2.7   Process flow of Inkjet printing

Following is the general process flow for inkjet printing (Fig. 24.7):

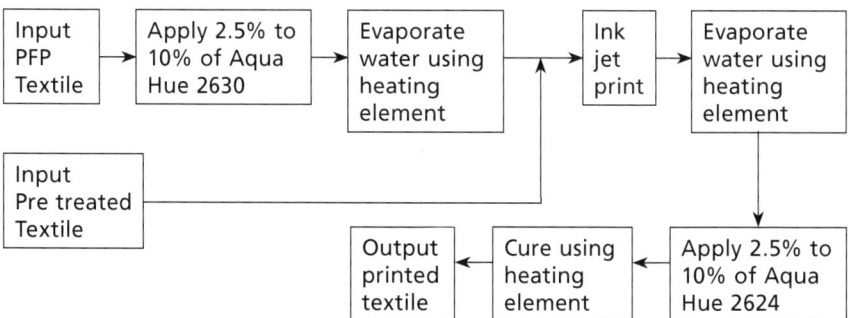

Figure 24.7   Process flow for inkjet printing.

24.3   Traditional styles of printing

Traditional textile printing techniques may be broadly categorized into following styles:

- **Direct printing**: In direct printing, colourants containing dyes, thickeners and the mordants or substances necessary for fixing the colour of the cloth are printed in the desired pattern. This is a process wherein the colours for the desired designs are applied directly to the white or dyed cloth as distinguished from discharge printing and resist printing.

- **Overprinting**: In the direct printing, a common approach is to apply a colour pattern onto a grey or bleached fabric. If done on coloured fabric, it is known as overprinting. The desired pattern is produced by pressing dye on the fabric in a paste form. To prepare the print paste, a thickening agent is added to a limited amount of water and dye is dissolved in it. Earlier starch was preferred as a thickening agent for printing. Nowadays, gums or alginates derived from seaweed are preferred as they allow better penetration of colour and are easier to wash out.

- **Mordant printing**: The mordant printing incorporates printing of a mordant in the desired pattern prior to dyeing cloth. The colour adheres only where the mordant is printed.

- **Resist dyeing**: In resist printing, a wax or other substance is printed onto fabric which is subsequently dyed. The waxed areas do not accept the dye, leaving uncoloured patterns on a coloured ground.

- **Discharge printing**: In discharge printing, a bleaching agent is printed onto previously dyed fabrics to remove some or the entire colour. In 'white' discharge printing, the fabric is piece dyed, then printed with a paste containing a chemical that reduces the dye and hence removes the colour where the white designs are desired. In 'coloured' discharge printing, a colour is added to the discharge paste in order to replace the discharged colour with another shade.

• **Resist printing**: Resist printing is a method in which the design can be produced either by applying a resist agent in the desired design, and then dyeing the fabric, or by including a resist agent and a dye in the paste which is applied for the design. In the first case, the design remains white although the rest of the fabric is dyed. In the second case, the colour of the design is not affected by subsequent dyeing of the fabric background. In this technique, a resist paste is fixed onto the fabric and then it is dyed. The dye affects only those parts that are not covered by the resist paste. After dyeing, the resist paste is removed leaving a pattern on the background of the fabric.

• **Pigment printing**: Printing is done by the use of pigments instead of dyes. The pigments do not penetrate the fibre but are affixed to the surface of the fabric by means of synthetic resins which are cured after application to make them insoluble. The pigments are insoluble, and application is in the form of water-in-oil or oil-in-water emulsions of pigment pastes and resins. The colours produced are bright and generally fat except for crocking. Most pigment printing is done without thickeners because the mixing up of resins, solvents and water produces thickening anyway.

24.4 Methods of printing

Printing is carried out with different instruments. Different methods are used to produce an impression on fabrics. Method of printing depends on the demand of the user and the quantity to be printed. It also depends on the type of material and the end use of the printed product. Following methods can be used for printing of a fabric:

1. Block printing
2. Burnout printing
3. Blotch printing
4. Digital printing
5. Duplex printing
6. Engraved roller printing

7. Electrostatic printing

8. Flock printing

9. Inkjet printing

10. Jet spray printing

11. Photo printing

12. Photographic printing

13. Screen printing—flat screen

14. Rotary screen printing

15. Stencil printing

16. Spray printing

17. Transfer printing

18. Warp printing

19. Special methods—tie and dye, batik printing, space dyeing and so on

24.4.1 Block printing

The block printing is used from time immemorial and is still being practiced in all Asian countries as it can be done even at homes by the people in their free time. The blocks are usually made of wood and the design is hand carved so that it stands out in relief against the background surface. The print paste is applied to the design surface on the block and the block is then pressed against the fabric. The process is repeated with different designs and colours until the pattern is complete.

Block printing is a slow and laborious process and is not suitable for high-volume commercial use (Fig. 24.8). This process is the most artistic and the earliest, simplest and slowest of all methods of printing. In this process, a design is drawn on, or transferred to, a prepared wooden block. A separate block is required for each distinct colour in the design.

A block-cutter carves out the wood around the heavier masses first, leaving the finer and more delicate work until the last so as to avoid any risk of injuring it during the cutting of the coarser parts. When finished, the block presents the appearance of flat relief carving, with the design standing out.

Fine details are very difficult to cut in wood, and, even when successfully cut, wear down very rapidly or break off in printing. They are, therefore, almost invariably built up in strips of brass or copper, bent to shape and driven edgewise into the flat surface of the block. This method is known as coppering.

Figure 24.8   Block printing.

To print the design on the fabric, the printer applies colour to the block and presses it firmly and steadily on the cloth, ensuring a good impression by striking it smartly on the back with a wooden mallet. The second impression is made in the same way, the printer taking care to see that it fits exactly to the first, a point which he can make sure of by means of the pins with which the blocks are provided at each corner and which are arranged in such a way that when those at the right side or at the top of the block fall upon those at the left side or the bottom of the previous impression, the two printings join up exactly and continue the pattern without a break. Each succeeding impression is made in precisely the same manner until the length of cloth is fully printed. When this is done, it is wound over the drying rollers, thus bringing forward a fresh length to be treated similarly.

If the pattern contains several colours, the cloth is usually first printed throughout with one, then dried, and printed with the second, the same operations being repeated until all the colours are printed.

The worker working on block printing should be highly skilled to place the blocks exactly in the place as needed.

24.4.2   Burnout printing

Burnout printing is a method of printing to obtain a raised design on a sheer ground. The design is applied with a special chemical onto a fabric woven of pairs of threads of different fibres. One of the fibres is then destroyed locally by chemical action. Burnout printing is often used on velvet. The product of this operation is known as a burnout print.

24.4.3 Blotch printing

Blotch printing is a process wherein the background colour of a design is printed rather than dyed. The result is that the reverse side of the fabric is typically white. This is a direct printing technique where both the background colour and the design are printed onto a white fabric, usually in one operation. Any methods such as block, roller or screen may be used. The ground colour is transferred from the cylinder and the motif retains the original hue of the cloth.

24.4.4 Digital printing

Digital printing is one of the most exciting developments in the textile industry (Fig. 24.9). It not only opens up endless opportunities for customization, small run printing, prototyping and experimentation but it also puts textile printing within the budget of average illustrator. Digital textile printing can reproduce unlimited colours and shades but—as with most forms of printing—what one sees on screen is not necessarily what he gets back.

Digital textile printing is described as any inkjet-based method of printing colourants onto fabric. Digital textile printing is normally referred to when identifying either printing smaller designs onto garments (direct to garment) or printing larger designs onto large format rolls of textile. The latter is a growing trend in visual communication, where advertisement and corporate branding is printed onto polyester media such as flags, banners, signs and retail graphics.

In this form of printing, micro-sized droplets of dye are placed onto the fabric through an inkjet printhead. The print system software interprets the data supplied by academic textile digital image file. The digital image file has the data to control the droplet output so that the image quality and colour control can be achieved.

The inks used in digital printing are formulated specifically for each type of fibre (cotton, silk, polyester, nylon, etc.). During the printing process, the fabric is fed through the printer using rollers and ink is applied to the surface in the form of thousands of tiny droplets. The fabric is then finished using heat and/or steam to cure the ink (some inks also require washing and drying). Digitally printed fabric will wash and wear the same as any other fabric, although with some types of ink you may see some initial fading in the first wash.

We shall discuss digital printing in detail in a separate chapter.

Figure 24.9   Digital printing.

24.4.5   Duplex printing

Duplex printing is a method of printing a pattern on the face and the back of a fabric with equal clarity. Printing is done on both sides of the fabric either through roller printing machine in two operations or a duplex printing machine in a single operation.

24.4.6   Engraved roller printing

In this method, engraved copper cylinders or rollers are used in place of hand-carved blocks (Fig. 24.10). When the rollers move, a repeat of the design is

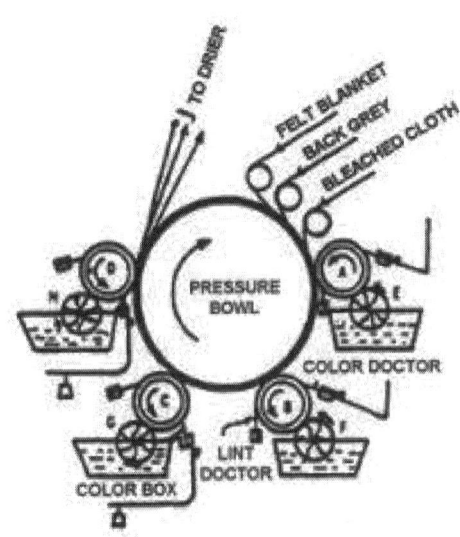

Figure 24.10   Roller printing.

printed on the fabric. The printed cloth is passed into a drying chamber and then in a steam chamber where the moisture and heat set the dye.

Roller printing is preferred for long production runs because of its very high speeds. It is also a versatile technique; up to a dozen different colours can be printed simultaneously. The basic roller printing equipment, shown in figure consists of a number of copper-faced rollers in which the design is etched. There is a separate printing roller for each colour being printed. Each of the rollers rotates over the fabric under pressure against an iron pressure roller. A blanket and backing cloth rotate over the pressure roller under the fabric and provide a flexible support for the fabric being printed. A colour doctor blade removes paste or fibres adhering to the roller after contact with the fabric. After the impression stage, the fabric passes to the drying and steaming stages.

24.4.7 Electrostatic printing

In electrostatic printing, a plate with electrostatic charge (to attract powdered dyes or ink into the fabric) is positioned behind the fabric and a stencil of the design to be printed is positioned between the fabric and the powder supply so the design is applied in the correct area (Fig. 24.11). A dye–resin mixture is spread on a screen bearing the design and the fabric is passed into an electrostatic field under the screen. The dye–resin mixture is pulled by the electrostatic field through the pattern area onto the fabric.

Electrophotographic printing system (electrostatic printing of fabrics) developed by Kyoto Municipal Institute of Industrial Technology and Culture in Kyoto in collaboration with the Denatex Workshop of Nagase & Co. in Amagasaki City, Hyogo prefecture is claimed to overcome the drawbacks of conventional inkjet printing systems. The principle behind the electrophotographic printing system is basically the same as that of a copier or a laser printer, 'A digital image composed of colours and designs is projected onto an electrostatically charged and revolving photosensitive drum, which is printed onto a fabric using cyan-, magenta-, yellow-, black-coloured dye toners through a printing paper'.

The amount of water used for dyeing and rinsing for a standard procedure in the printing of fabrics reaches approximately 75 kg in the traditional screen printing method or 20–30 kg for the conventional digital inkjet printing. In contrast, the electrostatic dye-sublimation and transcription system employing a new type of dye toner discharges neither water nor pollutants such as dyestuffs. Moreover, the new electrographic printing system achieved resolutions higher than 720 dpi allowing the printing of extra fine and colourful

Figure 24.11   Electrostatic printing—clothing printed using DENATEX, as exhibited in the JFW International Fashion Fair in Tokyo in July 2012 (the fabrics demonstrate the sort of vivid rainbow colours that cannot be realized without an electrostatic dye-sublimation printer).

designs of photograph quality, and at a printing speed of 10 m/min, which is about five times faster than the inkjet method of similar electrostatic dye-sublimation and transcription systems.

A textile material whose fibres have been coated, at least in part, with fluorocarbon particles is usable in an electrophotographic printing machine to clean toner particles off a fuser roll, and to supply a toner release agent to the fuser roll. The textile material can include woven goods, as well as non-woven felts and the like. The resultant product has reduced friction and decreased fibre shedding

24.4.8 Flock printing

Flocking is the technique of depositing many small fibre particles, called 'flock', onto a surface of a fabric to produce the design (Fig. 24.12). Instead of dyes, an adhesive is used to affix the flocks on the fabric. Then, roller printing produces design on its surface. Nowadays, this is done by the application of

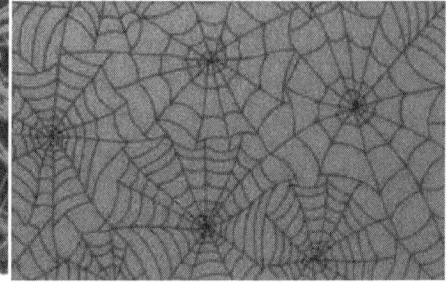

Figure 24.12   Flock printing.

high-voltage electric field as well. Flocks of cotton, wool, rayon, nylon and acrylic are all used for the purpose. Electrostatic flocking is applying flock to a textile fabric, covered or printed with adhesive in the presence of a strong electrostatic field. The field orients the flock fibres and accelerates them, and as they approach the fabric, they form a firmly attached pile perpendicular to the fabric surface.

In flock printing technique, very small particles of a fibre are stuck to the surface of the fabric to create a design or a pattern. These particles of fibre are stuck by means of a strong adhesive which is water-resistant so as to make it washable. Flock printing needs proper technological assistance for sticking the fibre on to the fabric. For this, techniques used are enclosing the fibres with the glue paste, dust it onto the glue paste or apply it electrostatically in order to hold it erect.

Floral prints are most appropriate for flock printing because the patterns are small and flock printing is ideal for small prints. Fabrics for textile flock printing are many but the most ideal and suitable type of fabric for flocking is lightweight or sheer fabric. Customization is possible for flocking as per the fibre used, the technique of flocking, colours, print and design and fabric used.

24.4.9 Inkjet printing

Inkjet printing is a non-impact printing. No printing gadgets such as block, screen or roller touch the fabric. Colour is sprayed from a distance. There has been considerable interest in the technology surrounding non-impact printing, mainly for the graphics market, but the potential benefits of reductions in the timescale from original design to final production have led to much activity in developing this technology for textile and carpet printing processes. The types of machines developed fall into two classes, drop-on-demand and continuous stream.

'Inkjet printing on fabric' is a way anyone can print on fabric using their home printer. Specially treated cotton, as well as various types of bamboo and silk fabric sheets, is available in various sizes. The fabric sheets have a paper backing which enables the fabric to go through the inkjet printer. Family photos printed on fabric are used to make memory quilts, pillows, notebook covers, wall hangings, ornaments and many other products. The printed fabric is dipped in water to set the ink after the inkjet ink dries, making it washable.

24.4.10 Jet spray printing

Jet printing is a non-contact application system originally developed for printing carpets, but now increasingly used in the textile sector. Designs are imparted to fabrics by spraying colours in a controlled manner through nozzles. Spray printing systems and first-generation jet printing methods cannot be controlled to produce a pre-specified pattern. Thus, the equipment must first be employed to produce a wide range of effects and only then can selections be made from these by the designer or marketing staff.

The first commercial jet printing machine for carpets was the Elektrocolor, followed by the Millitron. In the Millitron printing system, the injection of the dye into the substrate is accomplished by switching on and off a dye jet by means of a controlled airstream. As the carpet moves along, no parts of the machine are in contact with the face of the substrate. Airstreams are used to keep continuously flowing dye jets, deflected into a catcher or drain tray. This dye is drained back to the surge tank, filtered and recirculated. When a jet is requested to fire, the air jet is momentarily switched off, allowing the correct amount of dye to be injected into the textile substrate. The dye is supplied in continuous mode to the main storage tank to compensate for the amount of dye consumed.

An early improvement was made by the first digital carpet printers (Chromotronic and Titan by Zimmer and Tybar Engineering, respectively). These machines are based on the 'drop-on-demand principle'. The switchable electromagnetic valves placed in the dye liquor feed tubes are used to allow the jetting of discrete drops of dye liquor in a predetermined sequence according to the desired pattern. Although the amount of dye applied can be digitally controlled at each point of the substrate, further penetration of the dye into the substrate is still dependent on the capillary action of the fibre and fibre surface wetting forces. This can lead to problems of reproducibility (e.g. when the substrate is too wet) and means that it is still necessary to use thickeners to control the rheology of the dye liquor.

The latest improvement in the jet printing of carpet and bulky fabrics is by machines in which the colour is injected with precision deep into the face of the fabric without any machine parts touching the substrate. Here, the control of the quantity of liquor applied to the substrate (which may vary, for example, from lightweight articles to heavy quality fabrics) is achieved by varying not only the firing time but also the pumping pressure. This system is like an injection dyeing process. The name 'injection dyeing' is used as a commercial name to define the technology applied on the latest Milliken's Millitron machine. Another digital jet printing machine commercially available is Zimmer's ChromoJet. In the ChromoJet system, the printing head is equipped with 512 nozzles. These are magnetically controlled and can open and close up to 400 times/s. The carpet is accumulated into a J-box, and is then

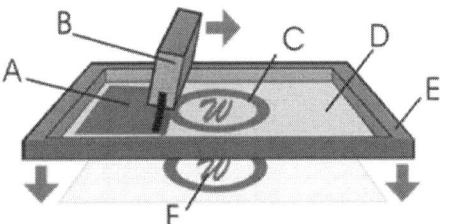

A – Rubber wiper blade
B – Handle for wiper
C – Design prepared on mesh (screen)
D – Nylon mesh
E – Frame of the screen
F – Design printed on the fabric

Figure 24.13 Screen printing.

steamed and brushed. When it reaches the printing table, it is stopped. The jets are mounted on a sliding frame that can itself be moved in the direction of the warp while the carpet remains stationary during the printing process.

24.4.11 Photo printing

In the photo printing, the fabric is coated with a chemical that is sensitive to light and then any photograph may be printed on it. The controlled light passes to the fabric through negative or a photo film, which allows the light to fall on the fabric as per the details in the photo.

24.4.12 Photographic printing

Photographic printing is a method of printing from photoengraved rollers. The resultant design looks like a photograph. The designs may also be photographed on a silk screen which is used in screen printing.

24.4.13 Screen printing

Screen printing is done either with flat or cylindrical screens made of silk threads, nylon, polyester or metal (Fig. 24.13). The printing paste or dye is poured on the screen and forced through its unblocked areas onto the fabric. Based on the type of the screen used, it is known as 'flat screen printing' or 'rotary screen printing'.

This type of printing has increased enormously in its use in recent years because of its versatility and the development of rotary screen printing machines which are capable of very high rates of production. An additional significant advantage is that heavy depths of shade can be produced by screen printing, a feature which has always been a limitation of roller printing because of the restriction to the amount of print paste which can be held in the shallow depth of the engraving on the print roller. Worldwide, about 61% of all printed textile fabric is produced by the rotary screen method and 23% by flat screen printing.

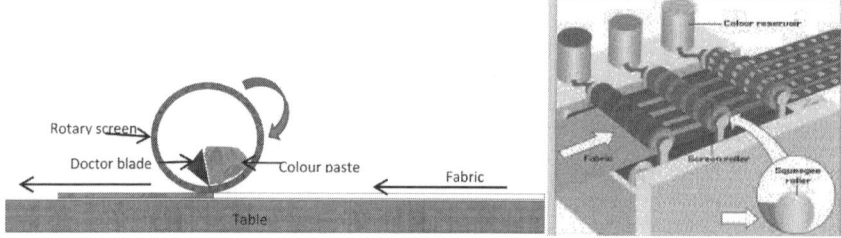

Figure 24.14   Rotary screen printing.

24.4.14   Rotary screen printing

The rotary screen is a screen in a cylindrical form (Fig. 24.14). The colour is applied from inside while the rotary screen is revolving.

There are three main advantages possessed by rotary printing over other flat screens: first, its high productivity, 10 000–12 000 yd being commonly printed in one shift of 8 h; second, by its capacity of being applied to the reproduction of every style of design; and third, the wonderful exactitude with which each portion of an elaborate multicolour pattern can be fitted into its proper place without faulty joints at its points of repetition.

The main disadvantage is that it is not economical if the orders are small and we need to change the design very frequently. It is very difficult to get orders in same design in lakhs of metres.

24.4.15   Stencil printing

The art of stencilling is not new. It has been applied to the decoration of textile fabrics from time immemorial by the Japanese. In stencil printing, the design is first cut in cardboard, wood or metal. The stencils may have fine delicate designs or large spaces through which colour is applied on the fabric. The pattern is cut out of a sheet of stout paper or thin metal with a sharp-pointed knife, the uncut portions representing the part that is to be reserved or left uncoloured. The sheet is now laid on the material to be decorated and the colour is brushed through its interstices.

The peculiarity of stencilled patterns is that they have to be held together by ties, that is to say, certain parts of them have to be left uncut so as to connect them with each other and prevent them from falling apart in separate pieces. For instance, a complete circle cannot be cut without its centre dropping out, and consequently, its outline has to be interrupted at convenient points by ties or uncut portions.

24.4.16   Airbrush (spray) printing

In this method, the dye is applied with a mechanized airbrush which blows or sprays colour on the fabric. A spray gun forces the colour through a screen; an electrocoating is used to apply a patterned pile.

24.4.17   Heat transfer printing

In heat transfer printing method, the design on a paper is transferred to a fabric by vaporization. There are two main processes for this: dry heat transfer printing and wet heat transfer printing. Various types of cylinders such as electrically heated cylinder, perforated cylinder and so forth are used for pressing a fabric against a printed paper which transfers the pattern to the fabric.

Transfer printing techniques involve the transfer of a design from one medium to another. The most common form used is heat transfer printing, in which the design is printed initially on to a special paper using conventional printing machinery. The paper is then placed in close contact with the fabric and heated, when the dyes sublime and transfer to the fabric through the vapour phase.

24.4.18   Warp printing

The printing of a design is done on the sheet of warp yarns before weaving. The filling is either white or neutral in colour, and a greyed effect is produced in the areas of the design.

24.4.19   Special methods

There are a number of special methods of dyeing apart from the ones discussed above. Following are some examples:

Tie dyeing: In tie dyeing, firm knots are tied in the cloth before it is immersed in a dye (Fig. 24.15). The outside portion of the immersed fabric is dyed but

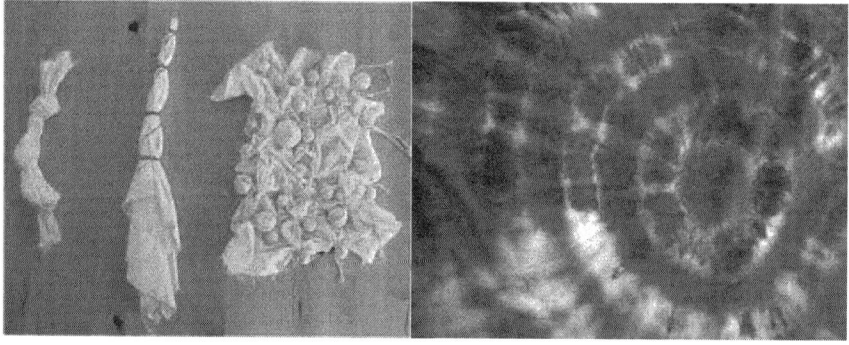

Figure 24.15   Tie dyeing.

Figure 24.16 Batik dyeing.

the colour does not penetrate the inside portions of the tied knots. There are various forms of tie dyeing such as ikat dyeing where bundles of warp and/or weft yarns are tie dyed prior to their weaving. In plangi dyeing, the gathered, folded or rolled fabric is usually held with stitching to form specific patterns.

Batik dyeing: It is a resist dyeing process (Fig. 24.16). Designs are made with wax on a fabric which is then immersed in a dye. The portion not having wax absorbs the colour.

Space dyeing: Space dyeing is a method of printing yarns using jet spray of colours. In space dyeing machines, 64 jet sprays are provided and 8 colours can be accommodated. There cannot be any uniformity or repetition of designs when these yarns are woven or knitted. It gives a special effect that is unique.

Kalamkari: Fabric is painted using a pen with dyes and mordants. Printing the outline of the design and filling inside with a pen (kalam) combine the printing and art with pen, that is, kalamkari. Kalamkari is an exquisite ancient craft of painted and printed fabrics practiced in Indian temples (Fig. 24.17).

Figure 24.17 Kalamkari.

24.5 Preparation of cloth for printing

Cotton goods intended for printing ought to be exceptionally well-bleached, otherwise stains and other serious defects are certain to arise during subsequent operations.

The chemical preparations used for special styles will be basic on the dyes and the fixing method employed. A general procedure employed for most colours that are developed and fixed by steaming only consists in passing the bleached fabrics through a weak solution of sulphated or Turkey red oil containing from 2.5 to 5% of fatty acid. Some colours are printed on pure bleached cloth but all patterns containing alizarin red, rose and salmon shades are considerably brightened by the presence of oil.

Apart from wet preparations, the cloth needs to be brushed always to free it from loose nap, flocks and dust that it picks up while being stored. Shearing by passing over rapidly revolving knives arranged spirally around an axle rapidly and effectually cuts off all filaments and knots, leaving the cloth perfectly smooth and clean and in a condition fit to receive impressions of the most delicate engraving. Some figured fabrics, especially those woven in checks, stripes and crossovers, require very careful stretching and straightening on a special machine, known as a stenter before they can be printed with certain formal styles of the pattern which are intended in one way or another to correspond with the cloth pattern. Finally, all descriptions of cloth are wound round hollow wooden or iron centres into rolls of convenient size for mounting on the printing machines.

24.6 Preparation of colours

The art of making colours for textile printing demands both chemical knowledge and extensive technical experience. The ingredients must not only be properly proportioned to each other but they must be specially chosen and compounded for the particular style of work in hand as well. For a pattern containing only one colour, any mixture may be used as long as it fulfils all conditions as to shade, quality and fastness; but where two or more colours are associated with the same design, each must be capable of undergoing without affecting the various operations necessary for the development and fixation of the others.

All printing pastes whether containing colouring matter or not are known technically as colours and are referred to as such in the sequence. Colours vary considerably in composition. The greater number of them contains all the elements necessary for the direct production and fixation of the colour lake.

Some contain the colouring matter alone and require various after-treatments for its fixation, and others are simply mordants thickened. A mordant is a metallic salt or other substance that combines with the colouring principal to form an insoluble colour lake, either directly by steaming or indirectly by dyeing.

All printing colours require thickening, for the twofold object of enabling them to transfer from colour box to cloth without loss and to prevent them from running or spreading beyond the limits of the pattern.

24.7 Selecting thickening agents

The printing thickeners used depend on the printing technique, the fabric and dyestuff used. Typical thickening agents are starch derivatives, flour, gum arabic, guar gum derivatives, tamarind, sodium alginate, sodium polyacrylate, gum senegal and gum tragacanth, British gum or dextrine and albumen.

Hot-water-soluble thickening agents as starch are made into pastes by boiling in double or jacketed pans, between the inner and outer casings of which either steam or water may be made to circulate, for boiling and cooling purposes. Mechanical agitators are also fitted in these pans to mix the various ingredients together and to destroy lumps and prevent the formation of lumps, keeping the contents thoroughly stirred up during the whole time they are being boiled and cooled to make a smooth paste. Most thickening agents used today are cold soluble and require only extensive stirring.

Starch paste: Starch paste is made from wheat starch, cold water and olive oil, and boiled for thickening.

Nonmodified starch was the most extensively used of all the thickenings. It is applicable to all but strong alkaline or strong acid colours. With the former it thickens up to a stiff unworkable jelly, while mineral acids or acid salts convert it into dextrin, thus diminishing its viscosity or thickening power. Acetic and formic acids have no action on it even at the boil. Nowadays, mostly, modified carboxymethylated cold-soluble starches are used which have a stable viscosity and are easier to rinse out of the fabric and give reproducible 'short' pasty rheology.

Flour paste is made in a similar way to starch paste. It is rarely used except for the thickening of aluminium and iron mordants. In the impressive textile traditions of Japan, several techniques using starch paste resists of rice flour have been perfected over several centuries.

Gums: Gum arabic and gum senegal are both very old thickenings but are expensive. Hence, they are not used for other shades except pale delicate tints.

They are especially useful thickenings for the light ground colours of soft muslins and satins. They dissolve completely out of the fibres of the cloth in the washing process after printing and have a long flowing, viscous rheology, giving sharp print and good penetration in the cloth. Today, guar gum and tamarind derivatives offer a cheaper alternative.

British gum or dextrin is prepared by heating starch. It varies considerably in composition sometimes being only slightly roasted and consequently only partly converted into dextrin, and at other times being highly torrefied, and almost completely soluble in cold water and very dark in colour. Its thickening power decreases and its gummy nature increases as the roasting temperature is raised. The lighter coloured gums or dextrins will make a good thickening with from 2 to 3 lb of gum to 1 gal of water, but the darkest and most highly calcined require from 6 to 10 lb/gal to give a substantial paste. Between these limits, all qualities are obtainable. The darkest qualities are very useful for strongly acid colours, and with the exception of gum senegal, are the best for strongly alkaline colours and discharges. Similar to the natural gums, neither light nor dark British gums penetrate into the fibre of the cloth so deeply as pure starch or flour, and are therefore unsuitable for very dark strong colours.

Gum tragacanth, or dragon, is one of the most indispensable thickening agents possessed by the printers. It may be mixed in any proportion with starch or flour and is equally useful for pigment and mordant colours. When added to the starch paste, it increases its penetrative power, adds to its softness without diminishing its thickness, makes it easier to wash out of the fabric and produces much more level colours than starch paste alone. Used by itself, it is suitable for printing all kinds of dark grounds on goods that are required to retain their soft clothy feel. Tragacanth mucilage may be made either by allowing it to stand a day or two in contact with cold water or by soaking it for 24 h in warm water and then boiling it up until it is perfectly smooth and homogeneous. If boiled under pressure, it gives a very fine, smooth mucilage (not a solution proper), much thinner than if made in the cold.

Starch always makes the printed cloth somewhat harsh (unless modified carboxymethylated starches are used) but are well suited to obtain very dark colours. Gum senegal, gum arabic or modified guar gum thickening are yielding beautifully clear and perfectly even tints comparing to starch, but give lighter colours and are washed away too much during the rinsing or washing of the printed fabric and are thus less suited for very dark colours. The gums are apparently preventing the colours from combining fully with the fibres. Therefore, a printing stock solution is mostly a combination of modified starch and gum stock solutions usually made by dissolving 6 or 8 lb of either in 1 gal of water.

Albumen is both a thickening and a fixing agent for insoluble pigments such as chrome yellow, the ochers, vermilion and ultramarine. Albumen is always dissolved in cold, a process that takes several days when large quantities are required. The usual strength of the solution is 4 lb/gal of water for blood albumen and 6 lb/gal for egg albumen. The latter is expensive and only used for the lightest shades. For most purposes, one part of albumen solution is mixed with one part of tragacanth mucilage, which is sufficient for the fixation of all ordinary pigment colours. In special instances, the blood albumen solution is made as strong as 50%, but this is only in cases where very dark colours are required to be absolutely fast to washing. After printing, albumen thickened colours are exposed to hot steam, which coagulates the albumen and effectually fixes the colours.

Printing thickeners and the dye system: Combinations of cold-water-soluble carboxymethylated starch, guar gum and tamarind derivatives are most commonly used today in disperse screen printing on polyester, for cotton printing with reactive dyes alginates are used, sodium polyacrylates for pigment printing and with vat dyes on cotton only carboxymethylated starch is used.

24.8   Printing paste preparation

Formerly, colours were always prepared for printing by boiling the thickening agent, the colouring matter and solvents and so forth, together, then cooling and adding the various fixing agents. At present, the concentrated solutions of the colouring matters and other adjuncts are simply added to the cold thickenings, of which large quantities are kept in stock.

Colours are reduced in the shade by simply adding more stock (printing) paste. For example, a dark blue containing 4 oz of methylene blue per gallon may readily be made into a pale shade by adding 30 times its bulk of starch paste or gum, as the case may be, similarly with other colours.

Before printing, it is essential to strain or sieve all colours in order to free them from lumps, fine sand and so forth, which would inevitably damage the highly polished surface of the engraved rollers and result in bad printing. Every scratch on the surface of a roller prints a fine line in the cloth. It is difficult to take care of all. As far as possible, all grit and other hard particles from every colour should be removed.

The straining is usually done by squeezing the colour through filter cloths as artisanal fine cotton, silk or industrial woven nylon. Fine sieves can also be employed for colours that are used hot or are very strongly alkaline or acid.

Digital printing of textile fabrics

25.1 Introduction

Digital printing is one of the most exciting developments in the textile industry. It not only opens up endless opportunities for customization, small run printing, prototyping and experimentation but it also puts textile printing within the budget of an average illustrator. Digital textile printing can reproduce unlimited colours and shades. It has one limitation, as with most forms of printing, what we see on a computer screen is not necessarily what we get back.

At present, the textile industry produces the majority of its (around 32 billion m^2) printed textiles by screen printing. However, gradually the developments in digital printing of paper are being adapted more and more for the textile market. Inkjet textile printing is growing while growth in analogue textile printing remains stagnant. As digital print technologies improve offering faster production and larger cost-effective print runs, digital printing is growing to become the technology that provides the majority of the world's printed textiles. The main advantage of digital printing is the ability to do very small runs of each design (even less than 1 yd) because there are no screens to prepare.

The inkjet printing technology used in digital printing was first patented in 1968. In the 1990s, inkjet printers became widely available for paper printing applications. The technology has continued to develop and there are now specialized wide-format printers which can handle a variety of substrates from paper to fabrics, vinyl and so forth.

'Digital' textile printing, an inkjet-based method of printing colourants onto fabric is referred normally when printing smaller designs onto garments abbreviated as DTG, which stands for 'direct to garment'; but now printing larger designs onto large format rolls of textile are also being done. The latter is a growing trend in visual communication, where advertisement and corporate branding are printed onto polyester media such as flags, banners, signs and retail graphics.

Digital textile printing can be divided into:

a) Direct to garment

b) Visual communication

c) Interior decoration

d) Fashion (the 'Como' industry)

25.2 Advantages of digital fabric printing

This technology uses large format digital inkjet printers similar to the technique used by desktop inkjet printers to print paper. Digital inkjet printing has major advantages over conventional textile printing methods as no screens or rollers are required. In digital inkjet printing, print heads, containing banks of fine nozzles, fire fine droplets of individual coloured inks onto a pretreated fabric. The print design is created digitally and the ink droplets are mixed together on the fabric surface to create the final colour. The digital textile printing has various advantages over traditional printing as listed below:

*   Photographic and tonal graphics with multiple shades as well as colours can be printed.

*   There is no limitation on a number of colours.

*   There is no limitation on repeat size.

*   It offers faster processing speed.

*   One can print as little as possible; there is no minimum order quantity as such.

*   High-precision printing is possible.

*   Overall cost of producing a sample is cheaper in comparison to other forms of printing.

*   Permits customers to control the design and printing process from remote locations.

*   Eliminates the expense and time of preparing screen/block/engraved roller/stencils and so forth.

*   The screens or blocks prepared for printing remains unused once the order is over, and people cannot scrap it or cannot keep without orders.

The investment becomes dead investment after the order is over. In digital printing, there is no such dead inventory.

- Makes 'just in time' delivery and quick response possible. Small runs of fabric can easily be printed for sampling purposes.

- Can virtually eliminate the threat of design theft before market release.

- Facilitates the increase in the number of fashion seasons.

- Prints directly from easily stored, transmitted and transported computer files.

- Reduces the space necessary for archiving art, films, plates and screens.

- Is more cost-effective for proof and short-run printing than analogue printing methods.

- Reduces proofing time from weeks to hours.

- Permits customization and personalization. Allows for design correction and modification at any time without significant schedule delays or cost increases.

- No design and process distortions as with on-contact analogue printing.

- It is cleaner, safer, and generally less wasteful, and less environmentally hazardous.

- The textile research and pattern design time can be reduced by over 50%.

- The designer can see how a particular piece of fabric or garment will look in different colours and shapes without having to commit to a final product.

- Every yard printed can be completely customized and personalized.

- Every item in line can come in a different colour.

- No need of fabric inventory—with an on-site printer, it is possible to print fabric on the same day that it will be cut. One can even have a pattern printed directly on the fabric.

25.3 Activity

The computer-aided design (CAD) system is used for the creation of original designs or the interpretation of artwork supplied by customers; the latter being either painted artwork, fabric samples or possibly black and white film negatives.

Design process: Designs can be created digitally with a graphic design software (Photoshop and Illustrator are the most popular). Alternatively, existing artwork or photographs can be scanned and then digitally manipulated to make a pattern. Usually, designs are created as a seamless pattern that is repeated (tiled) across the fabric. One can also create a design that fills an entire yard without repeating but may run into issues if the size of the file is too large for the printing service to process.

Scanning artwork: Generally, the use of a textile design system starts with work created on paper or another medium. Most companies making printed textiles use scanners to capture data in digital form. Scanner can also be used to scan in black and white, artwork and separates. One can also scan black and white film, and then merge them back into a coloured design in design station. The advantage is that a user can undo any element so that mistakes or unwanted effects can instantly be corrected. Other systems use digital cameras as a means of data capture.

Editing of designs: The first step in editing a design is to take the scan file and do some colour reduction to get the file as close as possible to the final number of colours that the finished design will have. Some of this will be done by the colour reduction function in the CAD system, where the computer picks the next closest colour left in the palette to replace the colour to be discarded or can manually choose which colour one wants to replace the colour being discarded. Then, CAD systems are used to manipulate the image, changing colours, adding, deleting, moving or copying elements and putting them overall into a suitable pattern repeat.

Normally, software packages have a wide range of effects such as watercolour and oil which enable different results to be created with the same input devices. The next thing is to put the pattern into repeat. If the design was painted in repeat, it is just a matter of trimming the design into the proper size of the repeat. If it was not in repeat, the CAD system has many functions to help create a repeat in a design. One can add areas around the existing design and fill these areas by copying motifs from the original design. The use of clone brushes is helpful in the blending of wash effect and backgrounds into the new areas of the design or the artist can draw new motifs to complete the repeat. Next step is the cleaning up of individual images or motifs. Here, use of combinations of reduce brushes, dot clear brushes, dot removal and drawing function are made.

Colour work on CAD system: The inkjet printer is used for three different purposes:

1. The proofing of designs in progress. At various stages of editing, inkjet proofs of the design are produced for the computer artist and the stylist to grow over and take decisions as to what still needs to be done to the design.

2. Full-size inkjet proofs are made of the design for customer or product manager approval. This allows one to make comments or changes before any engraving is done.

3. The colouring station is used for the creation of new colourways of patterns and for the customer's approval of these colourways before actual strike-offs done on a fabric.

Printings software: Softwares are developed for digital printing of textiles by different companies. There is been a continuous development, and the readers are suggested to discuss with the software manufacturers and select the best suited one for the purpose. Following are some examples.

Wasatch SoftRIP TX for textile printing

SoftRIP for textile printing developed by Wasatch is equipped with a suite of tools that allows one to produce digital fabric output without wasting media and time (Fig. 25.1). Colour Atlas Generator provided help in reproducing exact colours easily. Just pick a swatch and enter the red green and blue (RGB) value into SoftRIP, and the Colour Atlas Generator does the remaining work. The total solution package includes several pre-packaged colour profiles for use with Wasatch SoftRIP. Colourways can be managed as unlimited colourways can be stored for future use. Computer time, memory and money can be saved by efficiently printing endless repeats. Customize repeats with advanced controls is possible.

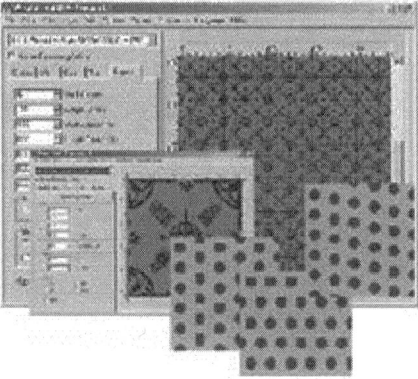

Figure 25.1   Wasatch SoftRIP TX for textile printing.

Figure 25.2   Adobe Photoshop.

Adobe Photoshop

Adobe Photoshop, or simply Photoshop, is a graphics editor developed and published by Adobe Systems (Fig. 25.2). It is claimed as the current market leader for commercial bitmap and image manipulation.

d.gen Textile Software Rip

d.gen Textile Software Rip is optimized textile software Rip for digital textile printing (Fig. 25.3). It is a professional solution which can digitally print colours with efficiency. It supports wide-format inkjet printing machines. This can be used with DTP programs such as Adobe, Photoshop, Illustrator, Corel Draw, Macromedia and Freehand. The claims are as follows:

- Suitable for design studios
- Printing system with compatibility and efficiency
- Easy to apply own creative idea
- Layout control before production
- The most similar and accurate colour reproduction by making colour library
- Able to create own digital designs
- Helps to match the image on a monitor and printout as much as possible.

Pointcarre Textile Software

With Pointcarre design software, it is possible to prepare artwork for any type of textile end use (Fig. 25.4); quickly modify, reduce colour and put all our patterns into repeat. It is provided with a variety of tools and features for a

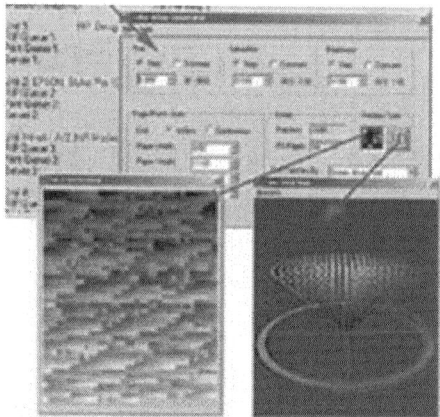

Figure 25.3   d.gen textile software rip.

Figure 25.4   Pointcarre textile software.

free expression of one's creativity. The tool icons are common graphics that are similar to those used in other popular graphics software.

Repeating design is possible by working directly in repeat mode. Any kind of repeat, such as straight, half drop, mirror, with or without overlap can be chosen.

Clean-up is easy (Fig. 25.5). Any scan of artwork, sketches or photographs can be quickly and easily cleaned up. Design can be enhanced by shading any texture over the design for realistic fabric renderings.

It is highly compatible. One can start working from photographs, scanned images or draw directly on screen; or even drag and drop images from popular graphics software such as Photoshop directly into Pointcarre. These images can be copied, modified or incorporated into other designs at any time with a click of the mouse.

By using colour reduction tool, flat artwork can be created quickly and effectively from a scan (Fig. 25.6).

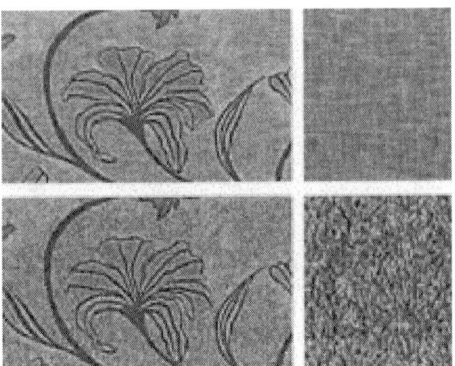

Figure 25.5   Clean-up.

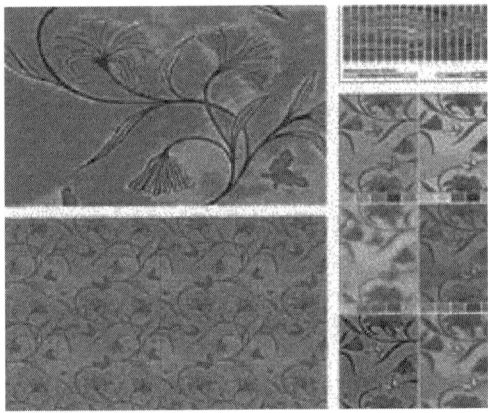

Figure 25.6   Flat artwork.

Changing colour in a design is easy. Just open the customized palettes and drag and drop a hue directly on the pattern. Time can be saved by building up the seasonal collections.

Pointcarre features familiar icons such as the pencil, paintbrush, rubber stamp, airbrush and eraser. With them, one can create a great variety of effects. The zoom, the customized grids and the instant screen preview guarantee an accurate and permanent control over the design.

The software CD-ROM also includes various libraries such as textures (woven, knit), knit and embroidery stitches. One can add his own for a greater variety of effects.

Multiple layer control helps selecting and moving, grouping, transforming and wrap objects more intuitively by clicking and dragging directly on the canvas. Aligning objects is easy with Smart Guides (Fig. 25.7).

Figure 25.7 Smart Guides.

It has non-destructive image correction; improve the colour, contrast and dynamic range of any image using a comprehensive set of professional correction tools and non-destructive adjustment layers, which display corrections while preserving the original.

'Step-and-repeat' is the term used for the process of duplicating an object and spacing.

Colour picker uses a colour wheel instead of swatches (Fig. 25.8). Painter's Picker allows one to quickly find and select any colour's complementary colour, and notes which colours one should stay away from, like those that clash. One can choose colours from five kinds of analogous schemes, five kinds of tetrad schemes, two kinds of triads and four complex complementary schemes.

Making colour is easy. Depending on the colour model, the printer uses (most often CYMK or Lab) the colours. The colours normally appear differently on the fabric than on the computer screen. Some colours such as deep, rich reds may be hard to reproduce. Large areas of solid colour may come out with bands of lighter and darker tones. Setting up design so that the colours can easily be changed (using layers or vector artwork) will save a lot of problems.

It is necessary to focus on the finish. It is easy to get caught up in the artistic aspect of creating a beautiful design and lose sight of the fact that fabric is never the end product—it is always a part of something else. It is necessary to make a habit of picturing the print as part of the finished product, especially concerning the size of the print. Have a ruler next to the computer—whenever you cannot quite decide if the scale is correct, hold the ruler up to the screen and zoom in or out until the size matches up.

The colour and texture of the fabric can have a noticeable effect on the print. Shiny fabrics like silk reflect light can make the print seem lighter. A

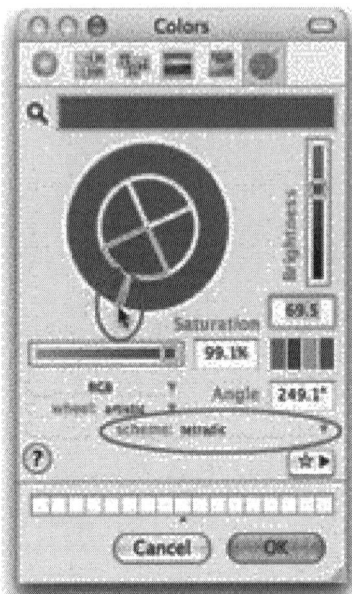

Figure 25.8   Colour picker.

thin fabric can be translucent and will make print look washed out. Most digital printing services offer affordable swatches; even if they only sell by the yard, one can gang up a couple of designs onto a single yard.

It may seem like a good idea to use digital printing to make a copy of a popular commercial print that is no longer available, but unlike clothing designs, print designs can be (and usually are) copyrighted by the artist or the manufacturer. It is best to stick with your own unique designs. If you are not artistically inclined, you can always hire a designer to make the perfect print for you.

25.4   Digital printing machines

Different types of digital printers are available. There are new entries in the market and are coming with various salient features. Let us have a look at some of them and the claims made by their manufacturers.

25.4.1   VARAD digital screen printing machines

M-Tex Machines Pvt Ltd, Mumbai, offers VARAD digital textile screen printing machine, which claims for its accuracy and best repeatability (Fig. 25.9). Its robust cast iron body ensures longer life.

Figure 25.9 VARAD digital textile screen printing machine.

Figure 25.10 Digital textile printer d.gen 740TX/C.

Other salient features include high production and power saving due to use of latest digital servo drives.

The screen printing machine includes an efficient and economic dryer. It is simple to operate with its user-friendly aspects, entailing less maintenance. It has special features for knit goods printing.

25.4.2 Digital textile printer d.gen 740TX/C

Digital textile printer, d.gen 740TX/C is able to print directly on most types of the fabrics including silk chiffon, stretchable fabrics such as knitted fabrics, and even some spandex (Fig. 25.10). It enables producing strike-off within a day with professional textile printing and high-profit business in high-end fashion industry, home furnishing and decorative fabric markets using garments, curtains, beddings, art walls, sofa and cushions. It is able to produce a sample within a day with much lower cost than the traditional way.

Figure 25.11   Silkohol digital printing machine.

Figure 25.12   Cromos of DUA Graphic systems.

25.4.3   Silkohol digital printing machine

The Silkohol digital printing technique, developed in the early 1990s, is large format digital printing (Fig. 25.11). Special inkjet or dye sublimation machines are used to print on material developed for the process. Although new inks and materials are in continuous development, digitally printed fabrics are somewhat less durable than traditional hand-painted or screen-printed fabrics. It has the advantage of producing multicolour and photographic images more cheaply than screen printing, particularly in smaller quantities.

25.4.4   Cromos of DUA Graphic Systems (DGS), Italy

Basic printer is Tx-2 of Mimaki, Japan, using piezoelectric print technology. Printing width is up to 1600 mm. The printing is done with the help of 8 heads with 16 colours (8×2) using Cromos of DUA Graphic systems (DGS; Fig. 25.12). Printing speed ranges from 6.6 to 28 m^2/h and dot size of fine, medium, large or variable can be selected during printing. Adhesive printing belt and integrated cleaning and drying mechanism make it possible to print all

Figure 25.13 Digital T-shirt printer.

Figure 25.14 DuPont Artistri.

types of fabrics even at a high speed. Usable inks are reactive, acid, disperse, pigment and sublimates. The DGS is equipped with a printer server and DGS software MatchPrint-II for controlling and calibrating the printer.

24.4.5 Texjet digital T-Shirt printer

Textile digital T-shirt printing can do direct pigment printing on to cotton T-shirts (Fig. 25.13). It also does inkjet direct printing on T-shirts, textile, leather and so forth. It consists of A2 size printer, table transport system, T-shirt holder and ink feeding bottle system. It uses water-based PixoInk's pigment inks and prints on almost any kind of fabric (white and light coloured), with very good wash fastness properties. It has a high productivity of up to 60 T-shirts/h at 720 dpi. Printer driver used is Windows XP and printing resolution is maximum 1440 dpi.

25.4.6 DuPont Artistri

DuPont Artistri is a fully integrated, production-capable digital printing system developed for printing on all types of fabrics (Fig. 25.14). The system was designed for a variety of applications, including printed textiles, accessories, apparel, home furnishings, table covers, flags and banners. High-speed, easy-to-use digital textile printing provides unparalleled design flexibility and greater cost-efficiency compared to traditional printing methods.

25.5   Typical procedure for digital printing

The final appearance of the fabric after printing can be visualized on the screen before starting printing. Here, an example of working step by step in digital printing is given. In this example, Spoonflower.com is used as printing service. There are many other print bureaus as well. The example of Spoonflower's colour library for Adobe Illustrator is given here (Fig. 25.15).

Step 1a—set the workspace: Depending on the design, set the area of the repeat tile (Fig. 25.16). It should be in a square form.

Step 1b—Set the colour swatches: When working with files for print, delete all unused swatches, gradients, brushes and graphic styles (Fig. 25.17). This makes it easier to manage and save colours later on. To delete all unused swatches click 'select all unused' in the Panel Menu and then click Delete Swatch. If any unused swatches remain, select and delete them.

Step 1c—Open the colour library: The easiest and cost-effective way to colour match a project is using a colour library (Fig. 25.18). All major printing companies have their own colour libraries which are available for download. These libraries are composed of swatches that fall within the printer's gamut range. If you do not have a colour profile, use colours that are not too bright or saturated. If you have downloaded a library, then load it by going to the

Figure 25.15   Spoonflower's colour library for Adobe Illustrator.

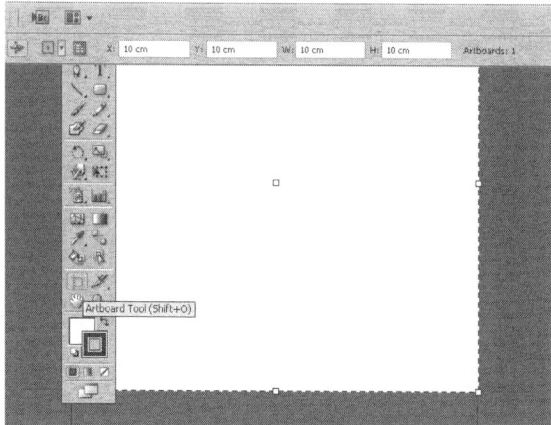

Figure 25.16 Workspace.

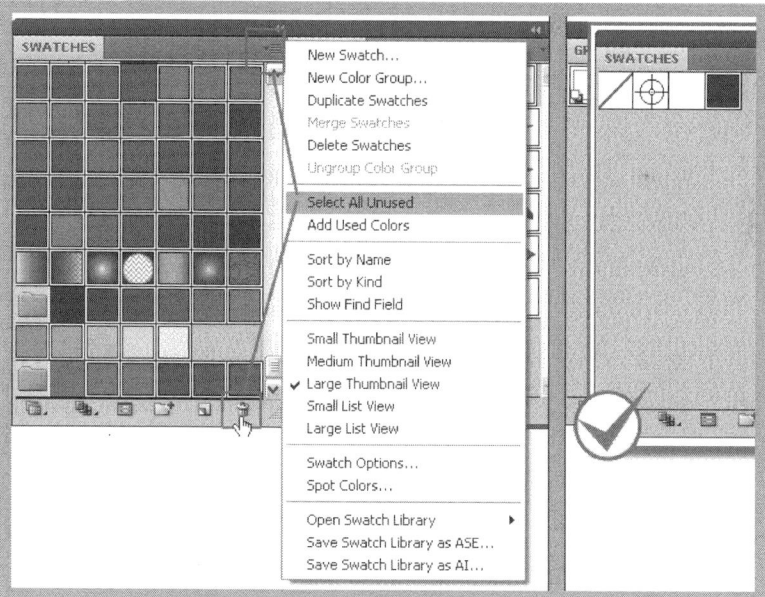

Figure 25.17 Colour swatches.

Panel Menu in the swatches pane and select Open Swatch Library > Other Library, then navigate to your file and click Open.

Step 1d—Make a colour group: Once the colour library is loaded into illustrator, choose the colours you want to use (Fig. 25.19). Brush the colours next to each other so you can get an idea of how they will sit together in an

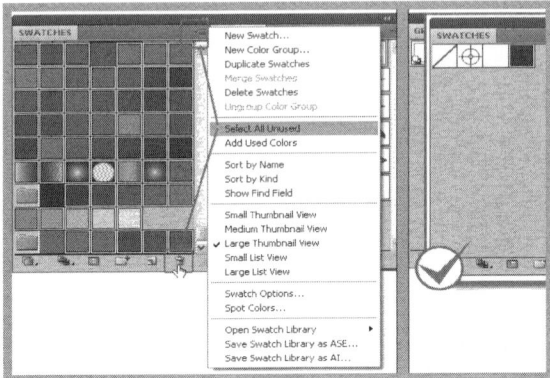

Figure 25.18   Colour library.

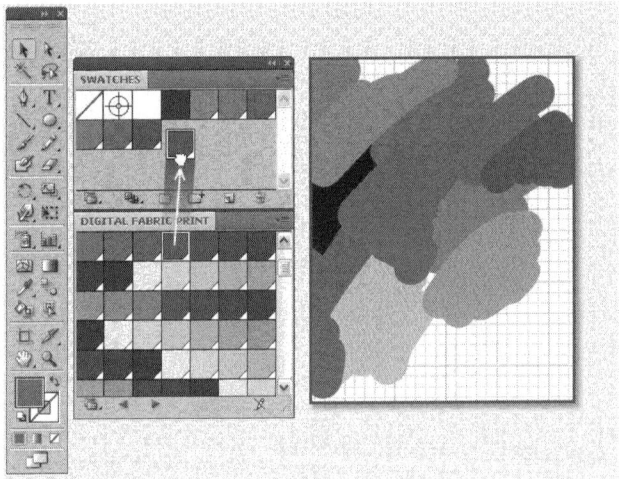

Figure 25.19   Colour group.

image. This is done by using the Blob Brush (Shift + B) but filled shapes work just as well. Once you have a selection of colours suitable for the design, drag the swatches from the colour library into the swatches pane. This is why we deleted all the swatches in Step 1b.

If the swatches used have a white corner on them, it means that they are Global colours. They are swatches linked to the fills and lines that use it. If a Global Colour Swatch is changed, then every colour in the document that uses that particular colour is also changed. This is extremely helpful for editing a document, such as a pattern, that is reliant on accurate colour management. To change a standard colour swatch to a Global Swatch, double-click on the

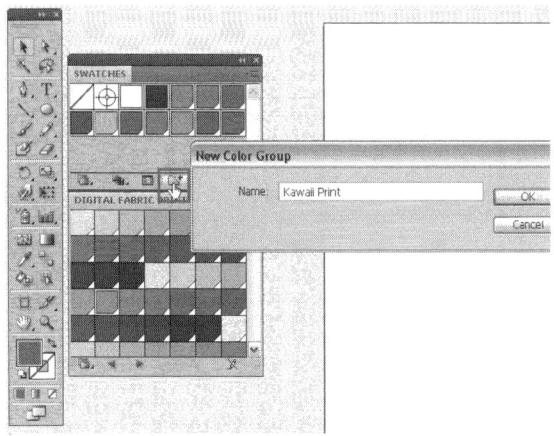

Figure 25.20 Adobe Swatch Exchange.

colour in the swatches pane (be careful not to have any objects selected) and select Global, then click OK.

Another way to keep track of colours is to make a colour group. To make a colour group, select the colours in your swatch pallet (command click) then click the 'new colour group' button at the bottom of the swatches pane and this will open a dialogue where you can name your colour group. You can also save these swatches for later use by going to the Panel Menu in the swatches pane and select Save Swatch Library Adobe Swatch Exchange (ASE; ASEswatches can be used with other Adobe programs such as Photoshop; Fig. 25.20).

If you are working from a pre-existing vector file, create a swatch pallet by selecting your image and clicking the New Colour Group button at the bottom of the Swatches pane; this will open the New Colour Group dialogue. Name your colour group and check Convert Process to Global, then click OK.

To change a colour swatch to make it suitable to print, double-click the swatch and change the CMYK values accordingly. If your colour is out of gamut (printable colour range), a warning shall appear in the Swatch Options dialogue—either manually change the colour values yourself or click the yellow alert triangle to change it to the nearest colour value that is within gamut.

If you are unsure of how your document will print or want greater accuracy of colour, replace each swatch with swatches suggested by your printing service. This can be done by loading the printers colour library and choosing the colours to use. Replace the old swatches by holding down Alt and dragging the new colour swatch on top of the colour to be changed. It is important to

Figure 25.21   Blob Brush.

have the swatches set to Global as each of the colours in your image will then change to be the colours you have chosen from the library.

Step 2a—Let's draw!—We now have a set of colours to use in our pattern—it is time to draw. For the absolute beginner, the quickest way to start drawing in Illustrator is to use the Blob Brush (Shift + B). The Blob Brush can be used to draw three different objects using the Colour Group I defined earlier (Fig. 25.21). To change the settings of the Blob Brush such as size and accuracy, double-click the Blob Brush icon in the Tools Pane.

The basic features of the Blob Brush are size, fidelity and smoothness. Size is self-explanatory. Fidelity is estimated according to the accuracy of the line—the higher the value, the smoother and less accurate the line will be. Smoothness is the amount of smoothing applied to your stroke—the higher the percentage, the smoother the path.

Note—Advanced users can either draw own pattern in any method you like and skip to Step 3 to learn how to test your repeat, or in case you already have a working repeat pattern, you can skip to Step 5 for instructions on exporting your file for print.

Step 2b—Copy and reflect objects: Here are four images drawn using the Blob Brush. They are fairly complex. To make the objects easier to work with, select the image with the Selection Tool (V), then press Command + G to group it together and repeat this for each object. As this is the quick and easy version of how to make repeating patterns, copy these four objects to save time to draw more. To do this, select the objects you wish to copy, select the Reflect Tool (O) and Command-click on a blank area of the art board—this

Figure 25.22   Reflect Tool.

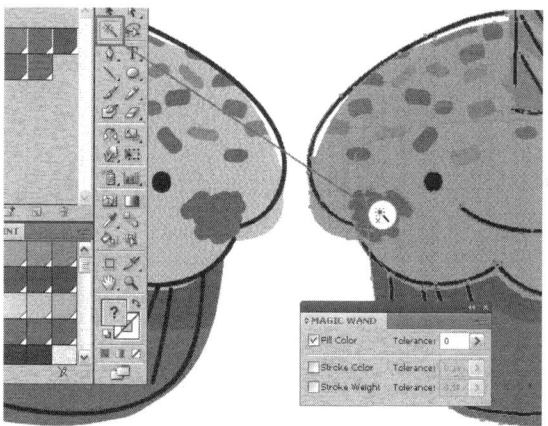

Figure 25.23   Magic Wand Tool.

will open the Reflect Dialogue (Fig. 25.22). Select Vertical and click Copy. You should now have a reflected copy of your four objects.

Step 2c—Recolour: Now recolour the copied objects by double-clicking the object you wish to recolour. This will take you to the object's isolation mode—from there you can recolour your object without accidentally changing the other objects. Since each object is grouped, this should be fairly easy.

To change every instance of a colour to another colour, one can use the Magic Wand Tool (Y) (Fig. 25.23). First, double-click the Magic Wand icon in the Tools Pane to bring up its settings, change the Tolerance to 0—this way you will only select exact colour matches. Once the colour you wish to

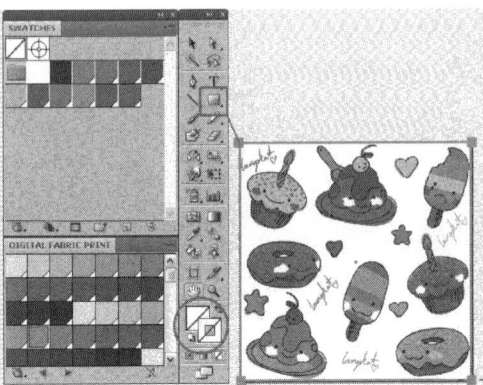

Figure 25.24   Arrangement of the objects.

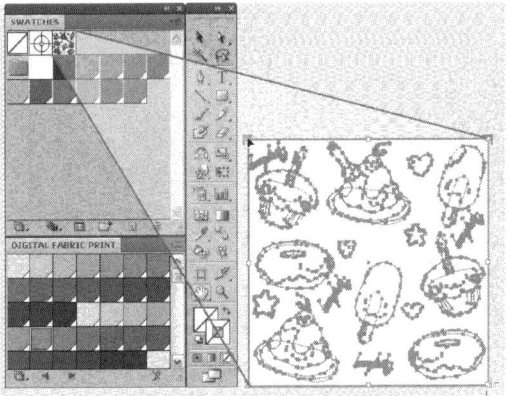

Figure 25.25   Transparent square.

change is selected using the Magic Wand, click the colour you wish to change it to in your colour Library. To exit isolation mode double-click outside of the isolated objects. Repeat this process with each mirrored object.

Step 3—Arrange: Now that we have our objects, it is time to arrange them on the art board—as you can see, we have added a few hearts, stars and signatures to fill the blank spaces (Fig. 25.24). This layout is looking pretty good, but before we export it for print, we need to test the pattern. To do this, first draw a square which has the size of the art board with the line and fill set to None, then select the square and send it to the back (Object > Arrange > Send to Back).

With the transparent square at the back of your image (Fig. 25.25), use the Selection tool (V) to click and drag a selection box around your art board.

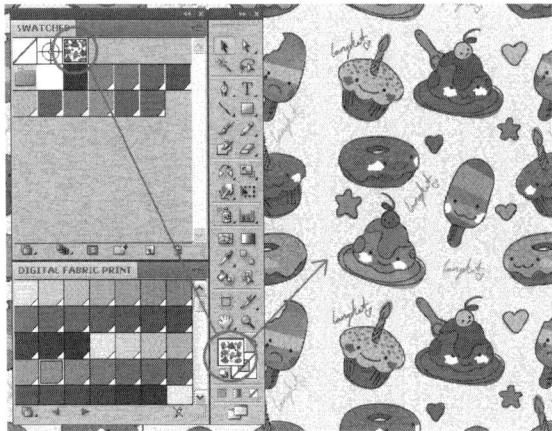

Figure 25.26 Rectangle pattern.

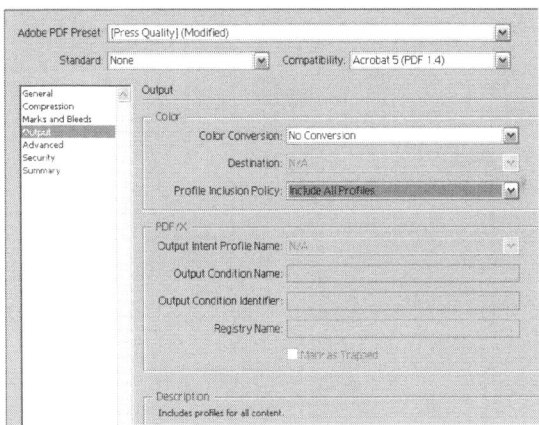

Figure 25.27 Saving the file as PDF.

Take the selection and drag it into the swatches pane. You should notice a new swatch is created, this is your pattern.

To test your pattern, draw a rectangle anywhere outside of the art board and fill it with the pattern swatch (Fig. 25.26). Remember to make the rectangle large enough to repeat the pattern a few times; this is better for detecting anything that needs to be changed.

Step 4—Save as PDF: Once you are happy with the positioning of the elements, save the file as a PDF (Fig. 25.27). One of the most popular file types (and recommended by Spoonflower) is an 8-bit, uncompressed TIFF in the LAB colour space. As it is impossible to export LAB colour directly from

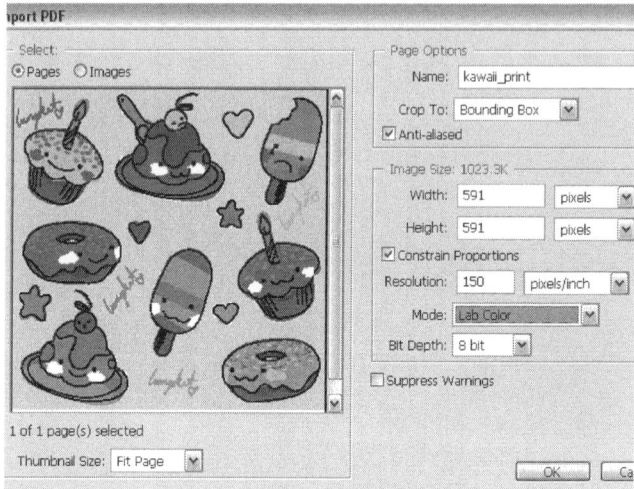

Figure 25.28   Opening PDF in Photoshop.

Illustrator, we have to save the image as a PDF and open it in Photoshop. Use PDF as it preserves colour values so that all the hard work choosing swatches is not ruined on export.

Go to File > Save As > PDF (Command + Shift + S)—this will open the Adobe PDF Dialogue Box. Go to Output and set the colour Conversion to No Conversion and then click OK.

Step 5—Open PDF in Photoshop: If you have not already done so, open the PDF in Photoshop (go to File > Open and select your saved PDF; Fig. 25.28). This will bring up the Import PDF Dialogue. Choose the following settings to import your image: Page Options as Crop to Bounding Box (this will import only what is inside the art board) Resolution as 150 px/in (this is the optimal resolution as specified by Spoonflower—other printers may differ, be sure to check.) colour Mode as LAB (LAB colour will be explained in the below sections) Bit Depth should be 8 (again, this is specified by Spoonflower) and then Click OK.

LAB colour, unlike RGB or CMYK has tones and colours separate so one adjusts without affecting the other. L is light, A and B are colour—so if A+B is red, you can subtract L to make maroon or add L to make pink.

LAB colour is also the only device-independent colour space in Photoshop, which makes it perfect for digital fabric printers as they work with colour differently to a standard commercial print machine.

Step 6—Save as a TIFF: Each printer has their own recommended file types, such as jpg and gif; similarly, Spoonflower asks for TIFF (8-bit, uncompressed).

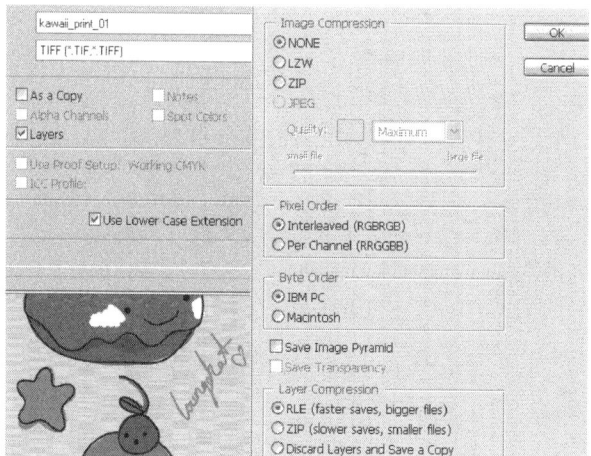

Figure 25.29   Saving the file as a TIFF.

It is recommended to use TIFF because it is a lossless file type, which means there is no quality loss due to compression (Fig. 25.29). To save your art as a TIFF, go to File > Save As > then choose TIFF—make sure the image compression is set to None, then click OK.

25.5.1   Conclusion

To have a print-ready file with (almost) foolproof colours, one needs to go to spoonflower.com, or the preferred printer, and upload the artwork to print.

25.6   Digital printing solutions

Various softwares are developed to give different effects of digital printing, which was impossible with other types of printing. For example, Digitex India, which offers textile digital printing solutions, give you the innovative and creative designs, such as concept of shadows, shimmering, vibration, reflection, optical, translucent, netting, blurring, layering, superimposing and more. Generally, the digital textile printing has been operated through transferring media, such as screens and rollers. Each transferring media is designed and allocated for a particular colour. A block printing and engraved copper printing are operated using these methods. Various types of wood blocks are used in textile printing solutions, which are designed in block printing and present a number of colours in designs. The advantage of digital textile printing is that it is possible to have three-dimensional effects as a result of its colour separation and layers. Digital textile printing solutions from Digitex India involves wide exploration in creating multiple fabric sheets with a wide range of traditional and modern designs.

Kornit Digital develops, manufactures and markets industrial and commercial printing solutions for the garment, apparel and textile industries. Leading the digital textile printing industry with revolutionary NeoPigment process including an integrated pretreatment solution, it caters directly to the needs of both designers and manufacturers. It has brought the benefits of pigment printing on textile to the digital world, driving this revolution. It has created a revolution with single-step solution that enables printing with one ink set on all types of fabric and with no additional finishing process. It is the first company in the world to introduce white digital ink for garment printing, enabling the garment industry to print on dark fabrics.

25.7 Problems/limitations associated with digital printing

The following are the problems normally associated with digital printing:

- Metallic colours cannot be printed by these machines.
- In case of flat colour printing, there can be a gamut of colours which the machine cannot produce.
- Maximum width of printing is 150 cm/58 in.
- Rejection level of printed fabric is much higher as compared to other forms of printing, especially where the print area is insignificant as compared to the entire fabric surface.
- Cost of printing for bigger production runs is much more as compared to other forms of printing.
- The major downside to digital printing is the cost. As with any new technology, the costs are always high when it first becomes available. As time goes on and the technology continues to develop, it will undoubtedly become more affordable.
- There is no sure-shot formula to achieve desired results in case of photographic files. Attaining good results in digital textile printing for photographic files is an art where good and desired results can be achieved only after correcting the file after various rounds of strike-off. It takes few attempts and trials before the optimum results are achieved. There is always a minor difference between the colours of the screen or artwork in comparison with the printed fabric.

26.1 Evolution of garment washing techniques

Jeans were created first for gold miners during the California Gold Rush. Dramatic changes have occurred in the function and design of jean garments since then. The evolution of the jeans' market led to the development of some unique and creative methods for the processing of denim garments. Originally, jeans were marketed and sold as workwear with primary emphasis on their durability and practicality. But when jeans were discovered and appreciated by consumers as general casual wear, they became fashionable and new techniques were developed to enhance denim garments and make them more unique. These techniques are popularly known as garment washing. Basically, all of these techniques involve the processing of garments in rotary drum machines

The first generation of indigo jeans was stiff and uncomfortable. Normally after weaving, greige denim is singed, finished with starch and a lubricant and then mechanically shrunk. This mechanical shrinking could make the hand somewhat soft, but no other processing techniques were employed to provide a soft handle. The consumers would take a newly purchased pair of jeans home and soften them by washing once or several times before wearing them for the first time. Denim fabric continues to be processed using the same basic finishing system, but after being cut and sewn, denim garments may undergo additional processing.

The second generation jeans incorporated pre-washing by the manufacturer. These jeans had a slightly faded appearance and a softer hand that felt comfortable, as though they had been laundered several times. This trend became fashionable and consumers were willing to pay the extra cost. Consumers no longer had to bother 'breaking-in' their jeans themselves. The added benefit was that the jeans were already shrunk to size with little or no residual shrinkage.

Not long after the introduction of pre-washed jeans, the idea of using abrasive stones to accelerate the aging process was developed and 'stone washing' was born, creating an even more 'broken-in' look. Next, chlorine bleach was incorporated in these wash techniques and a whole new paler blue denim family evolved. Ice washing was developed next, in which the porous stones were soaked in a bleaching agent and then tumbled with dry or slightly damp garments. This process has many names, including acid wash, snow wash, white wash, frosted, and so forth.

Cellulase wash procedure that was developed later used cellulase enzymes to accelerate the colour and fibre removal. A reduced quantity of stones could be used to create a desirable washed down appearance. This process was more efficient with fewer stones, larger load sizes and less of an abrasive effect on the inside of the rotary drum.

These garment washing techniques originally developed for denim garments, are now being used for a wide variety of different garment types. The mills and commission houses involved in garment processing continually search for ways to achieve unique new looks. Most of these garment processors have their own individual techniques, the details of which are not divulged. This chapter describes some of the basic wash techniques. Any of these procedures can be modified to fit a particular situation, depending upon the garment type (i.e., heavyweight denim versus lightweight chambray), available equipment and the process flow. Also, some of these procedures yield garments suitable for overdyeing, which may create a whole new look.

26.2 Types of garment washing

Different types or methods of garment washing are available. Primarily, garment washing is of two types. They are wet process (chemical process) and dry process (mechanical process). Following are some examples:

Wet process (chemical process)

- Normal wash/garment wash/rinse wash

- Pigment wash

- Caustic wash

- Enzyme wash

- Stone wash with or without bleach.

- Stone enzyme wash

- Tinting (Tie) and over dyeing (dip-dyeing)

- Super white wash

- Bleach wash
- Acid wash
- Silicon wash
- Soft wash
- Whitening
- Metal Wash

Dry process (mechanical process)

- Sand blasting
- Hands scraping
- Overall wrinkles
- Permanent wrinkle
- Broken and tagging
- Grinding and destroy
- Potassium permanganate (PP) spray and PP sponging and so forth.

26.3 Garment wash

The process which is used to transform the outlook appearance, ease and fashion application of the garments is called garment washing (Fig. 26.1). It is normally done after stitching. Wash types usually depend on the product nature and its usage. Based on the consumer demand and fashion trends, the buyer will fix the washing types of any product. For example, stone enzyme wash is required for denim items but light softener wash is perfect for knitted items. For the washing, apparel buyers mention exactly what types of washing they need for the order. Wash types mainly depend on the product types.

Indigo jeans were once the only item processed by the garment wash method, but now, a wide variety of garments with different types of woven and knit fabrics dyed by different systems are garment washed before retail distribution. Emphasis is on comfort and softness. Also, some fashion trends favour the broken-in look and worn/faded seams that can only be achieved through garment processing.

Depending on the garments' construction, different types of washing processes can be done. For example,

Figure 26.1   Garment washing machines.

- Twill/canvas/knitted/corduroy –normal wash, pigment wash, caustic, Si wash

- Denim/jeans/gabardine—enzyme wash, stone wash, bleach wash, acid wash

- Grey fabric— super white wash

26.4 Objects of garment washing

Garment washing is used for the following reasons:

a)   To modify the garment that shows up hard, feeling rough, stiff and not enough responsive for wearing by pre-washing

b)   To soften the garment hand feel and improve bulkiness

c)   To give a faded look or any other colour-tinted look to the garment

d)   Removing chemicals used during the printing and embroidery process

e)   Garment is often made bigger or becomes bigger and larger. Pre-wash returns those to the right size and make it dimensionally unstable

f)   After pre-wash, the garments become fit as they get rid of shrinkage; as a result, the garment become soft hand feel and becomes size free

g)   Some garments become more attractive, lucrative and lively after wash. To create a washed look appearance is a new trend in fashion

h)   Any dirt, spots, germs or oil stains that accumulate in the garments during manufacturing is removed due to washing

i)   Eliminating starch and chemicals used during fabric manufacturing and dying

j) Developing the market by attracting customers/buyers by fashionable washing

k) Making the garments ready to be worn directly after purchasing

l) Getting a faded, old, coloured or tinted effect

m) Preventing further shrinkage of the wash garments

26.5 Advantages of garment washing

1. Starch materials present in the new fabrics are removed, hence, they feel soft during use

2. Soft feeling of the garments could be further increased

3. Washed garment could be worn directly after purchase from the store

4. Fading effect is produced in the garment in regular or irregular pattern, and also in the specific area of the garment as per design

5. Different outlooks could be produced in the garment by different washing techniques

6. Similar outlook can be produced in the garments by different washing techniques

7. Initial investment cost to set up a garment washing plant is comparatively lower

8. Dirt and spots if present in the garment are removed

9. Shrinkage occurs in the garment washing, hence no possibility of further shrinkage

26.6 Different washing methods

26.6.1 Normal wash

Normal wash consists of washing garments in hot water with adequate detergent and softener, rinsing in plain water and drying in tumble dryer. Some soda ash is added to lend the garment a prominent washed look. Water temperature and the proportion of components of the wash are adjusted as per requirement of the wash and types of fabrics. The softener makes the fibre soft and tumble-drying makes the fabric fluffier. An expert technician can handle washing in an expedient way to solve many problems arising during washing.

26.6.1.1 Typical procedure for normal washing

Following is the general procedure followed for normal washing. The garments are to be inverted to minimize unwanted abrasion streaks (especially useful when preset creases are present).

1. Ensure that the machine is clean before loading the garments in the drum. Then load the machine with garments

2. Desize the garments by using a suitable enzyme like alpha-amylase enzyme and detergent

3. Drain and remove the desizing liquor

4. Rinse with clean water

5. Fill the machine with water and heat to 60°C. The liquor ratio can range from 10:1 to 20:1. A number of synthetic detergents can be used. Also, alkaline products such as soda ash or caustic soda can be added in amounts ranging from 0.5 to 2.0 g/l. Some chemical suppliers offer special products that accelerate the wash down process, depending on the particular dyestuff used

6. Wash/tumble action for 20–60 min, depending upon the desired effect

7. Drain and rinse

8. Apply softener

9. Tumble dry

10. Invert garments, if previously turned

11. Press, if required

26.6.2 Pigment wash

'Pigment wash' is similar to a normal wash but a bit costlier. The garment is solid colour dyed with a pigment dye. The requirement is that the colour should fade evenly to lend the garment a prominent washed look. Pigment wash requires a higher temperature of water than normal wash, normally 50–60°C. The tumble washer should not be loaded more than 70% of its capacity. This enables garments to smoothly move inside.

26.6.3 Bleach wash

In bleach wash, bleach chemical is used in water while washing in a tumble washer. Strict washing time is a requirement with such a wash, otherwise the garment may be overbleached and the colour cannot be reversed.

26.6.4 Stone-wash

Stone wash means washing garments with special stones so that garments achieve a very strong washed effect. To get a faded look on the garment surface, white stones are used with enzymes during washing. During washing, the fabric comes in contact with stones and the colour fades by the rubbing action.

Abrasive stones were introduced to the wash bath in order to accelerate the garment wash effect and to give garments an even more unique appearance and a softer hand. A variety of natural and synthetic stones are available for stone-washing; the most widely used being a pumice or volcanic rock. Volcanic stones, while washing, abrade the exposed parts of the garments. This idea of washing with porous volcanic stones is to give the garment a strong and rough wash to achieve the pronounced washed effect through abrasion on the exposed areas, such as the seams and pocket corners.

The stones, as they are used, slowly disintegrate reducing the severity of the stonewash effect over a period of time. The stones not only abrade the fabric but also gradually abrade the inside of the rotary drum. A machine used for stone-washing should not be used to dye delicate articles or when abrasion would be detrimental to the fabric.

Sometimes, bleach is added to the wash so that the colour fades in a more pronounced manner. This is done to turn navy blue jeans into a more faded light blue colour. Such a wash requires a lot of skill, experience, workmanship and expertise so that desired results are achieved. In stone wash, the following points should be carefully checked:

a) *Size of the stones:* Stone sizes have a varied effect on the garment being washed. Larger stones give tough abrasion while smaller ones lend less abrasion. Stones should be selected based on the required abrasion affect as well as the type of fabric of the garments. Larger stones may damage comparatively lightweight fabric

b) *Garment-stone ratio:* (Weight of the stones relative to the weight of garment) A wash with more stones may lead to a more apparent blue/white contrast on the fabric

c) *Washing time:* Washing time is also important in stone wash

d) *Quantity of the bleach:* Use of more bleach can shorten wash time, leading to more productivity. Bleach, however, cannot be used indiscriminately. Disproportionate amount of bleach may lead to loss of the desired blue/white contrast on the fabric. To get better results, one should cut a balance between the quantity of bleach, stone size and the amount of stone. Sometimes, one needs to use a normal quantity of stone and longer washing time to achieve the colour standard requirements.

26.6.4.1 A typical procedure for stone washing

1. Load stones into the machine

2. Load garments into the machine (ratio usually 0.5–3.0 part weight of stones for one part weight of garments)

3. Desize with alpha-amylase enzyme and detergent. Liquor ratio approximately 5–8:1

4. Rinse

5. Refill and tumble with stones for 30–90 min, depending upon the desired effect. Liquor ratio 5–8:1 at 50–70°C. Scouring additives can also be used

6. Drain. Separate the garments from stones (garments can be transferred to another machine)

7. Rinse

8. Apply softener (garments can be transferred to another machine for softening)

9. Extract and unload

10. De-stone and tumble dry

11. Press, if required

Softeners and/or lubricants can be added during steps 3 and 5 to reduce creasing potential. Steps 8, 9 and 10 may vary depending upon individual mill arrangement.

Pumice stone wash: Pumice has been used since the introduction of stone-washed jeans in the early 1980s. The jeans are washed with oval or round pumice stones which should all roughly have the same format. The pumice stones are very light with a rough surface. Freshly-dyed jeans are loaded into large washing machines and tumbled with stones. Adding pumice stones gives the additional effect of a faded or worn look. The pumice abrades the surface of the jeans like sandpaper, removing some dye particles from the surfaces of the yarn. In stone-washing, stones and denim are spun together in large industrial washing machines. The longer they are spun together, lighter is the colour of the fabric with better contrasts. The time duration of this procedure is set beforehand to avoid the tear and wear of the fabric. Thereafter, the fabric undergoes various processes of rinsing, softening and finally tumble-drying. These stone-washed fabrics are used for different purposes like garment making as well as for upholstery purpose.

Disadvantages of pumice stone usage: Stone-washing the denim with pumice stones has some disadvantages. The stones could cause wear and tear of the fabric. It creates the problem of environmental disposition of waste of the grit produced by the stones. High labour costs are to be borne as the pumice stones and their dust particles produced are to be physically removed from the pockets of the garments and machines by the labourers. Denim is required to be washed several times in order to completely get rid of the stones. The process of stone-washing also harms big, expensive laundry machines.

The quality of the abrasion process is difficult to control: Too little will not give the desired look. Too much can damage the fabric, particularly at the hems and waistbands. The outcome of a load of jeans is never uniform, with a significant percentage always getting ruined by too much abrasion. The process is also nonselective. Everything in the washing machines gets abraded, including the metal buttons and rivets on the jeans as well as the drum of the washing machine. This substantially reduces the quality of the products as well as the life of the equipment and increases production costs.

During the washing process, these stones will scrape off a thin layer of the denim, thus showing some of the white threads from the part of the cloth where the indigo dyeing stuff is not able to penetrate. It also creates the effect called brilliance. One may also encounter words like deep stone or super stone wash, which are an indication of how long the jeans have been stone-washed.

26.6.5 Alternate methods for stone-washing

To minimise the drawbacks explained earlier with pumice, stone-washing of denim is carried out with different methods as explained below.

Perlite stone wash: Perlite, a naturally-occurring silicon rock has the distinctive property of expanding 4–20 times its initial volume, when heated at a particular temperature. This happens because the raw perlite rock consists of 2–6% water content in it. The crude perlite rock, when heated at a temperature above 870°C, gets swollen up and tiny glass-sealed bubbles are formed. Its original colour is black or grey and it changes to greyish white or white. This heated form of perlite is used for stone washing.

Perlite does the same function of stone-washing as stones. This reduces the rate of harm caused to large washing machines compared to pumice stones and gives the denim better suppleness and softer finish. This also reduces the rate of wearing out of jeans when used. It gives a uniform worn-out and old

Figure 26.2   Dirty wash.

look to the denim. There are many grades of perlite differing in sizes that are used for giving a stone-wash finish to the denim right from the largest to the finest grades; some are very tiny just like ground earth.

Bio-stoning: At first, stone-washing involved using pea gravel, but pumice was discovered to float around with the jeans instead of lying in the bottom of the water and hence, manufacturers have switched. Turkish stone is commonly used for its porosity and cleanliness.

Rinse (water) wash: In rinse water wash, the jeans will be washed at about 50°C. There is a high risk of colour bleeding, so it is ideal to use old faded jeans for brightening up by washing them together. One should make certain to wash them separately from other garments the first few times. Some jeans brands will not even use Sanforized fabric, so you can shrink them to fit in a hot bath.

Dirty wash: After stone-washing the jeans or denim jackets, they will be dyed with special chemicals, thus creating a look in which the jeans will appear to be dirty (Fig. 26.2).

Destroyed/damaged/used/whiskers: There are several different techniques to make the jeans or denim jackets look old, worn and/or used. Most of these techniques involve actual sandblasting or abrading by some kind of power tool. Whiskers, which normally appear around the hip to crotch area of the pant, are usually made by using a grinder. Another popular way to make jeans appear damaged is to cut the edges at bottom, (back-) pockets, fly and knee area before the (stone) washing.

26.6.6 Stone-wash with chlorine

By incorporating chlorine in the stonewash procedure, a reduction of the indigo colour (or other chlorine-sensitive dyestuff) is obtained. It is very important that any residual chlorine be removed before drying, to prevent fibre degradation. This is accomplished by using an antichlor step with sodium bisulphite or hydrogen peroxide. The normal procedure is as follows:

- Load stones into the machine
- Load garments into the machine (ratio usually 0.5–3.0 part weight stones to one part garments).
- Desize with alpha-amylase enzyme and detergent (liquor ratio is approximately 10:1)
- Rinse
- Refill and add sodium or calcium hypochlorite
- Heat to 55°C
- Tumble for 15 min
- Add the second portion of sodium or calcium hypochlorite
- Tumble for 15 min, maintaining temperature of 55°C
- Drain
- Rinse well
- Antichlor with sodium bisulphite or hydrogen peroxide
- Drain
- Separate garments from stones (garments can be transferred to another machine)
- Rinse well
- Apply softener
- Extract and unload
- De-stone and tumble dry
- Press, if required

The amount of sodium or calcium hypochlorite required will vary depending upon the desired level of bleach down and the sensitivity of the colour to chlorine. Each addition can range from 0.075 to 0.225% available chlorine. The pH should be kept above 9.0, preferably 10.5–11.0. This is accomplished with the addition of soda ash with each addition of hypochlorite.

26.6.7 Acid wash

This is also a kind of stone wash. For acid wash, base colour of the garment is taken out by spraying acid on the specified areas. The wash is performed in two steps: in the first step, garment is washed without water and in the second step, with water. The normal procedure is as follows:

1. Soak volcanic stones in potassium permanganate solution. Stones absorb chemicals and become saturated. The stones are then dried in normal air or sun. The stones are ready for work.

2. Denim garments are now made ready for wash. They are desized and destarched in water in a tumble washer and dried in a spin dryer.

3. The garments are put in a separate tumble washer filled with treated stones. Water is not added. Now run the tumble dryer and wash the garments without water. Tumble washer is run to wash the garments without water. Stone will abrade the garments, especially the exposed parts. Hidden parts will not be abraded.

4. Thereafter, the garments are taken out of the tumble and transferred to another tumble filled with water for washing and rinsing. After rinsing is over, the prominent acid wash effect will show up.

The treated stones carry the chemical to bleach the exposed parts and bleach them to white. But the hidden parts remain untouched. Whitening agents are often added to water during rinsing, to make the white colour in the blue jeans whiter to display the acid wash.

26.6.8 Enzyme wash

Heavy enzyme or vintage wash is used to get the old or used appearance. Garments are washed inside a washing machine with enzymes. Fleece sweatshirts are washed with a heavy enzyme. Enzyme wash is performed with a kind of live cell, which can break some of the fibres of fabric and can give the fabric a special effect, desired on the garment. Enzyme wash provides the fabric a soft, sanded or 'peached' effect that is very desirable on many garments. Enzyme wash is also useful for indigo denims.

In this case, enzyme can replace stone but gives denim a stone-wash look, with a better and nicer blue and white contrast on the fabric. Enzyme wash is, however, costlier than stone wash.

Advantages of enzyme washing

• Soft handle and an attractive clean appearance is obtained without severe damage to the surface of yarn

• Inexpensive, low-grade fabric quality can be finished to resemble a top quality product by the removal of hairiness, fluff, pills and so forth.

• Simple process handling and minimum effluent problem

- Better feel to touch and increased gloss or lustre

- Prevents tendency of pilling after a relatively short period of wear

- Can be applied on cellulose and its blend

- Due to mild condition of the treatment, process is less corrosive

- Fancy colour-fenced surface can be obtained without or a partial use of stone

26.6.9 Ice wash

Either by accident or experimentation, a method was developed in which stones are used as a vehicle to deposit a chemical on garments to strip the colour. This surface deposit of chemical removes the colour only on the outer surface of the garment and produces a frosted appearance. Indigo and selected sulphur dyes are currently the most popular candidates for this procedure. The normal procedure is as follows.

- Soak stones in solutions of potassium permanganate for 1–2 h. Concentrations ranging from 1.5 to 5% are being used commercially. (5–10% sodium hypochlorite can be substituted.)

- Stones should be drained off excess liquor. This can be accomplished by placing stones in net or mesh fabric prior to soaking. Then the stones can be removed and the excess liquor can be drained off. Another alternative is to place the stones in a rotary tumble machine along with 'waste' fabric and tumbling for several minutes to remove the excess solution. A third alternative is to use any number of the pre-soaked stones or materials available from suppliers. These are available in many different shapes with varying levels of chemical and other additives that produce different effects. Trials should be conducted to determine the best method for achieving desired effects.

- Place stones and garments in the machine (garments should be scoured and/or desized and dry or slightly damp).

- Tumble for 10–30 min or until the desired effects are achieved. Results are dependent upon dyestuff, fabric, concentration of chemicals, stones, additives and equipment.

- In some cases, the stones can be reused for another load before resoaking, depending upon their porosity. It is advantageous to transfer the garments to another machine for washing, minimizing the number of machines used for the corrosive process of ice washing.

If potassium permanganate is used, manganese dioxide will form (a brown/ orange colour) and must be removed by treatment with sodium bisulphite, hydroxylamine sulphate or acidified hydrogen peroxide as the reducing agent. Fill the machine with water and add 1–5 g/L of the reducing agent. Heat to 50°C and run for 20 min. The process is normally repeated twice to ensure complete removal of the manganese dioxide. When sodium hypochlorite is used, the residual chlorine should be removed with sodium bisulphite or hydrogen peroxide. Adding jeans to a machine already charged with after-wash chemicals will increase contrast.

- Rinse well
- Repeat the step, if necessary
- Apply softener
- Tumble dry
- Press if required

The selection of sodium hypochlorite versus potassium permanganate depends upon the dyestuff and desired effect. Also, consideration must be given to the safety aspects of handling either of the chemicals.

26.6.10 Cellulase wash (Bio-stone-washing)

Cellulase is environmental friendly compared to pumice stones. Cellulase enzymes are natural proteins. It reduces the percentage of damage caused to denim by the rough effect of stones on them. Cellulase attacks primarily on the surface of the cellulose fibre, leaving the interior of the fibre as it is, by removing the indigo present in the surface layer of the fibre. This method is also known as bio-stone-washing. Enzymatic treatment has become another substitute for stones; also, the jeans stone-washed by this method have more shelf life. It ensures the same result with minimum amount of water, waste, time, volume and damage to machines.

Cellulase enzymes have gained acceptance in the garment wash industry as a means to achieve a washed down appearance without the use of stones or with reduced quantities of stones. These enzymes are different from the alpha-amylase enzymes used for starch removal in which they are selective only to the cellulose and will not degrade starch. Under certain conditions, their ability to react with cellulose (cotton) will result in surface fibre removal (weight loss). This will give the garments a washed appearance and soft hand.

Cellulase enzyme is classified into two classes, namely, acid cellulase and neutral cellulase.

- Acid cellulase: It works best in the pH range of 4.5–5.5 and exhibits optimum activity at 50°C

- Neutral cellulase: It works best at pH 6; however, its activity is not adversely affected in the range of pH 6–8 and shows maximum activity at 55°C.

As jeans are made up of cellulosic fibres, the use of cellulase enzyme is successful in giving a stone wash look. This enzyme breaks down the surface cellulose fibres and removes them without causing harm to the jeans. Better finishing and look is achieved even with indigo-dyed denim. In cellulase enzymatic wash, the denim is given an enzyme bath. Here, certain amount of indigo dye and cellulose fibres are removed from the surface of the fabric. As enzymes are like yeast in nature, they eat the cellulose present in denims. When the jeans get the preferred colour, enzymatic reaction is stopped by changing the alkalinity of the bath or else the water is heated. Thereafter, the fabric undergoes rinsing and softening process. The numbers of rinsing processes after the enzyme treatment are lesser than pumice stone-washing. There is a reduced amount of waste produced and the overall costs for stone-washing are also less.

The process flow is as follows

1. Load garments

2. Desize with alpha-amylase enzyme and detergent

3. Rinse

4. Add cellulase enzyme (amount, pH, temperature and cycle time dependent upon the type of fabric and desired effects; manufacturer's recommendations should be referred).

5. Adjust the pH as recommended

6. Tumble 30–90 min

7. Drain

8. Rinse well (70°C)

9. Drain

10. Rinse well (70°C)

11. Drain

12. Separate garments from the stones, if used (garments can be transferred to another machine)

13. Apply softener

14. Extract and unload

15. De-stone and tumble dry

16. Press, if required. After step 7, chlorine bleach may be used as described in stone-wash with chlorine.

The increase in temperature serves to deactivate the cellulase. Adjustment of pH to 9.0–10.0 with soda ash can also be incorporated. Some operations use both the increase in pH and temperature.

Disadvantages of cellulase treatment: There are certain disadvantages of the cellulase treatment. It could leave marks of back-staining, such as blue threads becoming more blue or white threads becoming blue. To get rid of such unwanted recolouration of threads, the jeans are rigorously washed by adding surfactants to it. This process could result in colour fading of jeans and there is added usage of water for the washing. Thus, wastage of water and certain amount of back-staining could be experienced. The primary target of stone-washing the denim with pumice stones or enzymes is to provide the garment a worn-out, old and aged look. Sometimes, both stones and enzymes are used for the purpose.

26.6.11 Mineral wash

In mineral wash, a chemical bleaching agent is sprayed on to the foam blocks (Fig. 26.3). The blocks are then washed with the garments. As the blocks hit the garments, the chemical removes the dye from the garment. Sampling is required to ensure that the desired outcome is achieved. Holes and amplified sewing issues can be found with this treatment. No two garments will look identical.

26.6.12 Caustic wash

Caustic wash is basically a preprinting wash. Caustic is a strong chemical with highly corrosive features. Prior to printing on cotton fabrics, grey goods are

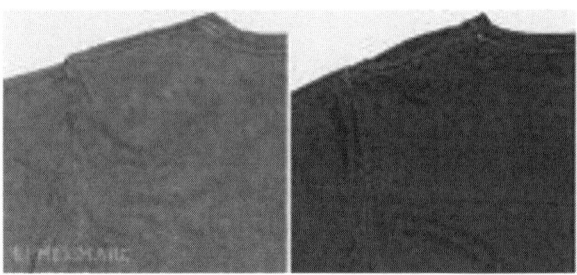

Figure 26.3 Mineral wash.

treated in boiling water with caustic, which also has strong cleaning power, especially for grease. This wash can remove all soil, dirt, grease, fine particles of cotton seeds as well as all foreign materials. As a result, only pure cotton fibre in the fabric for printing is left. This leads to stability of printing and a well cleaned fabric. However, when caustic wash is done on garments, just do the opposite of the above; fabric is not treated with caustic wash for cleaning prior to printing.

Printing is done on the raw and unclean fabric so that about 30% of the printing done on the surface may eventually fade away. Finally, printed garments are caustic washed. This leads to about 30% of the printing washing away along with the foreign materials–leaving about 70% of the printing on the fabric. This makes the design or stripe of the desired look. For this type of wash, the printing must be pigment printed with a binder. Baking treatment should also be performed so that the colour will stay on the fabric more or less securely to coincide with the caustic wash to be done later.

26.6.13 Garment wash and overdye

This is an additional dyeing after the jeans have been sewn. This ensures a very deep dark or black colour and makes the sewing thread blend in with the jeans colour. This is performed in the following way:

- Wash the denim garments with stone so that the double needle seams, pocket flaps and exposed parts get washed down to a light blue colour or white.

- Put the dye into the tumble to dye the garments to get the desired colour.

- A coat of new colour will appear on to the garment, especially, in areas where the garment has been washed to a light shade. This creates a special but different look. In this process of wash, the lining or pocketing will pick up the colour too. By this wash, direct dye or reactive dye, same as dyeing fabrics or yarn may be used. Direct dye may be used with concomitant use of a 'colour-fixing agent' after dyeing, to make the colour more stable. In case of a solid colour, fabric staining within the garment is not a problem. However, if garments of different colours are washed together by the consumers, the colour may transfer to other garments, so reactive dye is more preferable.

26.6.14 Whitening

Whitening agents are used to create a super white look in denims, where there are coloured warp threads and white weft threads. If such garments undergo 'stone wash and bleach', whitening powder is used in the final rinsing. This makes the white threads in the fabric whiter and generates a stronger contrast between the blue and white on the surface of the fabric.

After washing the denim, check the reverse side of the fabric to evaluate if adequate whitening agent has been used during rinsing. It is a common practice to wash garments having white parts with whitening powder at the time of rinsing. This generates a lively and desired look.

26.6.15 Denim bleaching

In this process, a strong oxidative bleaching agent, such as sodium hypochlorite or $KMnO_4$ is added during the washing with or without the stone addition. Discoloration is usually more apparent depending on the strength of the bleach liquor quantity, temperature and treatment time.

Limitations of denim bleaching are:

- Process is difficult to control, that is, it is difficult to reach the same level of bleaching in repeated runs

- When desired level of bleaching is reached the time span available to stop the bleaching is very narrow

- Due to harshness of the chemical, it may cause damage to the cellulose resulting in severe strength losses and/or breaks or pinholes at the seam, pocket, and so forth.

- Harmful to human health and causes corrosion of stainless steel

- Required antichlor treatment

- Problem of yellowing is very frequent due to residual chlorine

- Chlorinated organic substances occur as abundant products in bleaching and pass into the effluent where they cause severe environmental pollution

Bleaching with glucose, sulphinic acid derivatives and recently with laccase (enzyme) is being developed as an alternative. Laccase enzyme belongs to the oxidoreductase group. Laccase's oxidative effect is complex and it does not work independently. A mediator is necessary and a chemical mediator is employed between the enzyme and indigo.

26.6.16 Sand blasting

Sand blasting technique is based on blasting an abrasive material in granular, powdered or other forms through a nozzle at a very high speed and pressure onto specific areas of the garment surface to be treated to give the desired distressed/abraded/used look.

Advantages of sand blasting process:

- It is a purely mechanical process, not using any chemicals
- It is a water-free process, therefore, no drying required
- Variety of distressed or abraded looks possible
- Any number of designs could be created by special techniques

26.6.17 Mechanical abrasion

To give a worn-out effect, abraded look or a used look, some mechanical processes have been developed. These are based on mechanical abrasions by which the indigo can be removed. Some of these processes are sueding, raising, emerizing, peaching and brushing.

Advantages of these processes:

- Control on the abrasion
- Different look on the garment can be achieved
- All are dry processes
- Economical, ecological and environmentally-friendly.

26.6.18 Ozone fading

By using this technique, the garment can be bleached. Bleaching of denim garments is done in a washing machine with ozone dissolved in water. Denim garments can also be bleached or faded by using ozone gas in a closed chamber. The advantages associated with this process are:

- Colour removal is possible without losing strength
- This method is very simple and environmentally friendly because after laundering, ozonized water can easily be deozonized by ultraviolet radiation.

26.6.19 Water jet fading

Hydrojet treatment has been developed for patterning and/or enhancing the surface finish, texture, durability and other characteristics of the denim garment. Hydrojet treatment generally involves exposing one or both surfaces of the garment through Hydrojet nozzles.

The degree of colour washout, clarity of patterns and softness of the resulting fabric are related to the type of dye in the fabric and the amount as well as the manner of fluid impacts the energy applied to the fabric. Particularly good results are obtained with blue indigo dyed denim.

As this process is not involved with any chemical, it is pollution free. By using water recycling system, the technique can be used as an economic and environmental friendly denim processing method. Colour washout of dye in the striped areas produces a faded effect without blurring, loss of fabric strength or durability or excessive warp shrinkage.

26.6.20 Single-bath stone-washing and tinting technique

Tinting of denim garment is usually done after the stone wash process. In this, garment has been lightly coloured in order to give the final denim appearance a slight shift. This is not true about over dyeing but merely gives the impression of a change in the overall colour of the fabric. This process consumes large quantity of water and chemical. To make this process economical and ecologically friendly, some novel colour-based enzymes have been introduced in the market. By using this new technique, tinting and stone-washing effect can be achieved in a single bath.

The advantages of this system are:

• Less process time to achieve a tinted look.

• No extra chemical required, therefore, making process is more economical.

• Less water consumption.

• Less energy consumption.

• Less chance of patches or unevenness

26.6.21 Cloud wash

Cloud wash gives appearance of white patches on the garment surface that looks like clouds in the sky.

26.6.22 Laser marking/spray painting

It is a new development. The jeans laser marker and the jeans spray robot are used for this. However, the costs are relatively high for these machines.

26.7 Different types of washing faults/defects

Washing is one way of cleaning, namely with water and often some kind of soap or detergent. Washing is an essential part of wet processing. During the washing process, a number of faults occurred. Some occurred in the form of mechanical faults, while some were process faults.

Washing can also affect different garment parts differently. Shell fabric may be solid coloured, while pockets may be in the form of white pocketing piece. After washing, it may appear that pocketing piece has been stained. This may happen because too much softener may have been used or the fabric bleeding colour is of weak quality. Though softener makes the fibre soft, it also breaks the dye loose and may stain the fabric. Too much use of softener makes the fabric waxy. So, one should cut a balance in using softener and other components in a normal wash. In such cases, the garment should be washed with less softeners and should be checked if the staining problem has been solved or eliminated. If use of less softener cannot solve the problem of staining, it indicates poor colour quality of the fabric.

Colour change, colour staining and poor crooking (rubbing) are not accepted by the buyers. To improve this, following steps may be taken:

1. Have 40°C hot water in a tumble washer

2. Put 1% colour fixing agent and mix with hot water

3. Put the dry garment into the tumble washer and let them soak in the solution for 5 min and then lift them up for drying

4. Run the tumble washer a few times and stop

5. Do not run the washer too much; it may lead the dye to come off and stain

6. Just let the garments soak and absorb the chemicals. Move the garments inside the washer as gently as possible; when the garments are dried, the colour quality substantially improves. After treating the garments in the mentioned way, they can simply be imposed to a normal wash to give it the washed look.

Different types of washing faults/defects normally noticed are:

a) Over blasting/low blasting

b) Over grinding/low grinding

c) Bad smell due to poor neutralization

d) Poor hand feel

e) Too hairy

f) Poor brightness

g) Hole formation after washing

h) Very dark and very light

i) Bleach spot

j) Bottom hem and course edge destruction

k) Running shading

l) High or low affect/abrasion on garments

m) Spot on garments

n) Out of range pH value of garments.

o) Colour shade variation

p) Crease marks

Miscellaneous operations in value addition

27.1 General

In the previous chapters, we have discussed some value addition processes in textile fabric manufacture. There are some intermediate processes, may be value-adding or non-value-adding, but essential for the processing of fabrics. Let us try to understand their purpose and functions.

The activities of preparing batches from the grey fabric received, rotating the wet fabric rolls in order to complete reaction and to avoid settling of colours or chemicals at one place, opening the wet fabric rope and making it open width and folding the processed fabrics as per market requirements are some of the processes, which might not add value directly, but are essential.

27.2 Grey inspection

The activity of grey inspection may be done in the weaving section before dispatching them to value addition processes or done in the process house after grey fabrics are received (Fig. 27.1).

Fabric is checked to verify whether the grey fabrics are in conformity with the requirements and agreed standards, and all weaving faults are marked out. Fabric inspection involves three possible steps: perching, burling and mending. Perching is a visual inspection and the name derives from the frame, called a perch, of frosted glass with lights behind and above it. The fabric passes through the perch and is inspected. Flaws, stains or spots, yam knots and other imperfections are marked. Burling is the removal of yam knots or other imperfections from the fabric. The faults are then mended and any knots in the material are then pushed to the back. Mending is the actual repairing of imperfections. Knotting should be done carefully and thoroughly so that the repair or holes are not visible.

Figure 27.1   Grey inspection machine.

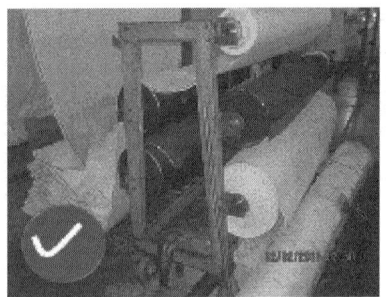

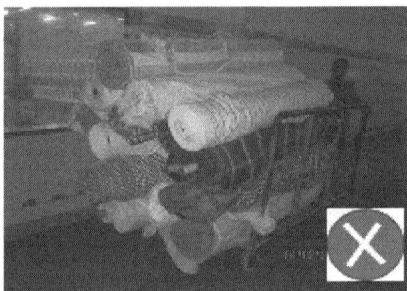

Correct handling of rolls   Too many rolls in one trolley

Figure 27.2   Transporting fabric rolls.

27.3   Bringing the rolls to process house

The fabrics inspected are brought to process house in roll trollies (Fig. 27.2). Care should be taken so that the rolls are properly identified as per sort number and loom number. The fabric should not be allowed to soil while handling.

27.4   Batching the fabrics

After the grey fabrics are inspected and checked, they are classed in the grey room, according to quality and stamped. Goods of similar weight, width and construction and the goods which will receive a similar treatment are sewn together, end to end, by sewing machines, specially constructed for this purpose, and each batch is given a number called lot number.

Figure 27.3   Fabric batch preparing machine.

The fabrics are usually sewn using a simple stitching machine. Stitching should be done in such a manner that the creases in fabric at the time of stitching are avoided. The use of proper stitching thread is necessary to avoid stitch marks during colour padding. For heavy fabrics intended for mercerizing and continuous operations, the seam should be wider (15 mm) and stronger.

The fabrics stitched may be rolled on a roller, which is normally referred to as a batch (Fig. 27.3). Alternately, the fabric can be put in a box trolley, where the length of fabric is much less and the fabric is fed to the next machine in a loose form.

27.5 Brushing

The precleaning of grey fabrics may be carried out in a separate unit just before cropping and shearing operations. The purpose of brushing is to remove the short and loose fibres from the surface of the cloth (Fig. 27.4). It

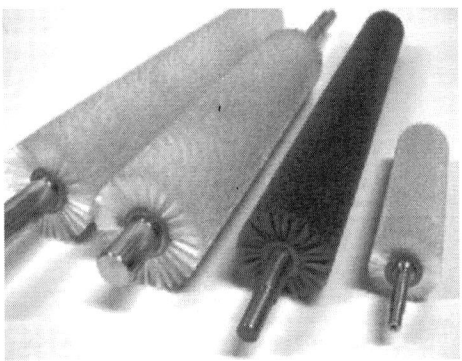

Figure 27.4   Brushes.

Figure 27.5   Fabric debatching machine.

also removes husk particles clinging to the cloth. Brushing is mainly done to fabrics of staple fibre content, as filament yams usually do not have loose fibre ends. Cylinders covered with fine bristles rotate over the fabric, pick up loose fibres and pull them away by either gravity or vacuum. The raised fibre ends are cut off during shearing operation. Brushing before cropping minimizes pilling.

27.6   Debatching

Once the required value addition processes are over on a fabric batch, it needs to be debatched. A fabric debatching machine with cutter helps in debatching the big fabric roll (Fig. 27.5). The fabrics are separated as per their original roll numbers taken from grey inspection. The stitches made for batching are either removed using a cutter or the fabric is cut just by the side of the stitches using a sharp cutter.

27.7   Rope opening

In processing, some operations are done in rope form and some in open width. For example, jet dyeing of fabrics, drum washing, dyeing in winch and so forth are done in rope form, whereas padding, stentering, chain mercerizing and singeing are done in open form. Rope opening machines are used to convert the wet fabric from rope form to open width (Fig. 27.6).

Figure 27.6 Rope opening machine.

27.8 Spot washing

The purpose of spot washing is to remove stains, if any, on the fabrics. It is done manually with the help of spotting auxiliaries, brush, water and solvent. The procedure involves following four steps:

a) Identifying the type of stain. It is essential to study the stain in detail before doing spotting.

b) Application of chemical as per stain seen. The use of appropriate chemical which does not harm fabric and the man is very important.

c) Tamping is the mechanical action which aids in the removal of stain. We should not allow harsh rubbing.

d) Applying water to dilute the chemical or solvent being used to avoid local spot. We should not allow deposition of chemical as it can damage the fabric later.

The spot washing is not real value addition activity but is a preventing activity of value loss. It is an indirect value addition activity.

The procedure normally adopted is as follows:

1. Clean the table and its surroundings before starting any work on the machines.

2. Bring the required fabric batch for spotting as per the instructions of the supervisor.

3. Bring spotting chemical, brush, pot and bucket full of water.

4. Verify the material brought against the programme given to you for that spotting. Check both the quality and the length in metres for each batch.

5. After verifying and ensuring that the materials are correct as per programme, start spotting the material.

6. Take the chemical for spotting as per the instruction given to you by the supervisor.

7. Refer the instructions given by your supervisor for spotting. If any problem is found in removing stain, then take instruction from supervisor.

8. Enter the details of the design number, batch number, lot number, quality number, the process done, the quantity of material processed and the problems faced in the production record.

9. If you find any irregularity or discrepancy, inform the supervisor immediately and take directions to correct the situation.

10. If you notice any problem in the spotting, may be of light or chemical, inform the concerned supervisor immediately and get it corrected. Enter complete details in the production report book.

11. Keep the material at the designated place as instructed by the supervisor.

27.9 Mending defects

The purpose of mending is to ensure that objectionable faults in the fabric are not passed on to end users. The fabrics made with fancy yarns or fancy designs and processes are high value-adding, but are highly dangerous when fails in the market because of unacceptable quality. Hence, special care is to be taken while dealing with uncommon and fancy products.

Mending is an operation of correcting minor defects/faults which are spoiling the appearance of the fabric. The faults are either removed by using a pincher or painted to make it not visible. The threads are realigned after removing a fault, and finally, the fabric is made acceptable to the customer. The faults are judged as mendable or non-mendable. The fabric is cut into short piece when a fault is non-mendable.

Mending is really not a value-adding process, but a preventer of value loss. If mending is not done, the value loss shall be very high compared to value added by various processes. It is, therefore, very necessary to prevent value losses. Mending is an indirect value-adding activity.

27.9.1 Stages of mending

Mending is done at three different stages. They are in the grey stage, in dyeing and after finishing.

a) Mending in the grey fabric stage is done, where the loose and entangled yarns, bunches, big knots and smashes are identified and removed. It is very important as any loose yarns, bunches of yarns, and so forth, when present, shall not allow the dye to penetrate in padding, and there are chances of getting white spec.

b) In the process house, mending stage is decided depending on the type of material and the style of dyeing being done. The stage of mending shall be decided in the flow chart in the dyeing depending on the type of fabric and the type of dyeing to be done.

c) In the finishing mending, the faults, such as contaminations, thick neps, colour spots and so forth, are cleaned.

Tools used in mending are mending scissors, pincher, round-tipped needle and painting pens.

1. The fabric checker shall mark the mendable defects with washable colour, either by using a tailor's chalk or a sketch pen.

2. The fabric after inspection is sent to mending section along with the job card.

3. The details are first entered in the stock register having columns of batch number, design number, user name and number of metres.

4. The respective heads of process house, such as cotton, wool, polyviscose (PV) and polywool (PW), and so forth, give the priority of their section referring to delivery schedules.

5. The in charge of mending allocates the skilled workers for different jobs of mending.

6. The mending operator checks the fabric and identifies the areas to be mended as marked by the checker.

7. The mender discusses with the in charge in case some faults are non-mendable or the frequency of such faults is very high, which may make mending unviable.

8. In case the faults are too many and are non-mendable, the fabric shall be downgraded or offered to the customer for approval. Only mendable faults will be mended.

9. In case of thick slubs, contaminations or knots, the operator uses a pincher and removes the defect.

10. In case of double ends or floats, the yarn is taken out by using round-tipped needle.

11. Special needles are used to darn the damaged woven fabrics.

12. Darning, also referred as reweaving, is the repairing of damaged woven fabric. This is done with the use of a special hand tool. Either a new section of material is taken from a hidden area of the fabric and used for the repair or replacement threads are directly woven into the damaged area (Fig. 27.7). Care is taken to select the threads of same material, same shade andsame depth. Once this new section is created, then it is woven into the fabric that surrounds the damaged area. The resulting repair, in most cases, is not perceivable to the naked eye. The new section is woven into the fabric. The repair begins to appear seamless (Fig. 27.8). Even with a fine weave, the result can be nearly perfect.

13. After mending, a seal is put on the job card indicating that the fabric is mended.

14. The materials mended are accounted against each mender's name, and mending registers are maintained separately for different sections of process house such as cotton, wool, PV, polycotton, PW and so on.

15. The columns in the mending register are batch number, lot number, design number, user name, number of pieces, total metres in each piece, received date, mending date, mender and remarks. The signature of the person receiving the report at user department is taken.

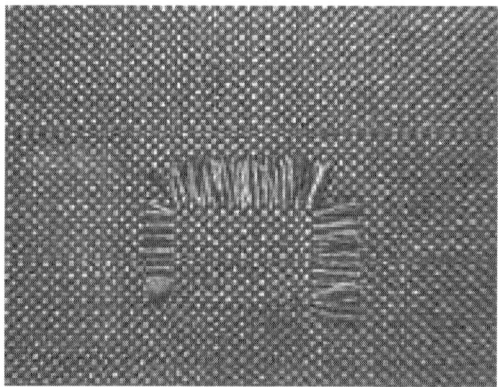

Figure 27.7 Preparing for darning.

Figure 27.8 After darning.

16. In case the damages are very high, they are entered in the high damage register with the columns of batch number, design number, metres and the type of damages seen. A sample is attached to indicate the type of faults and the signature is taken from the receiver of report in the user department.

17. Prefinish mending entry book is prepared after completing the mending job and giving the report to the user department for the purposes of accounting and making payment to the menders.

Mending fancyfabrics

Special care to be taken while dealing with fabrics having fancy yarns and dyed for a single component or cross-dyed:

1. Whenever a yarn is indented for special purposes, the Production Planning and Control section shall indicate the quality requirement of the yarns with specific terms.

2. If single component dyeing or cross-dyeing is adopted, the suppliers shall be informed that the yarns are going for single component dyeing, and yarns shall be specially prepared for that. No yarn shall be purchased from the market which is in stock.

3. A technical team shall visit the yarn manufacturers and approve the spinner for the supply of yarn for special purposes and fancy yarns. No yarn shall be purchased from the open market.

4. The supplier of yarn shall submit the test report of yarn when ordered specifically for special purposes.

5. Sample yarns shall be dyed and swatches are made to verify whether the yarn supplied is suitable for single component dyeing.

6. Marketing shall provide benchmark samples for deciding the acceptance levels for effects/defects to avoid overmending, over-rejections and accepting of substandard materials.

7. If marketing is unable to provide a benchmarking fabric sample, the Product Development Cell shall prepare three samples from different grades of available yarns and send for approval.

8. The marketing shall discuss with the customer and get the approval from the customer.

9. The approved sample shall be kept as benchmark. The approved sample shall be cut and made three reference samples. One sample shall be kept in Marketing, one with Quality Assurance and one with Product Development Cell.

10. A library of master samples shall be made by taking the benchmark yarns and the benchmark fabric samples given by marketing are compared with that. The approved master sample shall be preserved for reference.

11. The fabrics produced in bulk shall be compared with the approved master sample.

12. In the grey inspection, the loose and entangled yarns, bunches, big knots and smashes are identified. The loose yarns are trimmed off by using trimming scissors.

13. The checker shall mark on the grey fabric where mending is required. The grey fabric is mended at the greywarehouse as per the markings made by the checkers.

14. In process house, the fabrics are mended before mercerizing and after the vertical drying range drying.

15. In the intermediate fabric checking at process house, the checker shall verify the effects of mending to ensure that the mending is effective.

16. In the finished fabric inspection, the checker shall flag or put a sticker where mending is required.

Possible mending defects

Defect description	Reason
Double end	During mending, one extra end is not removed
Double pick	During mending, one extra pick is not removed
Loops	During mending, after filling ends or picks, they are not properly relaxed
Pinholes	During mending, if a knot on onside is relaxed unevenly, which is sheared in the shearing operation results as a pinhole after shrinking
Unmended	While mending, some defect misses the attention of mender. This normally happens if the checker has not marked the defect with a sticker or put a flag at the side of the fabric
Tight pick	While mending, if pick is tightly filled and not relaxed properly
Tails out	While mending, if tails are not properly cut or trimmed and not interlaced in tuck-in
Wrong mended	While mending, if end or pick is not properly filled as per the weave and pattern or if not relaxed properly after mending
Pincher marks	Pincher marks shall appear if the end or pick is tightly relaxed, creating abrasion while repairing. It can also happen because of using hard-quality rubber or relaxing directly with metal surface
Trailing picks	These are extra tuck-in length, more than 15 mm, which interlace in the fabric body, not pulled properly and cut during mending work
Mending soil mark	If the grey fabric is directly put on the floor or pushed from one place to another on a dusty floor. Wipe out the soil and dust periodically and do not keep fabric on the floor

27.10 Transporting fabrics for folding

Fabric, after completing all value addition processes, goes for folding. Horseback trollies (horsy) are more suitable for this operation as the length of fabric shall be less in each design (Fig. 27.9).

Figure 27.9   Horseback trolley.

27.11   Folding

The folding is done to ensure that the materials are folded as per the need of the customers for easy transportation and storing. The folding may be either by hand or machine with single fold or double fold. There are different requirements of the customer regarding the type of folding they need. It depends on the type of storage facility they have, the culture of the people and the ease of handling.

27.11.1   Fabric inspection and roll folding

Fabrics are inspected, cut and rolled using a roll folding machine (Fig. 27.10). A length counter is used to measure the actual length of fabric in each roll. Depending on the customer requirements, the fabrics are cut with predetermined lengths. The nonstandard lengths are packed separately with the label.

Hand folding—warkata (वारकाटा): A metal frame with a number of pins fixed at predetermined length is used for the purpose of hand folding (Fig. 27.11).

Figure 27.10   Fabric inspection, cutting and roll folding machine.

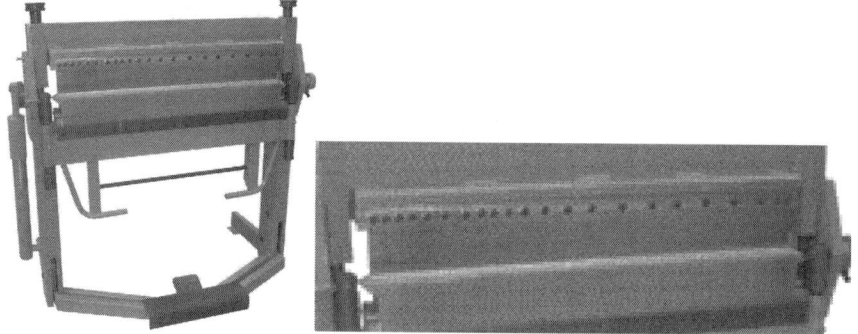

Figure 27.11 Hand folding.

Figure 27.12 Single folding.

One can select the pins depending on length needed in each fold so that it is easy to cut. Normally, 1 yd was very popular, and hence, the frame is called as 'warkata' in Hindi. *War* (वार) means yard and *kata* (काटा) means a measuring aid similar to a balance or stick.

Single folding: In single folding, the machine is adjusted to make folds of specified length, normally 1 m or 1 yd (Fig. 27.12). The machine can operate at 50 strokes/min. As 1 m is set per stroke, we can fold 50 m/h. However, depending on the length of fabric fed in each trolley, the time taken for set change shall vary.

Double folding: In double folding, the fabric is folded lengthwise by passing the fabrics on a 'V' plate and then folded for a given length, a metre or a yard (Fig. 27.13). Machines are available to make roll or book of double-folded

Figure 27.13   Double folding.

Figure 27.14   Book folding.

fabric. There are machines available which can be used for both single folding as well as double folding.

In case of machine folding, after folding, the number of folds is counted, and at the same time, the fabric is inspected. In hand folding, the folder himself counts the number of folds before removing the fabric from the stand.

Book folding: Book folding is the term used for folding the fabrics in small book size so that they can be put in the showcases within the space available like books kept in a library (Fig. 27.14). The size depends on where the fabrics are going to be shared. Fabrics of a *Thaan* are wrapped on either a thick hardboard or a wooden frame either by hand or by using a machine.

Figure 27.15 Roll packing.

27.12 Packing

The purpose of packing is to ensure that the fabrics inspected and sorted out are packed in suitable bale form (Fig. 27.15). The packing is carried out as per the guidelines given by marketing.

The packing of rolls into bales are done as per the following steps:

1. The rolls after shade sorting are grouped as suggested in the sorting report.

2. The sorted rolls shall be put in polythene bags and sealed with cello tape. Packer shall check the batch number on roll sticker and verify with the inspection report.

3. Number of rolls to be put in a bale is decided on the length of the rolls.

4. Data is fed into the system to generate barcode and barcode sticker is printed.

5. The barcode sticker of piece number is fixed on a roll after matching the metre and temporary roll number.

6. Bale marker will write down bale number at two places, design number, metre and number of pieces in the bag. Wherever specified, the shade number is also written.

7. The rolls as required are put in the bags by the helpers.

8. The stitcher shall stitch the bags.

9. Bale marker marks on the inspection report after rolls are put in the bale.

10. The inspection report after marking is given to record-keeping, where data is entered into the system.

27.13   Sample preparation and presentation

One may provide various properties and super finish for a fabric by adopting any of the value addition processes; however, unless it is sold and money is realized, the value addition has no meaning. It is necessary to educate the customer and show him your capabilities. Sample preparation and presentation help in this direction.

Samples of fabrics can be presented in a number of ways. Following are some examples.

*   Swatch cards—surface-mounted cards, sandwich cards, waterfall cards, pad cards, digital cards and presentation boards (Figs. 27.16–27.21).

*   Memo samples—ticketed memo samples, back-printed memo samples, labelled memo samples and chain memo samples.

*   Hanging samples—individual hanging samples and presentation hangers.

*   Sample books—stack books and waterfall books.

Figure 27.16   Surface mounted card.   *Figure 27.17*   Sandwich card.

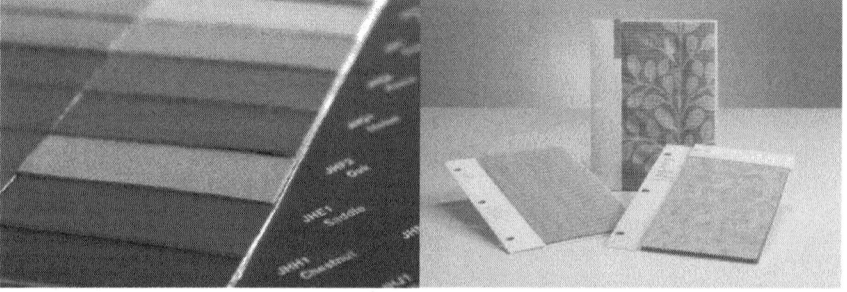

Figure 27.18   Waterfall card.   *Figure 27.19*   Pad card.

Figure 27.20 Digital card.   *Figure 27.21* Presentation board.

27.13.1 Swatch cards

Swatch card is the most common way of presenting material samples. Designers and consumers prefer this to get a quick, convenient overview of what is available in the materials catalogue. The variations in swatch cards include surface-mounted cards, sandwich cards, waterfall cards, pad cards, digital cards and presentation boards. The quality of swatch card is as important as the quality of fabric being presented. If the swatch card is not good, it is not possible to impress the customer and sell the products. It needs:

a) The right swatch in the right position on every card

b) Swatches that are straight, without stretching or wrapping

c) Sharp, clean die cuts

d) Clean, crisp printing

e) No smudges, smears or glue residue

The swatch cards are also presented online through the internet. The swatch card is scanned and put on the web so that the customers can select the samples. Once the samples are selected, orders may be placed online or else the customer may visit the supplier and see the sample personally and have a complete feel of the fabric.

Sample cards contain different samples of very small size of the same shade with different depth of dyeing in a row so that the customer can select the fabric and shade as per his choice (Fig. 27.22). A reference catalogue shall be given, indicating the sample number so that customer can quote that number in the orders placed. On the cards, only the sample identification number is given without any other details. Sample cards are given to regular customers, which can be kept in their offices and refer while placing orders.

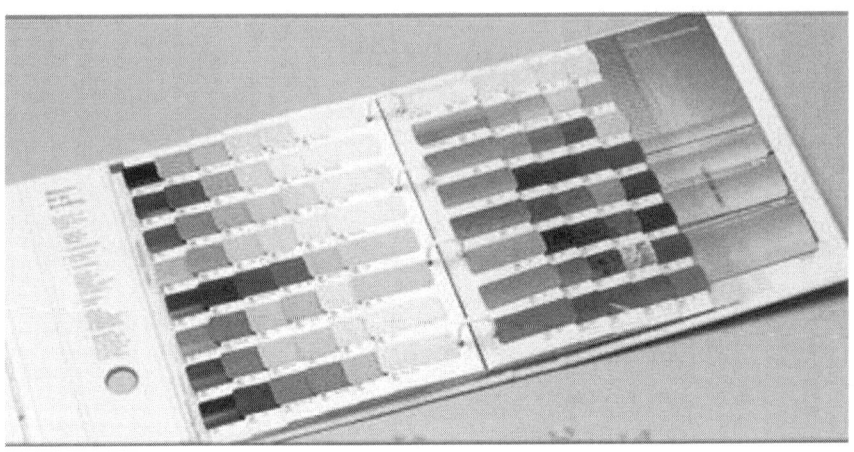

Figure 27.22   Sample cards.

27.13.2 Memo sample

A memo sample is an individual material sample used to confirm the designer's or consumer's initial material selection. Designers also use memo samples to present material recommendations to their clients. They are usually the final selection tool for the surface material and typically become the informal 'contract' specifying the material the end-consumer expects from their purchase.

In memo samples, one large-sized sample shall be attached to a card which contains all the details related to the samples. The details may be printed either on the back of the card or on the top or bottom of the face. This card shall be given to the fabric developers and also to inspectors, who may be internal auditors or final auditors.

Memo samples include ticketed memo samples, back-printed memo samples, labelled memo samples and chain memo samples (Figs. 27.23–27.26).

Back-printed memo samples are used only on surface materials that can accept ink directly on their backing. This form of memo identification is mostly limited to wall coverings, which are scrim, paper or spunback.

Labelled memo samples are used when graphic continuity with back-printed memos is wanted, but the quantity or material type does not allow for high-quantity back-printing.

A chain memo sample or peg sample is generally just a smaller version of a memo sample. The sample itself is usually smaller than the fold-over 'ticket' used for identification. Chain samples usually come in a set, secured with a

Figure 27.23 Ticketed memo sample. *Figure 27.24* Back-printed memo sample.

Figure 27.25 Labelled memo sample. *Figure 27.26* Chain memo sample.

ball chain such as those used for key rings. Peg samples are usually placed individually on wall display pegs in showroom scenarios.

27.13.3 Sample hangers

Sample hangers are kept on display in showrooms and marketing offices (Fig. 27.27). The samples are relatively of a larger size compared to samples presented in the sample cards. They are normally made of cardboard and, sometimes, plastic. It has a slot at the top so that it can be hanged on to a rod in the wardrobe or in a showcase.

Individual hanging samples are commonly used in retail or in a designer's showroom. Hanging samples are typically large enough to allow the user the ability to drape, fold or hang the sample in a way that matches its intended use, such as covering an upholstery frame or window.

Presentation hangers can also be used to show suggested colour and pattern correlations, while still allowing sufficient swatch size to effectively show the pattern design.

Hanging samples can be pinked or straight cut and serged. Serging is the most popular method, as these samples are subjected to a lot of use and handling. Hanging methods range from turned-edge cases with metal hanger units to plastic skirt hangers and fully incorporated die-cut cap units. Identifications can be applied directly to the incorporated hanger or they can be attached to the swatch with a label or fold-over ticket.

Figure 27.27   Sample hangers.

With hangers, the customers get a complete feel of fabric, the patterns, which help them in taking a decision, whereas with sample cards, as the sample size is very small, complete idea of the fabric cannot be obtained.

The hangers are easy to replace. The fabric is pinned with a stapler to the hanger. They can be arranged and rearranged depending on the theme and the occasion.

Individual hanging samples are the most common form of sampling in the retail environment (Fig. 27.28). Individual hanging samples are bound into a hanger-style cap and are usually identified with tickets or ID cards. The tickets can virtually be applied to any surface material, allowing for a consistent and contiguous graphic style.

Presentation hangers are ideal selling tools when you want to present a coordinated set of materials in a retail environment (Fig. 27.29). They are bound to a hanger-style cap and are usually identified with identification strips or individual labels.

27.13.4   Sample books

Sample books are made for taking different samples to the customers' place and explaining the salient features. The samples are bigger compared to cards but are smaller compared to hangers. In one book, a number of samples can be fixed and is easy to carry from place to place. In some showrooms, the

Figure 27.28 Individual hanging samples. *Figure 27.29* Presentation hangers.

Figure 27.30 Sample books.

books are also placed so that the customer can open the book and see the samples without taking the samples out.

Sample books are most often a permanently bound presentation (Fig. 27.30). This allows for a more controlled presentation to the viewer, encouraging materials to be seen together. This permanence also means that the entire presentation is dependent on the timeliness and longevity of the entire collection of materials shown. With smaller swatch sizes and manufacturing processes more suited to large quantities, sample books are a good way to make the product viewable in more locations. Sample books include stack books, waterfall books and specialized books (Figs. 27.31–27.33).

Stack books and stack pads are the most common sample books used in the residential and hospitality markets, which consist of individual swatches stacked on top of each other and permanently bound into a turned-edge case.

Figure 27.31 Stack book.   *Figure 27.32* Waterfall book.

Figure 27.33 Specialized book.

A waterfall book consists of multiple individual swatches assembled together onto a page before final assembly into the turned-edge case. The swatches stair-step over each other on a single page, allowing all colours to be viewed together. Waterfalling allows the viewer to see several swatches at the same time, providing quicker access to their desired colour.

Specialized books are highly customized ways to present material samples. Sometimes, the product is unique enough to warrant a little extra-special merchandizing to get the attention it deserves.

27.14 To conclude

There are a number of activities and processes which help in adding value to the fabrics by giving the required effects and feel to the fabric. There are numbers of subprocesses, which may not directly add value, but are essential. Other value-adding processes cannot be done without these processes. It is essential to identify all processes and understand their purpose. The processes need to be designed to achieve the purpose. When the working people understand the purpose and align their activities to achieve the same, there shall be real value addition.

28.1 General

In any process, there may be some problems. The competency of a technician is proved by the way he/she tackles the problems and solves them. There is no need for everyone to do experiments, but if they study and learn from the experiences of others, they can solve a number of problems. In this chapter, some normal problems are listed. However, each mill may have its own problems and it is needed to address them within the mill. It is practically impossible to list all the problems faced across the world relating to the value addition processes.

It is found that often, the root cause(s) of a problem in the dyed material can be traced as far back as to the raw materials used like the cotton and its consistency, the variations in blends, the variations in twist levels in the yarns used and the variations in fabric construction. However, the process control in value addition processes is also very important which can reduce the level of the problem.

28.2 Problems originating from the fibre and yarn

The problems originated from the fibres and yarn are as follows:

a) Problems caused by immature and/or dead cotton in which white- or light-coloured specs (neps) found in dyed material.

b) There are several stages in the spinning preparation where an attempt can be made to decrease the amount of neps of the immature and/or dead fibres that are usually clumped together. The process adopted should suit the cotton being processed.

c) Any mistake in setting the machines in spinning preparatory or spinning, the worn out machine parts can lead to more imperfections and twist variations which can result in streakiness after dyeing.

d)   Dyeability variations in cotton obtained from different sources can lead to shade variations in the fabrics. Therefore, the dyers normally demand spinners procuring cotton consistently from a single source. Since some dyestuffs are more sensitive to dyeability variations than others, those dyes should be selected for dyeing which are less sensitive to dyeability variation.

e)   The level of contamination in raw materials is another factor which affects the dyeing quality. For example, the level of contamination in cotton is affected by geology of cultivation area; soil constitution; weather conditions during the maturing period; cultivation techniques; chemicals, pesticides and fertilizers; as well as harvesting techniques. For the dyer, the elements that pose the greatest threat are alkaline earth and heavy metal contaminants such as calcium, magnesium, manganese and iron. Depending on its origin, raw cotton can exhibit widely different contents of alkaline earth and heavy metal ions.

f)   Variation in cotton colour grade has an effect on the colour yield of dyed goods.

g)   Studies by Hussein and his team indicate that as much as 25% of the faults responsible for downgrading cotton finished garments may be attributed to yarn.

h)   Variations in winding packages cause variations in dyeing in case of yarn dyeing, resulting in variation in the effects in dyed yarn woven fabrics.

28.3   Problems originating from fabric formation

There are problems that become more apparent after dyeing but are attributable to weaving. They include:

- Variation in the warp density of the cloth (wrong draw, missing end, double end)
- Selvedges thicker than the centre of the fabric
- Variation in size application on warp yarns
- Variation in drying of warp yarn after sizing
- Variation in warp tension during weaving
- Variation in weft density (missing pick, double pick)

- Variation in warp or weft yarns with respect to twist, twist direction, count, hairiness, colour, tensile properties, fibre composition and/or spinning batch

- Fly or foreign matter or fibre woven into the fabric

Major problems that become more apparent after dyeing but may be attributable to knitting include:

- Variation in course length

- Variation in wale density

- Variation in yarn with respect to count, twist, twist direction, hairiness, colour, tensile properties, fibre composition, lubrication and/or spinning batch

- Vertical lines of distorted loops, tuck stitches or cut stitches

- Fly or foreign matter knitted into the fabric

28.4 Problems due to water quality

The quality of textiles produced by any manufacturing operation which employs wet processes, such as preparation, dyeing and finishing, is generally affected by the water quality. The textile processes are influenced in different ways by the presence of impurities in the water supply and there are several major water use categories to be considered including water for processing, potable purposes, utilities and laboratory use. Each requires different water-quality parameters. Process water is to be mainly used for making concentrated bulk chemical stock solutions, substrate treatment solutions and washing. Potable water is used for drinking and food preparation. Utility use includes noncontact uses such as boiler use, equipment cleaning and so forth. Water from almost all supply sources contains impurities to some extent. The type and amount of impurities depend upon the type of water source.

The most common impurities that may be present in water are calcium and magnesium (hardness), heavy metals (such as iron, copper, manganese and aluminium), chlorine, miscellaneous anions (sulphide, fluoride, etc.), sediments, clay, suspended matter, acidity, alkalinity, and buffers, oil and grease and dissolved solids.

Contaminants from the water source are not the only ones found in textile water supplies. There are major internal contributions as well. Common sources of internal contamination are clear well (used for water storage), greige goods or other substrates, plumbing, valves and so forth, machinery and prior processes in the case of water reuse.

Water contaminants, especially metals, can have a substantial effect on many textile wet processes. The effects are not always adverse but even when a process is enhanced by water impurities, it is not desirable to have variance in processes and product quality due to water quality changes. Such variations in the quality of water make process and machinery optimization and control difficult.

Metallic ions in water can have a dramatic effect by either enhancing or inhibiting the action of many preparation processes. All of the wet preparation processes are affected in some way by metallic ion contaminants in water. In enzymatic desizing, the metallic ions may cause inactivation of the enzymes, resulting in poor size removal.

It is a common practice in some mills to use potable water for the laboratory supply while using non-potable water for production processing. Since potable water is usually chlorinated, it can alter the shade of dyeing and contribute to poor lab-to-bulk reproducibility. Moreover, most work in analytical laboratories is done with distilled and/or deionized water. However, many situations arising in textile wet processing laboratories will require the use of process water in order to correlate well with production. The laboratory technician must be able to decide when to use process water and when to use distilled or deionized water.

In scouring processes, calcium and magnesium ions (water hardness) cause the most problems. These ions precipitate soaps and form a sticky insoluble substance depositing on the substrate. Such deposits impair the fabric handle, resist dyeing, attract soil to the material and cause inconsistent absorbency in subsequent processes. Although most synthetic detergents used in scouring today do not precipitate in the presence of calcium and magnesium ions, the fatty acid hydrolysis products formed by the saponification of natural waxes, fats and oils in the fibres will precipitate. The formation of complexes with alkaline and alkaline earth salts drastically reduces the solubility and the rate of dissolution of surfactants, thus impairing the wash removal ability of the surfactants. Bleaching with hydrogen peroxide is greatly affected, even by trace quantities of metal ions in the water. The transition metal ions such as iron, copper, manganese, zinc, nickel, cobalt and chromium catalyse the decomposition of hydrogen peroxide, which is very rapid and frequently occurs before any bleaching can occur. In addition, the decomposition products attack cotton fibres leading to their degradation. Bleaching baths containing these ions will lead to a reduction in whiteness and high loss in fibre strength, as well as an increase in fluidity. The alkaline earth metal (magnesium), on the other hand, produces beneficial effects when present in peroxide bleaching solutions. These ions increase the stability of hydrogen peroxide under alkaline bleaching conditions, and as a result, increased whiteness and less fibre degradation are obtained. Electrolytes of other metals may have a harmful effect.

The most commonly observed dyeing problems caused by poor water quality include inconsistent shade, blotchy dyeing, filtering, spots, resists, poor washing-off and poor fastness.

Various measures and treatments may be employed in order to remove impurities from water and to avoid problems in textile processing, such as follows:

- Sedimentation and filtration treatments

- Softening treatments (such as cold lime-soda softening or zeolite softening)

- Reverse osmosis

- The use of sequestering agents

28.5   Problems in singeing

There are singeing faults that are optically demonstrable and are quite easily remedied during the actual working process. On the other hand, there are singeing faults that are not visible until after dyeing and that can no longer be repaired once they have occurred. Following are some examples.

Problem	Possible causes	Countermeasures
1. Incomplete singeing	• Low flame intensity • Very fast fabric speed • More distance between the fabric and the burner • Inappropriate singeing position (not severe enough) • Too much moisture in the fabric incoming for singeing	• Optimum flame intensity • Optimum fabric speed • Optimum distance between the fabric and the burner • Optimum singeing position • No excess moisture in the incoming fabric for singeing

Problem	Possible causes	Countermeasures
2. Uneven singeing (widthwise)	• Uneven moisture content across the fabric width • Uneven flame intensity across the fabric width • Uneven distance between the burner and the fabric	• Uniform moisture content across the fabric width • Uniform flame intensity across the fabric width • Uniform distance between the fabric and the burner
3. Uneven singeing (lengthways)	• Uneven moisture content along the fabric length • Uneven flame intensity along the fabric length • Change in fabric speed during singeing • Change in the distance between the fabric and the burner along the length	• Uniform moisture content along the fabric length • Uniform flame intensity along the fabric length • Uniform fabric speed during singeing • Uniform distance between the fabric and the burner along the length
4. Thermal damage or reduction in tear strength	• Very high flame intensity • Slow fabric speed • Very less distance between the fabric and the burner • Inappropriate singeing position (too severe)	• Optimum flame intensity • Optimum fabric speed • Optimum distance between the fabric and the burner • Optimum singeing position

There are some more practical problems during singeing, such as:

a) Uneven flame heights due to burnt fibres chocking the burners

b) Fluffs those have fallen on the fabric before singeing leading to excess burning

c) Improper brushing

28.6 Problems in desizing

The desizing procedure depends on the type of sizing materials used in sizing. It is, therefore, necessary to know what type of size is on the fabric before desizing.

Problem	Cause	Countermeasure
Incomplete desizing	• The pH of the bath is not appropriate • Inappropriate desizing bath temperature • Insufficient fabric pick-up • Insufficient digestion time • Poor enzyme activity • Deactivation of enzyme due to the presence of metals or other contaminants • Ineffective wetting agent • Incompatible wetting agent	• Optimum pH • Optimum temperature • Optimum squeeze pressure • Optimum use of wetting agent • Optimum digestion time • Use of good enzymes • Use of soft water • Use of appropriate sequestering agents • Use of good and effective wetting agent • Use of compatible wetting agent
Uneven desizing (widthways)	• Uneven pad pressure (across the width) • Nonuniform pad temperature • Nonuniform chemical concentration in the bath	• Uniform squeeze pressure • Uniform bath temperature • Maintaining uniform concentration of chemicals in the bath • Covering the batch with polythene or another suitable sheet • Keeping the batch rolling
Uneven desizing (lengthways)	• Uneven pick-up (along the length) • Preferential drying of outer layers of the batch • Temperature variation during digestion	• Uniform pick-up along the fabric length • Covering the batch with polythene or another suitable sheet • Keeping the batch rolling

Problem	Cause	Countermeasure
Uneven desizing (random)	• Poor wetting agent • Inappropriate bath temperature • Foaming in the bath • Improper use of defoamer • Uneven liquor distribution during padding • Nonuniform washing after desizing	• Use of effective and compatible wetting agent • Optimum bath temperature • Use of appropriate defoamers • Uniform liquor distribution during padding • Thorough and uniform washing after desizing • Covering the batch with polythene or another suitable sheet • Keeping the batch rolling

28.7 Problems in scouring

Depending on the amount of impurities and the reaction and wash conditions, the loss in weight of the raw cotton material due to boil-off change significantly. It can reach up to 7% or even higher in case of high-impurity cotton, whereas it may be just 2–2.5% in cleaner cotton. One has to decide the parameters and chemicals to be used depending on the impurities present in cotton, or else it results in improper scouring or increase the cost of scouring.

A very high concentration of caustic soda (e.g. >8% on weight of fabric) may result in a reduction in degree of polymerization as well as yellowing of the cotton fibre. The higher the concentration, the greater will be the fat removal which not only increases the absorbency but also introduces harshness in the handle of the material.

The inorganic polyphosphates in addition to sequestering most metals also aid in cleansing the fibres; however, they may hydrolyse at high temperature and lose their effectiveness. Oxalates and hydroxyl carboxylic acids (citrates, etc.) are excellent for sequestering iron but not effective for calcium and magnesium.

If the process temperature is above the cloud point of the surfactant, the surfactant may be ineffective and may actually be deposited on the substrate.

Higher scouring temperatures will reduce treatment time and remove fats and waxes completely, which will promote harsh handle of the material.

Selecting chemicals and deciding the process parameters is very important. Following table gives some guidance for actions to be taken for certain problems.

Problem	Possible cause	Countermeasure
Inadequate scouring or inadequate absorbency or high residual impurities (batch scouring of yarn or fabric)	• Very low concentration of scouring chemicals • Incompatible or ineffective surfactant or wetting agent • Very low scouring temperature • Inadequate scouring time • Inadequate washing after scouring	• Optimum concentration of scouring chemicals • Compatible and effective surfactant or wetting agent • Optimum scouring temperature • Optimum scouring time • Optimum washing after scouring
Inadequate scouring or Inadequate absorbency or high residual impurities (pad-steam scouring of fabric)	• Very low concentration of scouring chemicals • Incompatible or ineffective surfactant or wetting agent • Very low steaming temperature • Inadequate steaming time • Inadequate washing after scouring	• Optimum concentration of scouring chemicals • Compatible and effective surfactant or wetting agent • Optimum steaming temperature • Optimum steaming time • Optimum washing after scouring
Uneven scouring (random unevenness when scouring in fabric form)	• Poor stability of surfactant or wetting agent (cloud point below application temperature) • Water hardness or ineffective chelating agents • Nonuniform and/or ineffective washing after scouring • Improper use of defoamer	• Suitable selection and proper use of surfactant or wetting agent • Use of soft water or effective chelating agents • Uniform and thorough washing after scouring • Suitable selection and proper use of defoamer
Uneven scouring (random unevenness when scouring yarn in package form)	• Uneven package density • Yarn variations	• Maintain uniform package density • Control yarn variations (Good quality control of incoming yarn)

Problem	Possible cause	Countermeasure
Uneven scouring (widthways unevenness in pad-steam scouring)	• Uneven pad pressure • Nonuniform temperature across the bath • Nonuniform chemical concentration across the bath	• Maintain uniform pad pressure • Maintain uniform bath temperature • Maintain uniform chemical concentration in the bath
Uneven scouring (lengthways unevenness in pad-steam scouring)	• Variation in the concentration of scouring chemicals with time • Variation in the moisture content of the incoming fabric along the length	• Uniform concentration of scouring chemicals with time • Uniform moisture content in the incoming fabric along the length
Harsh handle	• Complete loss of natural oils and fats due to very high alkali concentration	• Optimum concentration of alkali during scouring
Resist marks	• Deposition of insoluble salts of surfactants • Redeposition of impurities	• Use of soft water or appropriate chelating agents • Careful selection of scouring auxiliaries • Thorough washing after scouring
Yellowing of the goods	• Very high alkali concentration • Very long dwell time	• Optimum alkali concentration • Optimum dwell time • Water purification • Use of appropriate complexing agent
Tendering or damage or loss in strength	• Presence of air in the machine, leading to the formation of oxycellulose • Contamination of iron	• Exclusion of air • Use of mild reducing agent • Demineralization (if iron present in the textile material)

28.8 Problems in bleaching

While bleaching with hydrogen peroxide, in order to get adequate bleach, there must be enough peroxide present from the start till the end of the process. The peroxide concentration based on the weight of the solution will determine the bleaching rate—the greater the solution concentration, the faster the bleaching.

Similar to any process using liquors, in peroxide bleaching system, its entire peroxide charge for active bleaching is not used, as some is always 'lost' during normal process. As the alkalinity in the system is primarily responsible for producing the desired scour properties, it is necessary to maintain a reasonably constant pH at the desired level throughout the bleaching cycle.

The quantity of the alkali to be added depends on the character of the goods, the finish required and the kind and quality of the other ingredients in the liquor. The pH value in peroxide bleaching is very important as it influences bleaching effectiveness, fibre degradation and peroxide stability in bleaching cotton fibres. With increasing pH, whiteness index increases to a maximum at a pH of 11.0 and then decreases. Fibre degradation is minimum at a pH of 9.0 but that which occurs at a pH of 10.0 is well within acceptable values. Above a pH of 11.0, fibre degradation is severe. Lower pH values can lead to decreasing solubility of sodium silicate stabilizer as well as lower whiteness due to less activation of the peroxide. By increasing the temperature, the degree of whiteness as well as its uniformity increase. However, at a very high temperature, there is a possibility of a decrease in the degree of polymerization of the cotton.

The fat removal action at high temperatures such as 110°C makes the handle of the material harsh and decreases the sewability of cotton fabrics.

Time, temperature and concentration of peroxide are all interrelated factors. At lower temperatures, longer times and higher concentrations are required. As the temperature of bleaching increases, shorter times and lower peroxide concentrations can be employed. The amount of peroxide decomposed is greatly reduced with increasing weight of cotton fibre in the bleach liquor. The raw fibre almost completely suppresses decomposition, while the scoured fibre is somewhat less effective.

The impurities such as magnesium and calcium have a good stabilizing effect when present in appropriate amounts. The impurities such as iron, copper and manganese have a harmful effect, resulting in catalytic decomposition of hydrogen peroxide leading to fibre damage. A good stabilizing system is indispensable in bleaching cotton with hydrogen peroxide.

The sodium silicate is one of the most commonly used stabilizers, but it results in a harsh handle of the fabric as well as resist spots leading to spotty dyeing. The alternatives to sodium silicate are organic stabilizers or a combination of silicate and organic stabilizers.

In addition to the normal ingredients of the bleaching recipe, namely hydrogen peroxide, caustic soda and the stabilizer, auxiliaries are used sometimes to aid the bleaching process. These may include surfactants and chelating agents. The type and concentration of these auxiliaries also play an important role in the bleaching effect obtained.

The most common problems in bleaching cotton with hydrogen peroxide are as follows:

- Inadequate mote removal

- Low degree of whiteness

- Uneven whiteness (or bleaching)

- Pinholes, tears, broken yarns, catalytic damage and loss of strength

- Resist marks

- Formation of oxycellulose

A summary of the possible causes of the problems and their countermeasures is given in the table below.

Problem	Causes	Countermeasures
Low degree of whiteness while bleaching yarn or fabric in batch form	• Inadequate concentration of hydrogen peroxide • Inadequate alkali concentration • Very low bleaching pH • Very short bleaching time • Very low bleaching temperature • Residual sodium acetate after neutralization	• Optimum concentration of hydrogen peroxide • Optimum alkali concentration • Optimum bleaching pH • Optimum bleaching time • Optimum bleaching temperature • Thorough rinsing after neutralization

Problem	Causes	Countermeasures
Low degree of whiteness while bleaching fabric by pad-steam process	• Inadequate concentration of hydrogen peroxide • Inadequate alkali concentration • Inadequate pick-up • Very low bleaching pH • Very short steaming time • Very low steaming temperature • Residual sodium acetate after neutralization	• Optimum concentration of hydrogen peroxide • Optimum alkali concentration • Optimum pick-up • Use of good wetting agents • Optimum bleaching pH • Optimum steaming time • Optimum steaming temperature • Thorough rinsing after neutralization
Uneven whiteness (random and/or lengthways) while bleaching of fabrics by pad-steam process	• Use of inappropriate surfactants • Water hardness • Irregular chemical feeding • Condensation or water marks • Foaming in the bath and inappropriate use of defoamer • Ineffective and/or nonuniform washing after bleaching	• Appropriate/compatible surfactants • Use only soft water or use of sequestering agents • Optimum chemical feeding • Maintain optimum steaming conditions • Appropriate use of defoamer • Thorough and uniform washing after bleaching
Uneven whiteness in widthways while bleaching of fabrics by pad-steam process	• Nonuniform pick-up with time • Variation in chemical concentration with time • Variation in steaming conditions with time • Variation in the fabric speed • Uneven pad pressure (across the fabric width) • Nonuniform bath temperature • Nonuniform chemical concentration	• Monitor uniform pick-up with time • Maintain uniform chemical concentration with time • Uniform steaming conditions with time • Uniform fabric speed • Maintain uniform pad pressure across the width • Maintain uniform bath temperature • Maintain uniform chemical concentration

Problem	Causes	Countermeasures
Harsh handle	• Silicate deposits • Very high concentration of alkali • Very high bleaching or steaming temperature	• Use of organic stabilizers • Optimum control of pH (low pH reduces silicate solubility) • Do thorough washing after bleaching • Maintain optimum concentration of alkali • Maintain bleaching and steaming temperature at optimum level
Fibre degradation or reduction in fibre strength	• Metal contaminants	• Demineralization to remove metals from the fibre • Treatment of water to remove metal contaminants • Use of appropriate complexing agents
Pinholes	• Localized fibre degradation usually due to the presence heavy metals	• Use of appropriate complexing agents for metal contaminants • Demineralization to remove metals from the fibre • Treatment of water to remove metal contaminants
Broken yarns	• Extreme condition of time, temperature and concentration of peroxide • Unstabilized hydrogen peroxide	• Use alternative stabilizer(s) • Appropriate ratio of NaOH • Proper stabilization of the bleaching liquor
Inadequate mote removal	• Very low bleaching pH/ alkalinity • Inadequate softening of motes during scouring	• Maintain optimum pH during bleaching and washing after bleaching • Adequate softening of motes during scouring
Resist spots	• Silicate deposits • Oxycellulose formation	• Use of appropriate stabilizer(s) • Optimum bleaching pH/ alkalinity

Problem	Causes	Countermeasures
Tears, loss in voluminous character of the material	• Extreme condition of time, temperature and concentration of peroxide • Very high alkalinity in the bleach liquor	• Optimum condition of time, temperature and concentration of peroxide • Optimum bleaching pH • Optimum alkalinity in the bleach liquor
Decrease in the elasticity of the material	• Very high bleaching pH • Very high alkalinity in the bleach liquor	• Optimum pH during bleaching and washing after bleaching • Optimum alkalinity in the bleach liquor
Low sewability of the material	• Very high bleaching temperature and/or alkalinity	• Optimum bleaching conditions • Optimum bleaching temperature and alkalinity

28.9 Problems in mercerizing

The main problems in mercerizing are (a) local mercerization leading to patchy dyeing, (b) low barium activity number and (c) width variations. The following table gives the possible causes and remedies for normal problems faced.

Problem	Possible cause	Countermeasure
Incomplete mercerization	• Low concentration of sodium hydroxide • Inappropriate wetting agent • Inappropriate temperature of the incoming fabric or the padder • Low pick-up • Insufficient contact time	• Maintain optimum concentration of sodium hydroxide • Use appropriate wetting agent • Maintain appropriate temperature of the incoming fabric or the padder • Optimum pick-up • Optimum contact time

Problem	Possible cause	Countermeasure
Low increase in lustre	• Low concentration of sodium hydroxide • Inappropriate temperature of the incoming fabric or the padder • Low pick-up • Insufficient contact time • Insufficient fabric stretching while on the frame • Too much caustic on the fabric as it comes off the frame	• Optimum concentration of sodium hydroxide • Appropriate temperature of the incoming fabric or the padder • Optimum pick-up • Optimum contact time • Optimum fabric stretching while on the frame • Removal of excess caustic from the fabric before it comes off the frame
Uneven mercerization (widthwise)	• Uneven pad temperature • Nonuniform bath temperature • Nonuniform alkali concentration in the bath • Nonuniform moisture in the fabric across the width	• Even pad temperature • Uniform bath temperature • Uniform alkali concentration in the bath • Uniform moisture in the fabric across the width
Uneven mercerization (lengthwise)	• Dilution of the bath with time • Increase in bath temperature with time • Lengthwise variation in the moisture content of the fabric • Variation in the pad pressure during the process • Variation in pick-up along the fabric length	• Uniform moisture content in the fabric • No variation in bath temperature with time • No lengthwise variation in the moisture content of the fabric • No variation in the pad pressure during the process • No variation in pick-up along the fabric length
Uneven mercerization (random)	• Ineffective and/or incompatible wetting agent • Low concentration of sodium hydroxide • Low pick-up • Poor shrinkage control	• Use of effective and compatible wetting agent • Optimum concentration of sodium hydroxide • Optimum pick-up

Problem	Possible cause	Countermeasure
Tearing of the fabric	• Insufficient fabric stretching while on the frame • Too much caustic on the fabric as it comes off the frame	• Optimum fabric stretching while on the frame • Removal of excess caustic from the fabric before it comes off the frame

28.10   Problems in dyeing

There are a number of problems of the reproducibility and difficulties in obtaining right-first-time dyeing. Two important aspects of dyeing, namely dye variables and system variables, along with important characteristics of dyeing such as exhaustion, migration and levelling, fixation and colour yield, and washing-off and fastness are responsible for this problem.

28.10.1   Problems with reactive dyes

In case of reactive dyes, there is a wide variety in terms of their chemical structure, and unless one understands the chemistry, he/she shall not be able to design the correct process to get the required results.

The two most important components of a reactive dye are the chromophore and the reactive group. Substantivity is more dependent on the chromophore as compared to the reactive system. A higher dye substantivity may result in a lower dye solubility, a higher primary exhaustion, a higher reaction rate for a given reactivity, a higher efficiency of fixation, a lower diffusion coefficient, less sensitivity of dye to the variation in processing conditions such as temperature and pH, less diffusion, migration and levelness, a higher risk of unlevel dyeing, and more difficult removal of unfixed dye. Substantivity is the best measure of the ability of a dye to cover dead or immature fibres. Covering power is best when the substantivity is either high or very low. An increase in the dye substantivity may be affected by lower concentration of the dye, higher concentration of the electrolyte, lower temperature, higher pH (up to 11) and lower liquor-to-goods ratio.

High dye reactivity entails a lower dyeing time and a lower efficiency of fixation. To improve the efficiency of fixation by reducing dye reactivity requires a longer dyeing time and is, therefore, less effective than an increase in substantivity.

There is a wider range of temperature and pH over which the dye can be applied. Reactivity of a dye can be modified by altering the pH or temperature, or both. By a suitable adjustment of pH and temperature, two dyes of intrinsically different reactivity may be made to react at a similar rate.

Dyes with higher diffusion coefficients usually result in better levelling and more rapid dyeing. Diffusion is hindered by the dye that has reacted with the fibre and the absorption of active dye is restrained by the presence of hydrolysed dye. Different types of dyes have different diffusion characteristics.

Dyes with better solubility can diffuse easily and rapidly into the fibres, resulting in better migration and levelling. An increase in dye solubility may be affected by increasing the temperature, adding urea and decreasing the use of electrolytes.

It is observed that a higher temperature in dyeing with reactive dyes results in a higher rate of dyeing, lower colour yield, better dye penetration, rapid diffusion, better levelling, easier shading, a higher risk of dye hydrolysis and lower substantivity.

While dyeing, the pH of the dyebath will not remain same and will reduce significantly as the dyeing proceeds. Different types of alkalis, such as caustic soda, soda ash, sodium silicate or a combination of these alkalis, are used in order to attain the required dyeing pH. The choice of alkali usually depends upon the dye used, the dyeing method as well as other economic and technical factors.

The addition of electrolyte results in an increase in the rate and extent of exhaustion, increase in dye aggregation and a decrease in diffusion. At lower liquor ratios, there is a higher exhaustion and higher colour strength. It is possible to enhance dye uptake on cellulosic fibres with the aid of suitable surfactants. The factors that affect the fastness of reactive dyes are the chromophore group, the stability of the dye–fibre bond and the completeness of the removal of the unfixed dye. To maximize wet fastness, particularly in deep shades, it is advisable to apply cationic agents for after-treatments.

It is, therefore, very necessary to understand the dyes and prepare the process sheet.

28.10.2 Problems in dyeing with direct dyes

Direct dyes represent an extensive range of colourants that are easy to apply and also are very economical. There are three common ways to classify direct dyes, namely, according to their chemical structure, dyeing properties and fastness properties.

Classification of direct dyes by the Society of Dyers and Colourists is based upon the compatibility of different groups of direct dyes with one another under certain conditions of batch dyeing; there are three classes of direct dyes: A, B and C. Class A consists of self-levelling direct dyes. Dyes in this group have good levelling characteristics and are capable of dyeing uniformly even when the electrolyte is added at the beginning of the dyeing operation. They may require relatively large amounts of salt to exhaust well. Class B consists of salt-controllable dyes. These dyes have relatively poor levelling or migration characteristics. They can be batch-dyed uniformly by controlled addition of electrolyte, usually after the dyebath has reached the dyeing temperature. Class C consists of salt- and temperature-controllable dyes. These dyes show relatively poor levelling or migration and their substantivity increases rapidly with increasing temperature. Their rate of dyeing is controlled by controlling the rate of rise of temperature as well as controlling the salt addition. The dyebath variables that influence the dyeing behaviour of direct dyes include temperature, time of dyeing, liquor ratio, dye solubility, and the presence of electrolyte and other auxiliaries.

Direct dyes can be applied by batch dyeing methods (on jigs, jet or package dyeing machines), by semi-continuous methods (such as pad-batch or pad-roll) and by continuous methods (such as pad-steam). Many direct dyes are suitable for application by combined scouring and dyeing. In this process, the usual practice is to employ soda ash and nonionic detergent. However, dyes containing amide groups are avoided because of the risk of alkaline hydrolysis. Direct dyes vary widely in their fastness properties and staining effects on various fibres. Most direct dyes have limited wet fastness in medium to full shades unless they are after-treated.

The fastness of selected direct dyes can be improved in several ways, such as treatment with cationic fixing agents, treatment with formaldehyde, treatment with copper salts such as copper sulphate, treatment with cationic agents and copper sulphate in combination, diazotization and development and treatment with cross-linking agents or resins.

An important consideration in dyeing with direct dyes is the ability of the dyes to cover the immature cotton fibre neps, which depends on both the molecular weight and hydrogen bond formation capacity of the dye molecules. Given a similar capacity to form hydrogen bonds, dyes having lower molecular weight show proportionately better neps coverage than those having higher molecular weight.

It is, therefore, necessary for the dyer to understand the dyes and the process.

28.10.3   Problems in sulphur dyeing

Sulphur dyes have been classified into four main groups: CI Sulphur dyes, CI Leuco Sulphur dyes, CI Solubilized Sulphur dyes and CI Condensed Sulphur dyes. They are available in various commercial forms such as powders, pre-reduced powders, grains, dispersed powders, dispersed pastes, liquids and water-soluble brands. The various steps in the application of sulphur dyes depend very much on their type and commercial form. The auxiliaries used in sulphur dyeing are reducing agents, antioxidants, sequestering agents, wetting agents, oxidizing agents and fixation additives.

Two special problems in dyeing with sulphur dyes are acid tendering and bronziness. In severe conditions of heat and humidity, some sulphur dyes, notably black, can generate a small amount of sulphuric acid within the cellulosic fibres, leading to tendering. Some sulphur-dyed textile material deteriorates under normal storage conditions. In the event of the dyeing needing subsequent correction, alkylated sulphur dyeings are difficult to strip and attempted removal will often entail the destruction of the dye chromogen. Fixation additives, such as alkylating agents based on epichlorohydrin, give dyeings of markedly improved washing fastness but often at the risk of some decrease in lightfastness.

The addition of copper sulphate to batch-wise oxidation baths of sodium dichromate/acetic acid improves the lightfastness but results in dulling of the shades, as well as harsher handle. It is not recommended with sulphur blacks, where the presence of copper promotes acid tendering. It is, therefore, necessary to understand the dyes and chemicals and decide the process.

28.10.4   Problems in dyeing with vat dyes

Vat dyes are preferred where the highest fastness to industrial laundering, weathering and light are required. Based on the temperature and the amount of caustic soda, hydrosulphite and salt used in dyeing, vat dyes can be classified into four main groups:

- IN dyes—require high temperature and a large amount of caustic soda and sodium hydrosulphite

- IW dyes—require medium temperature and a medium amount of caustic soda and sodium hydrosulphite with salt added

- IK dyes—require low temperature and a small amount of caustic soda and sodium hydrosulphite with salt added

- IN special dyes—require more caustic soda and higher temperature than IN dyes

Generally, vat dyes have a very rapid strike, a good degree of exhaustion and a very low rate of diffusion within the fibre. They have different chemical structures which differ in the solubility of their sodium leuco-vat, stability towards over-reduction, stability towards over-oxidation, substantivity and rate of diffusion.

Commercial competitive dyes have fairly equal particle sizes. Large particle sizes give dispersions of poor stability. For some vat dyes, colour yield decreases with increasing particle size. The effect is generally dye-specific.

The rate of reduction of vat dyes depends on various factors, such as the particle size of the dye, the temperature, time and pH during reduction and access to the reducing agent. The stability of alkaline solutions of reducing agents decreases with increased temperature, greater exposure to air, greater agitation and lower concentration of the reducing agent.

Vat dyes of the indanthrene type produce duller or greener shades at dyeing temperatures higher than 60°C due to over-reduction. Over-reduction can be prevented by the use of sodium nitrite if the reducing agent is hydrosulphite. In the case of thiourea oxide, over-reduction cannot be prevented by nitrite.

The factors influencing the rate of dyeing with vat dyes are the type of substrate, temperature, liquor ratio and concentration of dye and electrolyte. Mercerized cotton gives a higher rate of dyeing compared with unmercerized cotton, which in turn gives a higher rate than grey material. At low temperature, the rate of exhaustion is low, which might promote levelness but the rate of diffusion is also low.

At high temperature, the rate of exhaustion is high, which decrease levelness, but the rate of diffusion is high. Some dyes are not stable at very high temperatures so the stability of dyes to temperature must be taken into account. The reducing efficiency of sodium hydrosulphite in caustic soda solutions at high temperatures decreases rapidly in the presence of air. The higher the liquor ratio, the slower is the rate of dyeing. Most of the dyes exhaust more rapidly at low concentrations, increasing the risk of unlevel dyeing in light shades. Some have the same rate of dyeing irrespective of the concentration. The higher the concentration of the electrolyte, the higher is the rate of dyeing.

The very high pH and temperature during rinsing result in dulling of the shade. The ideal is to do rinsing thoroughly at low temperature at a rinsing bath pH value of 7.

Oxidation is done in vat dyeing to convert the water-soluble leuco form back into the insoluble pigment form. Normal variables in the oxidizing step are the type and concentration of oxidizing agent, the type of pH regulator and pH during oxidation, and temperature during oxidation. The oxidizing agent must provide a level of oxidation potential sufficient to oxidize the reduced vat dye into insoluble pigment, with no over-oxidation, that is, beyond the oxidation state of the original pigmentary form of the dye. Poor control of pH during oxidation may result in uneven oxidation and a lower temperature may result in slower oxidation.

At pH below 7.5, there is the possibility of formation of acid leuco forms of vat dyes. The optimum pH for oxidation is 7.5–8.5. The acid leuco form of vat dye is difficult to oxidize, has little affinity for fibre and is easily rinsed out. The higher the temperature, the faster is the oxidation, the optimum temperature being 120–140°F.

The soaping is done after oxidation to remove any dye that is not diffused into the fibre and to stabilize the final shade. This results in improved fastness properties and resistance to any shade change caused by a resin or other finish, or to consumer use. Important soaping parameters are time, temperature and type and concentration of soaping auxiliaries. After soaping, the fabric is rinsed and dried.

Both exhaust and continuous dyeing methods are used to apply vat dyes. Exhaust dyeing processes are mainly used for dyeing of loose stock, yarn and knitted fabrics. Woven fabrics can also be dyed by the exhaust method, but for large batch sizes, the continuous method is mostly used. Pad dyeing methods are preferred in the case of woven fabrics, particularly if these are in large batches. The commonly used pad dyeing methods are pad-jig, pad-steam and pad-thermosol. The most popular method for dyeing woven fabrics in a continuous manner is the pad-dry-pad-steam method, consisting of the following key steps:

a) Impregnating the fabric in a bath containing vat dye, dispersing agent, anti-migrant and a non-foaming wetting agent

b) Squeezing the impregnated fabric to a given pick-up level

c) Drying the fabric to achieve a uniform distribution of the vat pigment throughout the fabric

d) Impregnating the fabric with a solution of caustic soda and sodium hydrosulphite, with the optional use of salt

e) Expressing the impregnated fabric to a given pick-up level

f) Steaming the fabric to bring about reduction of the dye to the soluble leuco form and to promote diffusion of the dye into the cellulosic fibres

Rinsing, oxidation, soaping, rinsing and drying the fabric intermediately are the most important steps in the pad-dry-pad-steam process where the most common problem, 'migration', can take place. Important factors on which migration depends are as follows:

a) Dye constitution

b) Dye formulation

c) Pick-up

d) Additives in the dye padder

e) Residues of wetting agents and lubricants on the fabric

f) Fabric structure

g) Drying conditions

After drying, the fabric is padded with an alkaline solution of sodium hydrosulphite, and then the fabric undergoes steaming. Almost 40% of vat dyeing problems are related to improper steaming conditions. Ideal steaming conditions are controlled temperature and moisture, free from air, and sufficient dwell time.

After steaming, the fabric undergoes rinsing, oxidation and soaping. The most important control steps in vat dyeing are reduction, absorption and oxidation. The reduction and oxidation can best be controlled by metered addition of chemicals. The advantages of metered addition of hydrosulphite are as follows:

• Better levelling by slower vatting

• No need of levelling agent

• Protection from over-reduction

• Control of initial rate of dyeing (strike)

• Possibility of warm pre-pigmentation to give optimum fabric/liquor movement

• Good reproducibility

• Reduction of sulphite/sulphate effluent pollution

• Automatic monitoring of vat state and redox potential by means of measuring and regulating technology

The controlled dosage of hydrogen peroxide in the oxidation tank, together with the measurement and control of pH can result in obtaining sufficient peroxide for the oxidation of the dye as well as achieving an optimized dyeing procedure due to the control of speed of oxidation.

28.11   Factors affecting dyeing performance

28.11.1   Reproducibility and right-first-time dyeing

The reproducibility of dyeing can be affected by substrate, dyestuff, auxiliaries, water, steam, process conditions, machine and equipment, accuracy and calibration of gauges, methods and practices.

Substrate

    a)   Quality/characteristics of cotton

    b)   Quality/characteristics of yarn

    c)   Pretreatment

    d)   Absorbency

    e)   pH

    f)   Residual alkalinity

    g)   Residual peroxide

    h)   Whiteness/colorimetric coordinates

    i)   Dyeability

    j)   Validity with respect to database

    k)   Moisture content—conditioning

    l)   Weight

Dyestuff

    a)   Selection of dyes

    b)   Standardization of dyes

    c)   Source of the dye sample

    d)   Moisture content of dyes

    e)   Strength of dyes

    f)   Weight of dyes

    g)   Adulteration of dyes/impurities in dyes

    h)   Sensitivity of dyes to changes in process conditions

i) Compatibility of dyes

j) Reactivity of dyes

k) Distance of the dye colour from the target colour

l) Number of dyes in the recipe

m) Distance of the colour to be matched and the colour of the dye used in the recipe

n) Metameric index of the recipe

Auxiliaries

a) Types of auxiliaries

b) Strength of auxiliaries

c) Impurities in auxiliaries

d) Amount/weight of auxiliaries

Water

a) Impurities in water supply

b) Volume of dyebath

c) Ratio of water quantity to material quantity

Steam

a) Impurities in steam supply

b) Steam pressure variations

c) Condensates in steam

Process conditions

a) Liquor-to-goods ratio

b) Fill water temperature

c) Fixation temperature

d) Rate of rise of temperature/temperature gradient

e) Concentration of dye, electrolyte, alkali and other auxiliaries

f) Addition profile of dye

g) Addition profile of electrolyte/salt dosing

h) Conductivity

  i)   Addition profile of alkali/alkali dosing

  j)   Fixation pH

  k)   Addition profile of auxiliaries

  l)   Time (total time, before and after the addition of electrolyte, before and after the addition of alkali, before and after the addition of fixative or any other auxiliary)

  m) Load size

  n)   Liquor level

  o)   Machine flow and liquor reversal sequence

  p)   Method/conditions of washing-off

  q)   Method/conditions of drying

Machine and equipment

  a)   Leaking valves: steam, drain

  b)   Circulating pump or heat-exchanger performance at operating temperature

  c)   Location and integrity of temperature sensor

  d)   Location and evenness of steam injection for heating

Accuracy and calibration of gauges

  a)   Pressure indicators and controller

  b)   Flow indicators and controller

  c)   Level gauge

  d)   Temperature indicator and controller

  e)   Weighing balances

  f)   Spectrophotometer: inter- and intra-instrument calibration

  g)   Glassware such as pipettes, beakers and so forth

Methods and practices

  a)   Frequent change of suppliers

  b)   Spurious supply of dyes

  c)   Improper storage of dyes

  d)   Improper labelling of dyes

e) Accuracy of weighing

f) Improper location of balance where there is turbulence

g) Loss of the dye in the pan of the weighing balance

h) Spillage of solid dye prior to dissolution or after

i) Cross-contamination of vessels/materials

j) Age of the dye solutions

k) Selection of wrong method for dye strength evaluation

l) Blowing-out pipettes

m) Improper colour preparation

n) Calculation errors

o) Accuracy of dye recipe formulation

p) Dispensing methods for dyes and chemicals

q) Auxiliaries taken on the weight of the fabric

r) Improper substitution of Glauber's salt with common salt

s) Dye application method

t) Manner of drying the sample for colour assessment

u) Conditioning of the sample before colour assessment

v) Target shade very small or soiled

w) Target shade for textiles in paper/plastic

x) Dots/fluff in the area scanned

y) Colour judgement

z) Type of colorimeter and formula used

aa) Database preparation for computer colour matching

ab) Make-up and geometry of specimen—homogeneity, geometry and thickness

ac) Post-dyeing operations

ad) Poor housekeeping—lack of training/understanding, negligence, wrong attitude, wrong practice

28.12   Factors affecting dye selection and evaluation

Standardization

   a)   Homogeneity

   b)   Absorption in solution

   c)   Analysis and identification

Storage stability

   a)   Variation in moisture content

   b)   Storage conditions

Solubility and physical form

   a)   Aqueous solubility

   b)   Crystal modification

   c)   Particle size

   d)   Commercial form

Health and safety

   a)   Dustiness

   b)   Trace metals

   c)   Eye and skin irritation

   d)   Acute toxicity

   e)   Long-term hazards

   f)   Biodegradation

   g)   Sludge adsorption

   h)   Fish toxicity

Cost-effectiveness

   a)   Shade area

   b)   Colour value

   c)   Build-up reproducibility

Dye application properties

a) Levelling and migration

b) Substantivity and diffusion

c) Reactivity and fixation

d) Sensitivity to temperature

e) pH and redox potential

f) Compatibility

g) Cross-staining

h) Transfer and vapour pressure

i) Efficiency of wash-off

In-service requirements

a) Coverage

b) Penetration

c) Fastness

d) Tendering of substrate

e) Influence of finishes

28.13 To conclude

Value addition process adds value to the product, provided the processes are well designed by understanding the natural properties of materials and chemicals and worked as per the decision taken during designing, monitoring and controlling the processes by suitable checklists and analysing the variations found. If not done properly, the value addition processes shall turn out to be value-losing process. The shop floor technicians and chemists should keep their knowledge continuously enhanced by reading books, periodicals and by discussing with fellow technicians and chemicals, and share their experiences with juniors.

References and further reading

1. Sharup S A and Pagol B, An overview of textile scouring, Primeasia University: Available from: http://textilelearner.blogspot.com/2013/03/an-overview-of-textile-scouring-process.html#ixzz2rErlmlWd.

2. Cookson P G, Roczniok A F and Ly N G (1991), Measurement of relaxation shrinkage in woven wool fabrics, Text Res J, 61, 537–546. Available from: http://trj.sagepub.com/content/61/9/537.abstract

3. Tyndall R M (November 11–12, 1993), Relaxed Knit Fabric Finishing and Compacting (Tubular and Open Width), Cottech Conference.

4. Ströhle J and Benninger AG, (January 18, 2005), New process for the relaxation of knitwear containing elastane (spandex), CH-9240 Uzwil/Switzerland.

5. Choudhary A K R, Textile preparation and dyeing. Available from: http://books.google.co.in/books?id=0TamObsaaPQC&pg=PA450&lpg=PA450&dq=padding+process+textiles&source=bl&ots=7D47aebznX&sig=CZxwQlgIPHgetQE_QPgzzHP4OA0&hl=en&sa=X&ei=grbrUq6MNM3GrAfBhYGgBg&ved=0CCUQ6AEwAA#v=onepage&q=padding%20process%20textiles&f=false

6. Continuous weight reduction machine. Available from: http://hanayamakogyo.co.jp/en/genryoukakou/

7. Monga A, Radio frequency [RF] drying, Monga Strayfield Pvt Ltd. Available from: http://www.mongagroup.com/technical_paper.html

8. Purushothama B (2007), Guidelines for process management in textiles, CVG Publications.

9. Purushothama, B (2011), Training and development of technical staff in textile industry, India, Woodhead Publishing.

10. A Beginners Guide to Digital Textile Printing by LoungeKat5 Oct 2009. Available from: http://design.tutsplus.com/tutorials/a-beginners-guide-to-digital-textile-printing--vector-3189.

11. AIRO Fabrics Softening Washing and Drying Machine by Biancalani S.P.A. Available from: http://www.biancalani.com/index.php?option=com_content&view=article&id=69&Itemid=70&lang=en

12. Rotary Drum Washer by CAPTO Engineering Limited.

13. Textile Machinery by CAPTO Engineering Limited.

14. Cizopan W R, Product information sheet, Ciess Texaux Specialty Chemicals Pvt Ltd Available from: www.ciesstexaux.com/pdf/.../Cizopan%20WR%20(CHART)

15. Industrial Drum Washers by Colo Fab, (India) Surat

16. Relax Dryers, Colour soft, Exolloys Engineering works. Available from: http://www.exolloys.com/relax-dryers.html

17. Seven roller calendering machine, Cumins (Hongkong) Equip and Engineering Co ltd. Available from: http://www.textilefinish.com/

18. Technology of Shearing, Danti paolo Textile finishing machines. Available from: http://www.danti.it/010981.asp

19. Batch Process. Available from: http://www.definetextile.com/2013/05/batch-process.html

20. Multi Nozzle Soft Flow Dyeing Machines (A). Available from: http://www.devrekha.com/multi-nozzle-soft-flow-dyeing-machine.html

21. Digital Textile Printing. Available from: http://www.digitexindia.net/digital-textile-printing.html

22. Dollfus & Muller decatizing machine. Available from: http://www.dollfus-muller.com/en/decatizing-machine/

23. Hussain T, Pad dyeing methods with reactive dyes. Available from: https://www.academia.edu/Download

24. Shenoy V A, Technology of fabric Finishing.

25. Pad Dyeing Machines; Padding Mangles. Available from: http://dyeingworld1.blogspot.in/2010/01/pad-dyeing-machines.html

26. Yarn Mercerizing. Available from: http://dyeingworld1.blogspot.in/2009/12/yarn-mercerizing.html

27. Dyeing Process » Semi-continuous dyeing. Available from: http://dyes-pigments.standardcon.com/semi-continuous-process.html

28. Dyeing Machinery. Available from: http://dyes-pigments.standardcon.com/machinery.html

29. Overflow dyeing machine. Available from: http://dyes-pigments.standardcon.com/overflow-dyeing-machine.html

30. Air flow dyeing machines. Available from: http://dyes-pigments.standardcon.com/airflow-machine.html

31. Soft flow dyeing machine. Available from: http://dyes-pigments.standardcon.com/soft-flow-dyeing-machine.html

32. Best practice: Hot air drying—stenters. Available from: http://www.e4s.org.uk/textilesonline/content/6library/report5/15_hot_air_drying_stenters.htm

33. Drying in textile industry, From Efficiency Finder. Available from: http://wiki-ze.elas-calculator.eu/index.php/Drying_in_textile_industry

34. Shearing. Available from: http://encyclopedia2.thefreedictionary.com/Shearing

35. Dyeing and Processing, Fabric Mercerizing. Available from: http://dyeingworld1.blogspot.com.ar/2010/02/fabric-mercerising-process.html

36. Introduction to Digital Fabric Printing. Available from: http://www.fashion-incubator.com/archive/introduction-to-digital-fabric-printing/

37. Radio Frequency Drying: New trend in drying, Fibre2fashion. Available from: http://www.fibre2fashion.com/industry-article/3/205/radio-frequency-drying2.asp

38. Surface Finishing Gains New Precision, Trends in the latest surface finishing technology, Textile World. Available from: http://www.textileworld.com/Issues/2000/May/Textile_News/Surface_Finishing_Gains_New_Precision

39. Gargo, Rotary Drum Washers, Gargo Corporation.

40. Technical Bulletin, Garment Washing Techniques for Cotton Apparel- 6399 Weston Parkway, Cary, North Carolina, 27513.

41. Textile Learner, Garment Washing/Objects of garment washing/Types of garment wash/Advantages of Garment washing. Available from: http://textilelearner.blogspot.in/2012/04/garment-washing-objects-of-garment.html

42. Textile Learner, What is garment wash? Types of garment washing. Procedure for garment washing. Available from: http://textilelearner.blogspot.in/2012/05/what-is-garment-wash-types-of-garment.html

43. Types of garments wash, By apparel merchandising. Available from: http://goldnfiber.blogspot.in/2013/04/types-of-garments-wash.html

44. MELMARC, Mineral wash. Available from: http://www.melmarc.com/mineral-wash/

45. The Indian Textile Journal, Processing (February 2010), dyeing and finishing, types of stone wash and their effects on the denim fabric. Available from: http://www.indiantextilejournal.com/articles/FAdetails.asp?id=2683

46. Gopalakrishnan D and Karthik T, Development of anti-shrink treatment on cellulosic knits. Available from: http://www.fibre2fashion.com/industry-article/38/3771/development-of-antishrink-treatment-on-cellulosic-knits-part--21.asp

47. Gore S, Alkaline weight reduction.

48. Public Relation Office, Government of Japan (2012), Toward zero emission fabric printing. Available from: http://www.gov-online.go.jp/eng/publicity/book/hlj/html/201211/201211_07.html

49. Ujiie H (2002), Digital printing of textiles - Centre of Excellence in Digital Ink Jet Printing, Philadelphia University, USA, Woodhead Publishing Series in Textiles No. 53.

50. What actually is meant by Decatizing? Available from: http://www.hocks.de/decatising_wrappers/applications/decatizing/decatizing.html

51. Kier decatizing. Available from: http://www.hocks.de/decatising_wrappers/applications/decatizing/kier_decatizing/kier_decatizing.html

52. Finishing process. Available from: http://www.teonline.com/knowledge-centre/finishing-processes.html

53. Flock printing. Available from: http://www.indiamart.com/harsh-enterprises/flock-printing.html.

54. Textile printing. Available from: http://textiles.indianetzone.com/1/techniques_textile_decoration.htm

55. Dyeing fabric in rope form. Available from: http://www.ineris.fr/ippc/sites/default/interactive/bref_text/breftext/anglais/bref/BREF_tex_gb55.html

56. Modi J R and Mehta P C, Chemical Processing Tablet, I, Mercerization, The Textile Association (India).

57. Operation manual of Suprema, KD Biella Shrink Process.

58. Biella Shrunk Process, KD SUPREMA 95. Available from: http://www.lpbatson.com/

59. KD Gigante, Machine for "KD" permanent finishing and for atmospheric decatizing. Available from: http://www.kd-biella.com/kd_gigante.html

60. Continuous-decatizing-machines, Superfinish GFP 800, system Kettling + Braun. Available from: http://www.m-tec-gmbh.com/en/produkte/trockenausruestung/kontinue_dekatier_maschinen.htm

61. Continuous pressing and setting machines, Contipress GPP 400, system Kettling & Braun. Available from: http://www.m-tec-gmbh.com/en/produkte/trockenausruestung/kontinue_pressmaschinen.htm

62. Mercerizing of knit fabrics. Available from: http://dyeingworld1.blogspot.in/2010/02/mercerizing-of-knit-fabrics.html, Feb 2, 2010.

63. Knit Mercerizer by Swastik Machinery. Available from: http://www.swastiktextile.com/merceriser_for_tubular_fabric.html

64. About Kornit. Available from: http://www.kornit.com/about/

65. Lafer SpA, Raising machine manual

66. Lafer SpA, Shearing machine catalogue

67. Operation Manual–Lakshmi 'Insta' RF dryer

68. LoungeKat, A Beginners Guide to Digital Textile Printing, October 5, 2009. Available from: http://design.tutsplus.com/tutorials/a-beginners-guide-to-digital-textile-printing–vector-3189

69. Lucy Engineering WORKS, Relax dryer. Available from: http://www.lucyengineering.com/relax-dryer.html

70. Lucy Engineering Works Pvt Ltd, Hot air stenter. Available from: http://www.lucyengineering.com/hot-air-stenter.html

71. Lucy Engineering Works, Relax dryer. Available from: http://www.lucyengineering.com/relax-dryer.html

72. CombiSoft, Cloth rope finishing machine by Mat Di A, Bertoldi S P A. Available from: http://machinery.fibre2fashion.com/productDetail.aspx?refno=5866

73. Kiron M I, Dyeing process of jigger dyeing machine. Available from: http://www.fibre2fashion.com/industry-article/49/4898/dyeing-process-of-jigger-dyeing-machine1.asp

74. Morrison Chain Mercerizing Range. Available from: http://www.morrisontexmach.com/chain_merc.cfm

75. Mehta N, "Energy saving in textile processing", Nandish Mehta, Harish Enterprise Pvt. Ltd. Available from: http://www.energymanagertraining.com/textiles/pdf/Energy%20Saving%20in%20Textile%20Processing.pdf

76. Cloth finishing, (Originally Published Early 1900s). Available from: http://www.oldandsold.com/articles04/textiles22.shtml

77. Pad steam machine. Available from: http://www.scribd.com/doc/61408787/PAD-STEAM-MACHINE

78. Latest developments for finishing of woven and knitted fabrics, Pak Text J. Available from: http://ptj.com.pk/Web%202003/8-2003/fleissner.htm

79. Fabric printing methods. Available from: http://www.pinterest.com/emilykeener01/fabric-printing-methods/

80. Kholiya R, Jahan S and Raghuvanshi R, Implementation of CAD in printing. Available from: http://www.fibre2fashion.com/industry-article/14/1328/implementation-of-cad-in-printing1.asp

81. Karmarkar S R, Chemical Technology in the Pre-treatment process of textiles, Google books (1999). Available from: books.google.co.in/books/isbn=0080539475

82. Le, C V, Tester D H, Ly N G and De Jong S (1994), Changes in Fabric Mechanical Properties after Pressure Decatizing as Measured by FAST, Text Res J, 64, 61–69. CSIRO. Available from: http://trj.sagepub.com/content/64/2/61.abstract

83. Shakti Textile Engineers, Multi cylinder drying range. Available from: http://shaktitextileengrs.com/machine_multicylinderverticaldryingrange.html

84. IBENA Shanghai Technical Textiles Co. Ltd, Decatizing wrappers. Available from: http/ cn.ibena.cn:9999/en/decatizing-wrappers-sh-en.html

85. The Smart Time, Chain mercerising range. Available from: http://thesmarttime.com/pretreatment/chain-mercerizer.html

86. Stabila of Sperotto Rimer, Italy, Colourage; Mar 2000, 47(3), 57.

87. RF Drying Technology, Stalam. Available from: http://www.stalam.it/it/applications/rf-drying-technology.html

88. Sueder Machine. Available from: http://textilelearner.blogspot.in/2013/01/sueding-machine-specification-of.html

89. Sueding machines, Over 800 Ultrasoft and Microsand sueding machines supplied to 45 countries in 5 years! Available from: www.lafer.com

90. Tensionless dryer for fabric relaxing dryer. Available from: http://sunwinwuxi.en.alibaba.com/product/386447420-218367441/Tensionless_Dryer_for_fabric_relaxing_dryer.html

91. Stenter, Optima 2510 By Swastik Processing Machinery. Available from: http://www.swastiktextile.com/stenter.html

92. Relax dryers, Swastik Group of Companies. Available from: http://www.swastiktextile.com/relax_dryers_for_open_width_and_tubular_fabrics.html

93. Single Door Drum Washer by Swastik Engineering & Fabrication Vadodara

94. Printing. Available from: http://www.teonline.com/knowledge-centre/printing-textile-services.html

95. Tex Rolls India, Industrial drying machine. Available from: http://www.indiamart.com/texrollsindia/industrial-drying-machine.html

96. Texshrink Relax dryer. Available from: http://thakoreexports.com/Texshrink_Relax_dryer.html.

97. Textile Learner, Different methods of reactive dye application. Available from: http://textilelearner.blogspot.in/2012/01/different-methods-of-reactive-dye.html

98. Textile Learner, Textile Dryer. Available from: http://textilelearner.blogspot.in/2013/02/textile-dryer-working-principle-of.html

99. Textile Learner, Available from: http://textilelearner.blogspot.com/2013/01/stenter-machine-function-of-stenter.html#ixzz41uiN7pDP

100. Textile Learner, Raising or Napping. Available from: textilelearner.blogspot.com/.../raising-or-napping-finishing-working.htm

101. Textile Learner, Calendering. Available from: http://textilelearner.blogspot.in/2012/10/objectivesaims-of-calendering.html

102. Textile Learner, Printing Methods. Available from: http://textilelearner.blogspot.in/2011/07/printing-method-method-of-printing_5617.html

103. Textile Learner, Washing Faults. Available from: http://textilelearner.blogspot.com/2012/08/washing-faultsdefects-of-fabric.html#ixzz25D3PeWRT

104. Textile School, Fabric washing techniques. Available from: http://www.textile-school.com/articles/358/fabric-washing-techniques

105. Singeing and desizing. Available from: http://textilebe-dyeing.blogspot.in/2011/05/singeing-and-desizing.html

106. Textile Technology, Wet processing. Available from: http://textechdip.wordpress.com/contents/wet-processing/

107. Textile Fashion Study, Printing flow chart of 100% synthetic fabrics. Available from: http://textilefashionstudy.com/printing-process-flow-chart-of-100-synthetic-fabric/

108. Textile Fashion Study, Printing flow chart for 100% cotton fabrics. Available from: http://textilefashionstudy.com/process-flow-of-printing-process-for-100-cotton-fabric/

109. Textile Fashion Study, Printing process flow of plastic solution (High Density Printing Process). Available from: http://textilefashionstudy.com/printing-process-flow-of-plastic-solution-high-density-printing-process/

110. Textile Fashion Study, Sequence of discharge printing process on cotton, Discharge printing style - Available from: http://textilefashionstudy.com/sequence-of-discharge-printing-process-on-cotton-discharge-printing-style/

111. Textile Fashion Study, Process flow of emboss printing, Pub printing process. Available from: http://textilefashionstudy.com/process-flow-of-emboss-printing-pub-printing-process/

112. Textile Fashion Study, Sequence of glitter printing process on textile materials. Available from: http://textilefashionstudy.com/sequence-of-glitter-printing-process-on-textile-materials-fabric-printing/

113. Textile Fashion Study, Printing flow chart for 100% cotton fabrics. Available from: http://textilefashionstudy.com/process-flow-of-printing-process-for-100-cotton-fabric/

114. Fire up your textile knowledge, Dyeing Process by Jigger Dyeing Machine with limitation. Available from: http://textilefire.blogspot.in/2013/04/dyeing-process-by-jigger-dyeing-machine.html

115. Textile Learner, Jet dyeing machine. Available from: http://textilelearner.blogspot.in/2012/01/jet-dyeing-machine-working-process-of.html

116. Textile Printing, Jet printing. Available from: http://textilesindepth-textileprinting.blogspot.in/2011/05/jet-printing_01.html

117. Fluorocarbon particle coated textiles for use in electrostatic printing machines, US Patent US 6054399 A. Available from: http://www.google.st/patents/US6054399

118. Vijaya Shanbhag, Latest textile processing machineries, Fires2fashion. Available from: http://www.fibre2fashion.com/industry-article

119. Prepare for EU Compliance, new seven roll calender machine for many materials. Available from: http://www.weiku.com/products/10421372/new_seven_roll_calender_machine_for_many_materials.html

120. Sanforization, From Wikipedia, the free encyclopedia. Available from: http://en.wikipedia.org/wiki/Sanforization

121. Desizing. Available from: http://en.wikipedia.org/wiki/Desizing

122. Wikipedia Free Encyclopaedia. Available from: http://en.wikipedia.org/wiki/Kier_(industrial)

123. Fulling, From Wikipedia, the free encyclopaedia. Available from: https://en.wikipedia.org/wiki/Fulling

124. Decatising, From Wikipedia, the free encyclopaedia. Available from: http://en.wikipedia.org/wiki/Decatising

125. Textile printing. Available from: http://en.wikipedia.org/wiki/Textile_printingView shared post

126. Textile Fashion Study, Textile printing/Definitions, Styles and method of printing. Available from: http://textilefashionstudy.com/textile-printing-definition-styles-and-methods-of-printing/

127. Calendering, From Wikipedia, the free encyclopedia. Available from: http://en.wikipedia.org/wiki/Calender

128. History of textile printing. Available from: http://wftprintpm.wikispaces.com/Textile+%26+Screen+Printing

129. Drying in textile industry, Hot air drying. Available from: http://wiki-ze.elas-calculator.eu/index.php/Drying_in_textile_industry

130. Additional information: Finishing of wool From Efficiency Finder. Available from: http://wiki-ze.elas-calculator.eu/index.php/Additional_information:_Finishing_of_wool

131. Xetma Multisystem XR, Manual. Available from: http://www.xetma.com/en-multisystem-xr.html

132. Xetma Multisystem XRE, Manual.

133. Xetma Multisystem XREB, Manual.

134. Xetma History. Available from: www.xetma.com/en-company.html

135. Yamuna Hot Air stenters. Available from: http://www.fibre2fashion.com/machinery-yearbook/yamuna/products.asp

136. Yamuna Textile Processing Machines. Available from: http://www.fibre2fashion.com/machinery-yearbook/yamuna/products.asp

137. Air-Flow Softening Finishing Machine for Washing, Drying and Softening (LD-4) Zhangjiagang Polygee Environmental Protection Technology Co., Ltd. Available from: http://polygee.en.made-in-china.com/product/aenEfPMXOCRc/China-Air-Flow-Softening-Finishing-Machine-for-Washing-Drying-and-Softening-LD-4-.html

138. Machine catalogue and operation manuals provided by the machine manufacturers.

139. Manuals and leaflets given by the dyes and chemical manufacturers from time to time.

140. Dyes and chemicals catalogue given by dyestuff manufacturers

141. Drying | Drying Systems Used in Sizing | Cylinder Drying/Hot Air Drying/Infrared Drying/Combined Drying. Available from: http://textilelearner.blogspot.in/2012/05/drying-drying-systems-used-in-sizing.html

142. Stenter Machine | Function of Stenter Machine | Working Process of Stenter Machine. Available from: http://www.swastiktextile.com/stenter.html

143. Process Recommendation for 100% polyester processing. Available from: www.fibre2fashion.com/product_launch/rossera_dec/macro.htm

144. Moire fabric, Wikipedia. Available from: http://en.wikipedia.org/wiki/Moire_(fabric)

145. The elegant art of embossing. Available from: www.fibre2fashion.com › Knowledge › Article

146. Shenoy V A., The Technology of Textile Finishing. Available from: http://noteboi.net/main/index.php?option=com_content&view=category&layout=blog&id=74&Itemid=160&limitstart=62

147. Machine catalogues and operation manuals supplied by machine manufacturers.

148. International Finance Corporation (April 30, 2007), Environmental, Health and Safety Guidelines for Textile Manufacturing.

149. Shroff P, Problems and remedies in wet processing of 100% cotton (woven).

150. Madhu C R, Quality Control in Wet Processing Part I.

151. Periyasamy A P, Recent technologies in textile wet processing, SSMITT Komarapalayam.

152. Washing Faults. Available from: http://textilelearner.blogspot.com/2012/08/washing-faultsdefects-of-fabric.html#ixzz25D3GydQc

153. Mazharul Islam, Effects of Pretreatment Process in Textile Wet Processing: Available from: http://www.textilehelpline.com/effects-of-pretreatment-process-in-textile-wet-processing/

154. Flowchart of wet processing. Available from: http://textilelearner.blogspot.in/2011/08/flow-chart-of-wet-processing-process.html

155. Machinery for preparation of textiles. Available from: http://nptel.ac.in/courses/116102016/39, IIT Delhi

156. Wirote Sarakarnkosol, Textile Pretreatment, Right First Time. Available from: www.academia.edu/.../Textile_Chemistry_Textile_auxiliaries

157. Batching. Available from: http://textilelearner.blogspot.in/2013/01/process-flow-chart-of-batching-section.html

158. Ben-Colour Continuous dyeing system developed by Benninger. Available from: www.benningergroup.com/uploads/tx_userdownloads/BEN-COLOUR_English.pdf

159. Sahinli M (2005), Continuous processes, cold pad-batch dyeing process and tailing problem, Huntsman Textile Effects.

160. Athalye A, Shah D, Sequiera J, Controlling continuous dyeing by vat dyes, Novatic MD, Technical Service, Atul Ltd (Colors Division), Valsad, Gujarat, India.

161. Teli M D (March 1996), 'New developments in dyeing process control', Indian J Fibre Text Res, 21.

162. Mishra R R, 'Some of the process sequence with dyeing shade discussion for Continuous Dyeing Range machine (CDR)' Available from: http://www.fibre-2fashion.com/industry-article/3580/some-of-the-process-sequence-with-dyeing-shade?page=1

163. Sheshir M H, Textile dyeing machinery, Southeast University, Bangladesh.

164. Apparatus for continuously reducing the weight of fabrics of polyester fibers with alkali, Europe Patent EP 0069436 B1. Available from: http://www.google.com/patents/EP0069436B1?cl=en

165. The Textile Institute, Textile progress, doi:10.1533/tepr.2005.0001©.

166. Shamey R and Husein T (2007), 'Critical solutions in the dyeing of cotton textile materials', Textile Progress, 37, 1–84.

167. Abhinav and Nathan Y, Textile chemical processing. Available from: http://www.slideshare.net/abhinav_nathany/wet-processing

168. Wikipedia, Wet processing engineering. Available from: https://en.wikipedia.org/wiki/Wet_processing_engineering

169. Hiemac Machinery. Available from: http://www.hiemac.com.au/portfolio-view/textile-machinery/

170. Purushothama B (2016), Handbook on Fabric Manufacturing, India, Woodhead Publishing.

171. Franklin Beech, The Project Gutenberg eBook of The Dyeing of Woollen Fabrics, www.basiccarpentrytechniques.com